Electromagnetic Materials: Properties and Applications

Electromagnetic Materials: Properties and Applications

Edited by
Brooklynn Murphy

www.willfordpress.com

Published by Willford Press,
118-35 Queens Blvd., Suite 400,
Forest Hills, NY 11375, USA

ISBN: 978-1-64728-451-0

Cataloging-in-publication Data

Electromagnetic materials : properties and applications / edited by Brooklynn Murphy.
p. cm.
Includes bibliographical references and index.
ISBN 978-1-64728-451-0
1. Electromagnetism. 2. Metamaterials. 3. Materials--Magnetic properties. I. Murphy, Brooklynn.
QC760 .E44 2023
537--dc23

For information on all Willford Press publications
visit our website at www.willfordpress.com

Contents

Preface

An electromagnet is a type of magnet in which electric current produces a magnetic field. The electromagnets are used in electrical devices such as motors, generators, electrochemical solenoids, relays, and MRI machines. A solenoid is an electromechanical device that generates a mechanical motion when electric current is passed through it. The principle of solenoid is used in a device called as a relay to open and close light-current electrical circuits. Electromagnetic materials have a wide range of applications. They are used in various fields such as communications, research, and magnetic recording. Modern telephone systems are based on the reed relay along with solid-state circuits for complex routing of connections. This book outlines studies that have been carried out with respect to the properties and applications of electromagnetic materials. It will provide comprehensive knowledge to the readers.

This book unites the global concepts and researches in an organized manner for a comprehensive understanding of the subject. It is a ripe text for all researchers, students, scientists or anyone else who is interested in acquiring a better knowledge of this dynamic field.

I extend my sincere thanks to the contributors for such eloquent research chapters. Finally, I thank my family for being a source of support and help.

Editor

1

The Influence of the Dielectric Materials on the Fields in the Millimeter and Infrared Wave Regimes

Zion Menachem

Abstract

This chapter presents the influence of the dielectric materials on the output field for the millimeter and infrared regimes. This chapter presents seven examples of the discontinuous problems in the cross section of the straight waveguide. Two different geometrical of the dielectric profiles in the cross section of the straight rectangular and circular waveguides will be proposed to understand the behavior of the output fields. The two different methods for rectangular and circular waveguides and the techniques to calculate any geometry in the cross section are very important to understand the influence of the dielectric materials on the output fields. The two different methods are based on Laplace and Fourier transforms and the inverse Laplace and Fourier transforms. Laplace transform on the differential wave equations is needed to obtain the wave equations and the output fields that are expressed directly as functions of the transmitted fields at the entrance of the waveguide. Thus, the Laplace transform is necessary to obtain the comfortable and simple *input-output* connections of the fields. The applications are useful for straight waveguides in the millimeter and infrared wave regimes.

Keywords: wave propagation, dielectric profiles, rectangular and circular waveguides, dielectric materials

1. Introduction

Various methods for analysis of waveguides have been studied in the literature. The review for the modal analysis of general methods has been published [1]. The important methods, such as the finite difference method and integral equation method, and methods based on series expansion have been described. An analytical model for the corrugated rectangular waveguide has been extended to compute the dispersion and interaction impedance [2]. The application of analytical method based on the field equations has been presented to design corrugated rectangular waveguide slow-wave structure TH_z amplifiers.

A fundamental technique has been proposed to compute the propagation constant of waves in a lossy rectangular waveguide [3]. An important consequence of this work is the demonstration that the loss computed for degenerate modes propagating simultaneously is not simply additive. The electromagnetic fields in rectangular conducting waveguides filled with uniaxial anisotropic media have been characterized [4].

A full-vectorial boundary integral equation method for computing guided modes of optical waveguides has been proposed [5]. The integral equations are used to compute the Neumann-to-Dirichlet operators for sub-domains of constant refractive index on the transverse plane of the waveguide. Wave propagation in an inhomogeneous transversely magnetized rectangular waveguide has been studied with the aid of a modified Sturm-Liouville differential equation [6].

An advantageous finite element method for the rectangular waveguide problem has been developed [7] by which complex propagation characteristics may be obtained for arbitrarily shaped waveguide. The characteristic impedance of the fundamental mode in a rectangular waveguide was computed using finite element method. The finite element method has been used to derive approximate values of the possible propagation constant for each frequency. A new structure has been proposed for microwave filters [8]. This structure utilizes a waveguide filled by several dielectric layers. The relative electric permittivity and the length of the layers were optimally obtained using least mean square method.

An interesting method has been introduced for frequency domain analysis of arbitrary longitudinally inhomogeneous waveguides [9]. In this method, the integral equations of the longitudinally inhomogeneous waveguides are converted from their differential equations and solved using the method of moments. A general method has been introduced to frequency domain analysis of longitudinally inhomogeneous waveguides [10]. In this method, the electric permittivity and also the transverse electric and magnetic fields were expanded in Taylor's series. The field solutions were obtained after finding unknown coefficients of the series. A general method has been introduced to analyze aperiodic or periodic longitudinally inhomogeneous waveguides [11]. The periodic longitudinally inhomogeneous waveguides were analyzed using the Fourier series expansion of the electric permittivity function to find their propagation constant and characteristic impedances.

Various methods for the analysis of cylindrical hollow metallic or metallic with inner dielectric coating waveguide have been studied in the literature. A review of the hollow waveguide technology [12, 13] and a review of IR transmitting, hollow waveguides, fibers, and integrated optics [14] were published. Hollow waveguides with both metallic and dielectric internal layers have been proposed to reduce the transmission losses. A hollow waveguide can be made, in principle, from any flexible or rigid tube (plastic, glass, metal, etc.) if its inner hollow surface (the core) is covered by a metallic layer and a dielectric overlayer. This layer structure enables us to transmit both the TE and TM polarizations with low attenuation [15].

A transfer matrix function for the analysis of electromagnetic wave propagation along the straight dielectric waveguide with arbitrary profiles has been proposed [16]. This method is based on the Laplace and Fourier transforms. This method is based on Fourier coefficients of the transverse dielectric profile and those of the input-wave profile. Laplace transform is necessary to obtain the comfortable and simple input–output connections of the fields. The transverse field profiles are computed by the inverse Laplace and Fourier transforms.

The influence of the spot size and cross section on the output fields and power density along the straight hollow waveguide has been proposed [17]. The derivation is based on Maxwell's equations. The longitudinal components of the fields are developed into the Fourier-Bessel series. The transverse components of the fields are expressed as functions of the longitudinal components in the Laplace plane and are obtained by using the inverse Laplace transform by the residue method.

These are two kinds of different methods that enable us to solve practical problems with different boundary conditions. The calculations in all methods are based on using Laplace and Fourier transforms, and the output fields are computed by the inverse Laplace and Fourier transforms. Laplace transform on the differential wave equations is needed to obtain the wave equations (and thus also the output fields) that are expressed directly as functions of the transmitted fields at the entrance of the waveguide at $z = 0^+$. Thus, the Laplace transform is necessary to obtain the comfortable and simple *input-output* connections of the fields.

All models that are mentioned refer to solve interesting wave propagation problems with a particular geometry. If we want to solve more complex discontinuous problems of coatings in the cross section of the dielectric waveguides, then it is important to develop for each problem an improved technique for calculating the profiles with the dielectric material in the cross section of the straight waveguide.

This chapter presents two techniques for two different geometries of the straight waveguide. The two proposed techniques are very important to solve discontinuous problems with dielectric material in the cross section of the straight rectangular and circular waveguides. The proposed technique relates to the method for the propagation along the straight rectangular metallic waveguide [16]. The examples will be demonstrated for the rectangular and circular dielectric profiles in the straight rectangular waveguide.

In this chapter, we present seven dielectric structures as shown in **Figure 1(a)–(g)**. **Figure 1(a)–(e)** shows five examples of the discontinuous dielectric materials in the cross section of the rectangular straight waveguide. **Figure 1(f)–(g)** shows two examples of the discontinuous dielectric materials in the cross section of the circular straight waveguide.

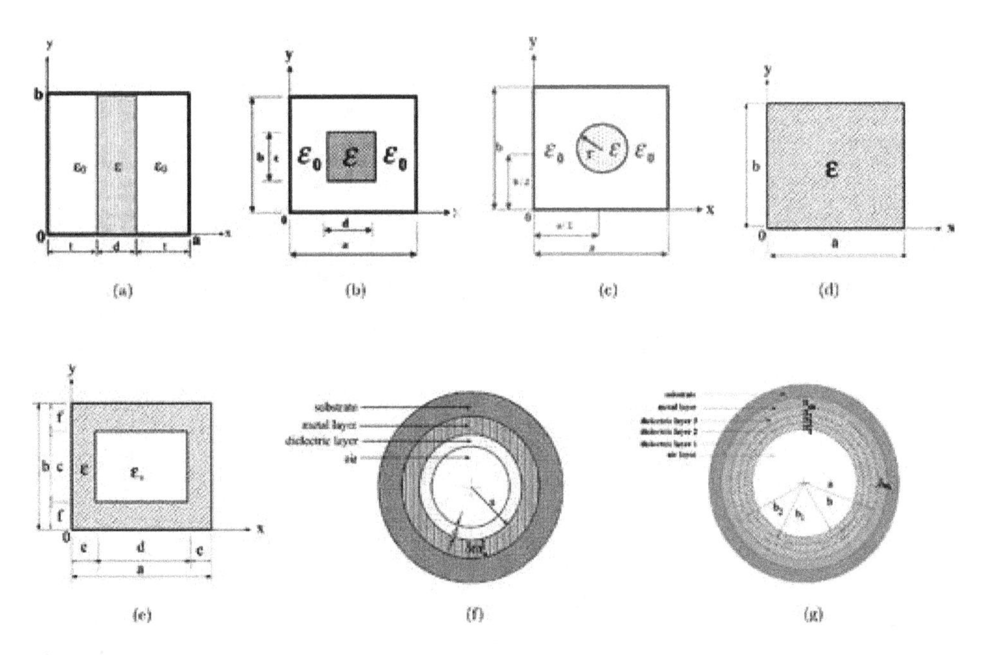

Figure 1.

Seven examples of dielectric materials in the cross section of the straight waveguides. (a) Slab profile in the cross section of the straight waveguides. (b) Rectangular profile in the rectangular waveguide. (c) Circular profile in the rectangular waveguide. (d) Full dielectric material in the rectangular waveguide. (e) Hollow rectangular waveguide with one dielectric material between the hollow rectangle and the metal. (f) Hollow waveguide with one dielectric coating. (g) Hollow waveguide with three dielectric coatings.

2. Two proposed techniques for discontinuous problems in the cross section of the rectangular and circular waveguides

The wave equations for the components of the electric and magnetic field are given by

$$\nabla^2 E + \omega^2 \mu \varepsilon E + \nabla \left(E \cdot \frac{\nabla \varepsilon}{\varepsilon} \right) = 0 \tag{1}$$

and

$$\nabla^2 H + \omega^2 \mu \varepsilon H + \frac{\nabla \varepsilon}{\varepsilon} \times (\nabla \times H) = 0 \tag{2}$$

where ε_0 represents the vacuum dielectric constant, χ_0 is the susceptibility, and g is its dielectric profile function in the waveguide.

Let us introduce the dielectric profile function for the examples as shown in **Figure 1(a)–(g)** for the inhomogeneous dielectric materials.

3. The derivation for rectangular straight waveguide

The wave Eqs. (1) and (2) are given in the case of the rectangular straight waveguide, where

$$\varepsilon(x,y) = \varepsilon_0(1 + \chi_0 g(x,y)),\ g_x = [1/\varepsilon(x,y)][\partial\varepsilon(x,y)/\partial x], \text{ and}$$
$$g_y = [1/\varepsilon(x,y)][\partial\varepsilon(x,y)/\partial y].$$

3.1 The first technique to calculate the discontinuous structure of the cross section

Figure 2(a) shows an example of the cross section of the straight waveguide (**Figure 1(a)**) for $gx()$function. In order to solve inhomogeneous dielectric profiles, we use the ω_ε function, with the parameters ε_1 and ε_2 (**Figures 1(a)** and **(b)**).

The ω_ε function [18] is used in order to solve discontinuous problems in the cross section of the straight waveguide. The ω_ε function is defined according to **Figure 2(b)** as

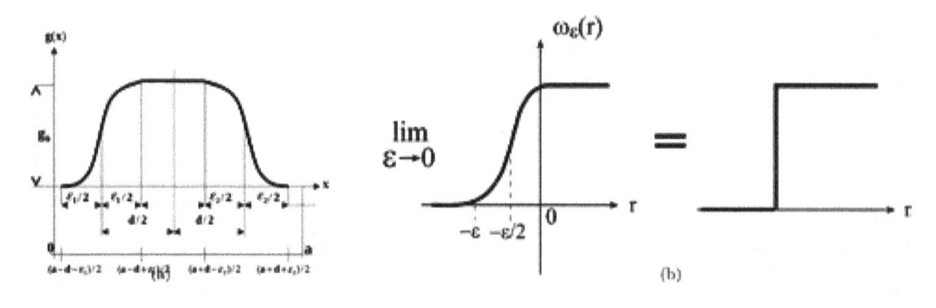

Figure 2.
(a) An example of the discontinuous problem of the slab dielectric profile in the straight rectangular waveguide.
(b) The ω_ε function in the limit $\varepsilon \to 0$.

$$\omega_\varepsilon(r) = \begin{cases} C_\varepsilon e^{-\frac{\varepsilon^2}{\varepsilon^2 - |r|^2}} & |r| \le \varepsilon \\ 0 & |r| > \varepsilon \end{cases}, \tag{3}$$

where C_ε is a constant and $\int \omega_\varepsilon(r)dr = 1$.

In order to solve inhomogeneous dielectric profiles (e.g., in **Figure 1(a)–(b)**) in the cross section of the straight waveguide, the parameter ε is used according to the ω_ε function (**Figure 2(b)**), where $\varepsilon \to 0$. The dielectric profile in this case of a rectangular dielectric material in the rectangular cross section (**Figure 1(b)**) is given by

$$g(x) = \begin{cases} 0 & 0 \le x < (a-d-\varepsilon)/2 \\ g_0 \exp\left[1 - \dfrac{\varepsilon^2}{\varepsilon^2 - [x-(a-d+\varepsilon)/2]^2}\right] & (a-d-\varepsilon)/2 \le x < (a-d+\varepsilon)/2 \\ g_0 & (a-d+\varepsilon)/2 < x < (a+d-\varepsilon)/2 \\ g_0 \exp\left[1 - \dfrac{\varepsilon^2}{\varepsilon^2 - [x-(a+d-\varepsilon)/2]^2}\right] & (a+d-\varepsilon)/2 \le x < (a+d+\varepsilon)/2 \\ 0 & (a+d+\varepsilon)/2 < x \le a \end{cases}, \tag{4}$$

and

$$g(y) = \begin{cases} 0 & 0 \le y < (b-c-\varepsilon)/2 \\ g_0 \exp\left[1 - \dfrac{\varepsilon^2}{\varepsilon^2 - [y-(b-c+\varepsilon)/2]^2}\right] & (b-c-\varepsilon)/2 \le y < (b-c+\varepsilon)/2 \\ g_0 & (b-c+\varepsilon)/2 < y < (b+c-\varepsilon)/2 \\ g_0 \exp\left[1 - \dfrac{\varepsilon^2}{\varepsilon^2 - [y-(b+c-\varepsilon)/2]^2}\right] & (b+c-\varepsilon)/2 \le y < (b+c+\varepsilon)/2 \\ 0 & (b+c+\varepsilon)/2 < y \le b \end{cases}. \tag{5}$$

The elements of the matrices are given according to **Figure 1(b)**, in the case of b $\neq$ c by

$$\begin{aligned} g(n,m) = \frac{g_0}{ab} & \left\{ \int_{(a-d-\varepsilon)/2}^{(a-d+\varepsilon)/2} \exp\left[1 - \frac{\varepsilon^2}{\varepsilon^2 - [x-(a-d+\varepsilon)/2]^2}\right] \cos\left(\frac{n\pi x}{a}\right) dx \right. \\ & \left. + \int_{(a-d+\varepsilon)/2}^{(a+d-\varepsilon)/2} \cos\left(\frac{n\pi x}{a}\right) dx + \int_{(a+d-\varepsilon)/2}^{(a+d+\varepsilon)/2} \exp\left[1 - \frac{\varepsilon^2}{\varepsilon^2 - [x-(a+d-\varepsilon)/2]^2}\right] \cos\left(\frac{n\pi x}{a}\right) dx \right\} \\ & \left\{ \int_{(b-c-\varepsilon)/2}^{(b-c+\varepsilon)/2} \exp\left[1 - \frac{\varepsilon^2}{\varepsilon^2 - [y-(b-c+\varepsilon)/2]^2}\right] \cos\left(\frac{m\pi y}{b}\right) dy \right. \\ & \left. + \int_{(b-c+\varepsilon)/2}^{(b+c-\varepsilon)/2} \cos\left(\frac{m\pi y}{b}\right) dy + \int_{(b+c-\varepsilon)/2}^{(b+c+\varepsilon)/2} \exp\left[1 - \frac{\varepsilon^2}{\varepsilon^2 - [y-(b+c-\varepsilon)/2]^2}\right] \cos\left(\frac{m\pi y}{b}\right) dy \right\}, \end{aligned} \tag{6}$$

where

$$\int_{(a-d+\varepsilon)/2}^{(a+d-\varepsilon)/2} \cos\left(\frac{n\pi x}{a}\right) dx = \begin{cases} (2a/n\pi)\sin((n\pi/2a)(d-\varepsilon))\cos((n\pi)/2) & n \neq 0 \\ d-\varepsilon & n = 0 \end{cases}$$

and

$$\int_{(b-c+\varepsilon)/2}^{(b+c-\varepsilon)/2} \cos\left(\frac{m\pi y}{b}\right) dy = \begin{cases} (2b/m\pi)\sin((m\pi/2b)(c-\varepsilon))\cos((m\pi)/2) & n \neq 0 \\ c-\varepsilon & n = 0 \end{cases}.$$

The elements of the matrices are given according to **Figure 1(a)**, in the case of b = c by

$$g(n,m) = \frac{g_0}{ab}\left\{\int_{(a-d-\varepsilon)/2}^{(a-d+\varepsilon)/2} \exp\left[1 - \frac{\varepsilon^2}{\varepsilon^2 - [x-(a-d+\varepsilon)/2]^2}\right]\cos\left(\frac{n\pi x}{a}\right) dx\right.$$
$$\left. + \int_{(a-d+\varepsilon)/2}^{(a+d-\varepsilon)/2} \cos\left(\frac{n\pi x}{a}\right) dx + \int_{(a+d-\varepsilon)/2}^{(a+d+\varepsilon)/2} \exp\left[1 - \frac{\varepsilon^2}{\varepsilon^2 - [x-(a+d-\varepsilon)/2]^2}\right]\cos\left(\frac{n\pi x}{a}\right) dx\right\}$$
$$\left\{\int_0^b \cos\left(\frac{m\pi y}{b}\right) dy\right\},$$

where

$$\int_0^b \cos\left(\frac{m\pi y}{b}\right) dy = (b/m\pi)\sin(m\pi) = \begin{cases} b & m = 0 \\ 0 & m \neq 0 \end{cases}.$$

3.2 The second technique to calculate the discontinuous structure of the cross section

The second technique to calculate the discontinuous structure of the cross section as shown in **Figure 1(a)** and **(b)**.

The dielectric profile $g(x,y)$ is given according to $\varepsilon(x,y) = \varepsilon_0(1+g(x,y))$. According to **Figure 3** and for $g(x,y) = g_0$, we obtain

$$g(n,m) = \frac{g_0}{4ab}\int_{-a}^{a} dx \int_{-b}^{b} \exp[-j(k_x x + k_y y)] dy$$
$$= \frac{g_0}{4ab}\left\{\int_{x_{11}}^{x_{12}} dx \int_{y_{11}}^{y_{12}} \exp[-j(k_x x + k_y y)] dy + \int_{-x_{12}}^{-x_{11}} dx \int_{y_{11}}^{y_{12}} \exp[-j(k_x x + k_y y)] dy\right.$$
$$\left. + \int_{-x_{12}}^{-x_{11}} dx \int_{-y_{12}}^{-y_{11}} \exp[-j(k_x x + k_y y)] dy + \int_{x_{11}}^{x_{12}} dx \int_{-y_{12}}^{-y_{11}} \exp[-j(k_x x + k_y y)] dy\right\}. \quad (7)$$

If y_{11} and y_{12} are functions of x, then we obtain

$$g(n,m) = \frac{g_0}{abk_y}\int_{x_{11}}^{x_{12}} [\sin(k_y y_{12}) - \sin(k_y y_{11})]\cos(k_x x) dx$$
$$= \frac{2g_0}{am\pi}\int_{x_{11}}^{x_{12}} \sin\left[\frac{k_y}{2}(y_{12} - y_{11})\right]\cos\left[\frac{k_y}{2}(y_{12} + y_{11})\right]\cos(k_x x) dx. \quad (8)$$

The dielectric profile for **Figure 1(b)** is given by

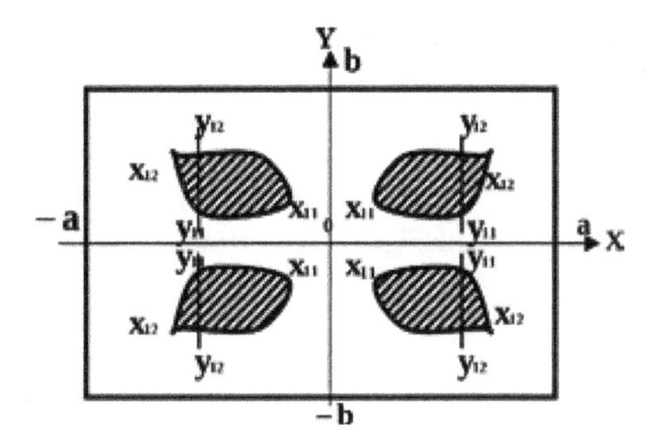

Figure 3.
The arbitrary profile in the cross section.

$$g(n,m) = \begin{cases} \frac{g_0}{4ab}(4cd) & n=0, m=0 \\ \frac{g_0}{4ab}\left(\frac{8d}{k_{0y}m}\sin\left(\frac{k_{0y}mc}{2}\right)\cos\left(\frac{k_{0y}mb}{2}\right)\right) & n=0, m\neq 0 \\ \frac{g_0}{4ab}\left(\frac{8c}{k_{0x}n}\sin\left(\frac{k_{0x}nd}{2}\right)\cos\left(\frac{k_{0x}na}{2}\right)\right) & n\neq 0, m=0 \\ \frac{g_0}{4ab}\left(\frac{16}{k_{0x}k_{0y}nm}\sin\left(\frac{k_{0x}nd}{2}\right)\cos\left(\frac{k_{0x}na}{2}\right)\sin\left(\frac{k_{0y}mc}{2}\right)\cos\left(\frac{k_{0y}mb}{2}\right)\right) & n\neq 0, m\neq 0 \end{cases} \tag{9}$$

3.3 The dielectric profile for the circular profile in the cross section

The dielectric profile for the circular profile in the cross section of the straight rectangular waveguide is given by (**Figure 1(c)**)

$$g(x,y) = \begin{cases} g_0 & 0 \leq r < r_1 - \varepsilon_1/2 \\ g_0 \exp\left[1 - q_\varepsilon(r)\right] & r_1 - \varepsilon/2 \leq r < r_1 + \varepsilon/2 \end{cases}, \tag{10}$$

where

$$q_\varepsilon(r) = \frac{\varepsilon^2}{\varepsilon^2 - \left[r - (r_1 - \varepsilon/2)\right]^2},$$

else $g(x,y) = 0$. The radius of the circle is given by $r = \sqrt{(x-a/2)^2 + (y-b/2)^2}$.

3.4 The dielectric profile for the waveguide filled with dielectric material in the entire cross section

The dielectric profile (**Figure 1(d)**) is given by

$$g(n,m) = \begin{cases} g_0 & n=0, m=0 \\ \frac{g_0}{4ab}\left(\frac{8a}{k_{0y}m}\sin\left(\frac{k_{0y}mb}{2}\right)\cos\left(\frac{k_{0y}mb}{2}\right)\right) & n=0, m\neq 0 \\ \frac{g_0}{4ab}\left(\frac{8b}{k_{0x}n}\sin\left(\frac{k_{0x}na}{2}\right)\cos\left(\frac{k_{0x}na}{2}\right)\right) & n\neq 0, m=0 \\ \frac{g_0}{4ab}\left(\frac{16}{k_{0x}k_{0y}nm}\sin\left(\frac{k_{0x}na}{2}\right)\cos\left(\frac{k_{0x}na}{2}\right)\sin\left(\frac{k_{0y}mb}{2}\right)\cos\left(\frac{k_{0y}mb}{2}\right)\right) & n\neq 0, m\neq 0 \end{cases} \tag{11}$$

3.5 The hollow rectangular waveguide with the dielectric material between the hollow rectangle and the metallic

The dielectric profile of the hollow rectangular waveguide with the dielectric material between the hollow rectangle and the metal (**Figure 1(e)**) is calculated by subtracting the dielectric profile of **Figure 1(b)** from the dielectric profile of **Figure 1(d)**.

The matrix **G** is given by the form

$$G = \begin{bmatrix} g_{00} & g_{-10} & g_{-20} & \cdots & g_{-nm} & \cdots & g_{-NM} \\ g_{10} & g_{00} & g_{-10} & \cdots & g_{-(n-1)m} & \cdots & g_{-(N-1)M} \\ g_{20} & g_{10} & \ddots & \ddots & \ddots & & \\ \vdots & g_{20} & \ddots & \ddots & \ddots & & \\ g_{nm} & \ddots & \ddots & \ddots & g_{00} & \vdots & \\ \vdots & & & & & & \\ g_{NM} & \cdots & \cdots & \cdots & \cdots & \cdots & g_{00} \end{bmatrix}. \tag{12}$$

Similarly, the G_x and G_y matrices are obtained by the derivatives of the dielectric profile. These matrices relate to the method that is based on the Laplace and Fourier transforms and the inverse Laplace and Fourier transforms [16]. Laplace transform is necessary to obtain the comfortable and simple input-output connections of the fields. The output transverse fields are computed by the inverse Laplace and Fourier transforms.

This method becomes an improved method by using the proposed technique and the particular application also in the cases of discontinuous problems of the hollow rectangular waveguide with dielectric material between the hollow rectangle and the metal (**Figure 1(e)**), in the cross section of the straight rectangular waveguide. In addition, we can find the thickness of the dielectric layer that is recommended to obtain the desired behavior of the output fields.

Several examples will demonstrate in the next section in order to understand the influence of the hollow rectangular waveguide with dielectric material in the cross section (**Figure 1**) on the output field. All the graphical results will be demonstrated as a response to a half-sine (TE_{10}) input-wave profile and the hollow rectangular waveguide with dielectric material in the cross section of the straight rectangular waveguide.

4. The derivation for circular straight waveguide

The wave Eqs. (1) and (2) are given in the case of the circular straight waveguide, where

$$\varepsilon(r) = \varepsilon_0[1 + \chi_0 g(r)] \text{ and } g_r(r) = [1/\varepsilon(r)][\partial\varepsilon(r)/\partial r].$$

The proposed technique to calculate the refractive index for discontinuous problems (**Figure 1(f)** and **(g)**) is given in this section for the one dielectric coating (**Figure 1(f)**) and for three dielectric coatings (**Figure 1(g)**).

4.1 The refractive index for the circular hollow waveguide with one dielectric coating in the cross section

The cross section of the hollow waveguide (**Figure 1(f)**) is made of a tube of various types of one dielectric layer and a metallic layer. The refractive indices of the air, dielectric, and metallic layers are $n_{(0)} = 1, n_{(AgI)} = 2$, and $n_{(Ag)} = 10 - j60$, respectively. The value of the refractive index of the material at a wavelength of λ = 10.6 μm is taken from the table performed by Miyagi et al. [19]. The refractive indices of the air, dielectric layer (AgI), and metallic layer (Ag) are shown in **Figure 1(f)**.

The refractive index (n(r)) is dependent on the transition's regions in the cross section between the two different materials (air-AgI, AgI-Ag).

The refractive index is calculated as follows:

$$n(r) = \begin{cases} n_0 & 0 \le r < b - \varepsilon_1/2 \\ n_0 + (n_d - n_0)\exp\left[1 - \dfrac{\varepsilon_1{}^2}{\varepsilon_1{}^2 - [r - (b + \varepsilon_1/2)]^2}\right] & b - \varepsilon_1/2 \le r < b + \varepsilon_1/2 \\ n_d & b + \varepsilon_1/2 \le r < a - \varepsilon_2/2, \\ n_d + (n_m - n_d)\exp\left[1 - \dfrac{\varepsilon_2{}^2}{\varepsilon_2{}^2 - [r - (a + \varepsilon_2/2)]^2}\right] & a - \varepsilon_2/2 \le r < a + \varepsilon_2/2 \\ n_m & else \end{cases}$$

where the internal and external diameters are denoted as 2b, 2a, and 2(a+δ_m), respectively, where δ_m is the metallic layer. The thickness of the dielectric coating (d) is defined as [a – b], and the thickness of the metallic layer (δ_m) is defined as [(a+δ_m) – a]. The parameter ε is very small [ε = [a – b]/50]. The refractive indices of the air, dielectric, and metallic layers are denoted as n_0, n_d, and n_m, respectively.

4.2 The refractive index for the circular hollow waveguide with three dielectric coatings in the cross section

The cross section of the hollow waveguide (**Figure 1(g)**) is made of a tube of various types of three dielectric layers and a metallic layer. The internal and external diameters are denoted as 2b, 2 b_1, 2 b_2, 2a, and 2(a + δ_m), respectively, where δ_m is the thickness of the metallic layer. In addition, we denote the thickness of the dielectric layers as d_1, d_2, and d_3, respectively, where $d_1 = b_1 - b$, $d_2 = b_2 - b_1$, and $d_3 = a - b_2$. The refractive index in the particular case with the three dielectric layers and the metallic layer in the cross section of the straight hollow waveguide (**Figure 1(g)**) is calculated as follows:

$$n(r) = \begin{cases} n_0 & 0 \le r < b - \varepsilon/2 \\ n_0 + (n_1 - n_0)\exp\left[1 - \dfrac{\varepsilon^2}{\varepsilon^2 - [r - (b + \varepsilon/2)]^2}\right] & b - \varepsilon/2 \le r < b + \varepsilon/2 \\ n_1 & b + \varepsilon/2 \le r < b_1 - \varepsilon/2 \\ n_1 + (n_2 - n_1)\exp\left[1 - \dfrac{\varepsilon^2}{\varepsilon^2 - [r - (b_1 + \varepsilon/2)]^2}\right] & b_1 - \varepsilon_2/2 \le r < b_1 + \varepsilon_2/2 \\ n_2 & b_1 + \varepsilon/2 \le r < b_2 - \varepsilon/2 \quad , \\ n_2 + (n_3 - n_2)\exp\left[1 - \dfrac{\varepsilon^2}{\varepsilon^2 - [r - (b_2 + \varepsilon/2)]^2}\right] & b_2 - \varepsilon/2 \le r < b_2 + \varepsilon/2 \\ n_3 & b_2 + \varepsilon/2 \le r < a - \varepsilon/2 \\ n_3 + (n_m - n_3)\exp\left[1 - \dfrac{\varepsilon^2}{\varepsilon^2 - [r - (a + \varepsilon/2)]^2}\right] & a - \varepsilon/2 \le r < a + \varepsilon/2 \\ n_m & else \end{cases}$$

where the parameter ε is very small $[\varepsilon = [a - b]/50]$. The refractive indices of the air, dielectric, and metallic layers are denoted as n_0, n_1, n_2, n_3, and n_m, respectively. In this study we suppose that $n_3 > n_2 > n_1$.

The proposed technique to calculate the refractive indices of the dielectric profile of one dielectric coating (**Figure 1(f)**) or three dielectric coatings (**Figure 1(g)**), and the metallic layer in the cross section relate to the method that is based on Maxwell's equations, the Fourier-Bessel series, Laplace transform, and the inverse Laplace transform by the residue method [17]. This method becomes an improved method by using the proposed technique also in the cases of discontinuous problems of the hollow circular waveguide with one dielectric coating (**Figure 1(f)**), three dielectric coatings (**Figure 1(g)**), or more dielectric coatings.

5. Numerical results

Several examples for the rectangular and circular waveguides with the discontinuous dielectric profile in the cross section of the straight waveguide are demonstrated in this section according to **Figure 1(a)–(g)**.

Figure 4(a)–(c) demonstrates the output field as a response to a half-sine (TE_{10}) input-wave profile in the case of the slab profile (**Figure 1(a)**), where a = b = 20 mm, c = 20 mm, and d = 2 mm, for ε_r = 3, 4, and 5, respectively. **Figure 4 (c)** shows the output field for ε_r = 3, 4, and 5, respectively, where y = b/2 = 10 mm.

By increasing only the value of the dielectric profile from ε_r = 3 to ε_r = 5, the width of the output field decreased, and also the output amplitude decreased.

Figure 5(a)–(e) demonstrates the output field as a response to a half-sine (TE_{10}) input-wave profile in the case of the rectangular dielectric profile in the rectangular waveguide (**Figure 1(b)**), where a = b = 20 mm and c = d = 2 mm, for ε_r = 3, 5, 7, and 10, respectively. **Figure 5(e)** shows the output field for ε_r = 3, 5, 7, and 10, respectively, where y = b/2 = 10 mm.

By increasing only the dielectric profile from ε_r = 3 t o ε_r = 5, the width of the output field increased, and also the output amplitude increased.

The output fields are strongly affected by the input-wave profile (TE_{10} mode), the location, and the dielectric profile, as shown in **Figure 4(a)–(c)** and **Figure 5(a)–(e)**.

Figure 6(a)–(e) shows the output field as a response to a half-sine (TE_{10}) input-wave profile in the case of the circular dielectric profile (**Figure 1(c)**), for ε_r = 3, 5, 7 and 10, respectively, where a = b = 20 mm, and the radius of the circular dielectric profile is equal to 1 mm. **Figure 6(e)** shows the output field for ε_r =3, 5, 7, and 10, respectively, where y = b/2 = 10 mm. The other parameters are z = 0.15 m, k_0 = 167 $1/m$, λ = 3.75 cm, and β = 58 $1/m$.

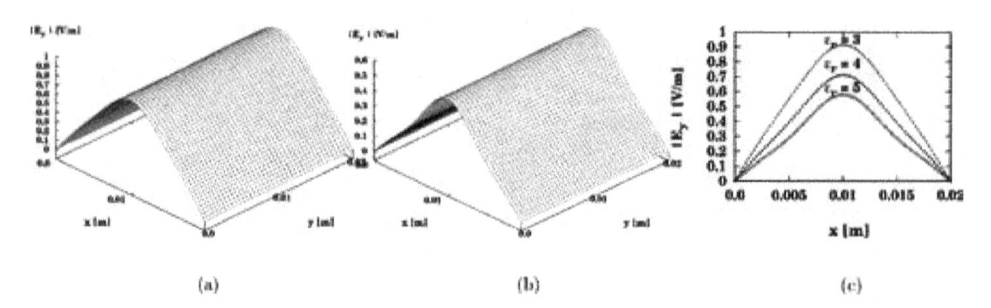

Figure 4.
*The output field as a response to a half-sine (TE_{10}) input-wave profile in the case of the slab dielectric profile (**Figure 1(a)**), where a = b = 20 mm, c = 20 mm, and d = 2 mm where (a) ε_r = 3 and (b) ε_r = 5. (c). The output field for ε_r =3, 4, and 5, respectively, where y = b/2 = 10 mm.*

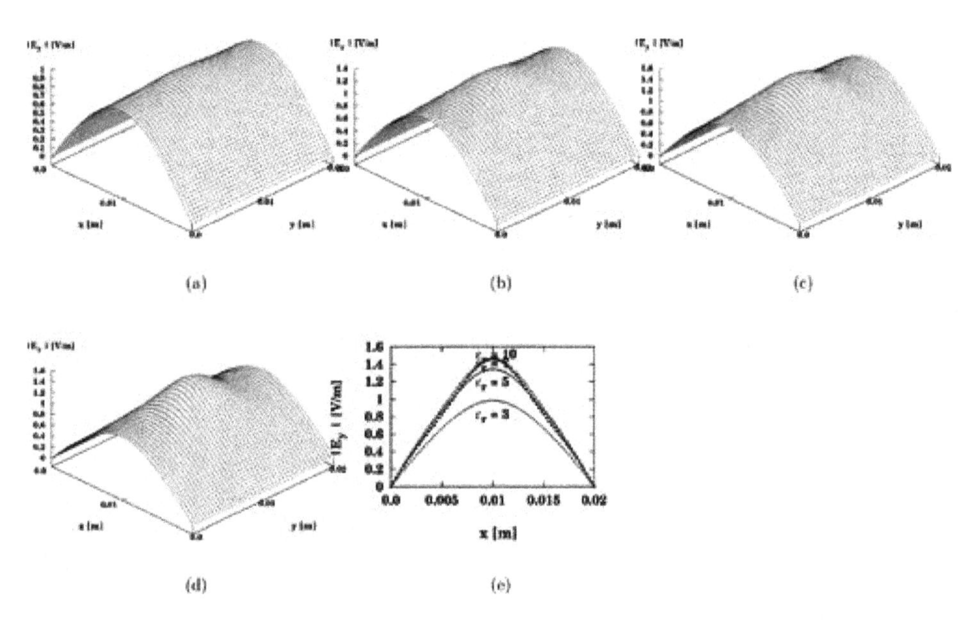

Figure 5.
*The output field as a response to a half-sine (TE_{10}) input-wave profile in the case of the rectangular dielectric material (**Figure 1(b)**), where a = b = 20 mm and c = d = 2 mm: (a) $\varepsilon_r = 3$, (b) $\varepsilon_r = 5$, (c) $\varepsilon_r = 7$, and (d) $\varepsilon_r = 10$. The other parameters are a = b = 20 mm, z = 0.15 m, $k_0 = 167$ 1/m, $\lambda = 3.75$ cm, and $\beta = 58$ 1/m.(e) The output field for $\varepsilon_r = 3, 5, 7$, and 10, respectively, where y = b/2 = 10 mm.*

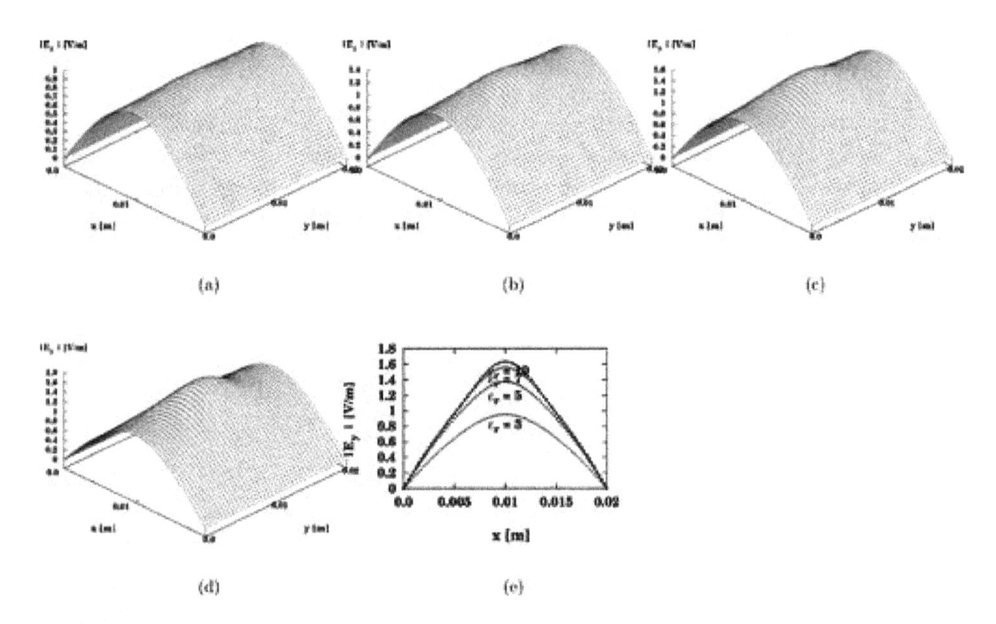

Figure 6.
*The output field as a response to a half-sine (TE_{10}) input-wave profile in the case of the circular dielectric profile (**Figure 1(c)**), where a = b = 20 mm, and the radius of the circular dielectric profile is equal to 1 mm: (a) $\varepsilon_r = 3$, (b) $\varepsilon_r = 5$, (c) $\varepsilon_r = 7$, and (d) $\varepsilon_r = 10$. The other parameters are z = 0.15 m, $k_0 = 167$ 1/m, $\lambda = 3.75$ cm, and $\beta = 58$ 1/m. (e) The output field for $\varepsilon_r = 3, 5, 7$, and 10, respectively, where y = b/2 = 10 mm.*

The proposed technique in Section 3.3 is also effective to solve discontinuous problems of periodic circular profiles in the cross section of the straight rectangular waveguides, and some examples were demonstrated in Ref. [20].

The behavior of the output fields (**Figures 5(a)–(e)** and **6(a)–(e)**) is similar when the dimensions of the rectangular dielectric profile (**Figure 1(b)**) and the circular profile (**Figure 1(c)**) are very close . The output field (**Figure 5(a)–(e)**) is shown for c = d = 2 mm as regards to the dimensions a = b = 20 mm. The output field (**Figure 6(a)–(e)**) is shown where the radius of circular profile is equal to 1 mm (viz., the diameter 2 mm), as regards to the dimensions a = b = 20 mm.

Figure 7(a)–(e) shows the output field as a response to a half-sine (TE_{10}) input-wave profile in the case of the circular dielectric profile (**Figure 1(c)**), where a = b = 20 mm, and the radius of the circular dielectric profile is equal to 2 mm for ε_r = 3, 5, 7, and 10, respectively. The other parameters are z = 0.15 m, $k_0 = 167\ 1/m$, λ = 3.75 cm, and $\beta = 58\ 1/m$. **Figure 7(e)** shows the output field for ε_r =3, 5, 7, and 10, respectively, where y = b/2 = 10 mm.

By changing only the value of the radius of the circular dielectric profile (**Figure 1(c)**) from 1 mm to 2 mm, as regards to the dimensions of the cross section of the waveguide (a = b = 20 mm), the output field of the Gaussian shape increased, and the half-sine (TE_{10}) input-wave profile decreased.

The dielectric profile of the hollow rectangular waveguide with the dielectric material between the hollow rectangle and the metal (**Figure 1(e)**) is calculated by subtracting the dielectric profile of the waveguide with the dielectric material in the core (**Figure 1(b)**) from the dielectric profile according to the waveguide entirely with the dielectric profile (**Figure 1(d)**).

Figure 8(a)–(c) shows the output field as a response to a half-sine (TE_{10}) input-wave profile in the case of the hollow rectangular waveguide with one dielectric material between the hollow rectangle and the metal (**Figure 1(e)**), where a = b = 20 mm, c = d = 14 mm, and d = 14 mm, namely, e = 3 mm and f = 3 mm. **Figure 8(a)–(b)** shows the output field for ε_r = 2.5 and ε_r = 4, respectively. **Figure 8(c)** shows the output field for ε_r = 2.5, 3, 3.5, and 4, respectively, where y = b/2 = 10 mm. The other parameters are z = 0.15 m, $k_0 = 167\ 1/m$, λ = 3.75 cm, and $\beta = 58\ 1/m$.

Figure 9(a)–(c) shows the output power density in the case of the hollow circular waveguide with one dielectric coating (**Figure 1(f)**), where a = 0.5 mm. **Figure 9(a)–(b)** shows the output power density for w_0 = 0.15 mm and w_0 = 0.25 mm, respectively. The output power density of the central peak is shown

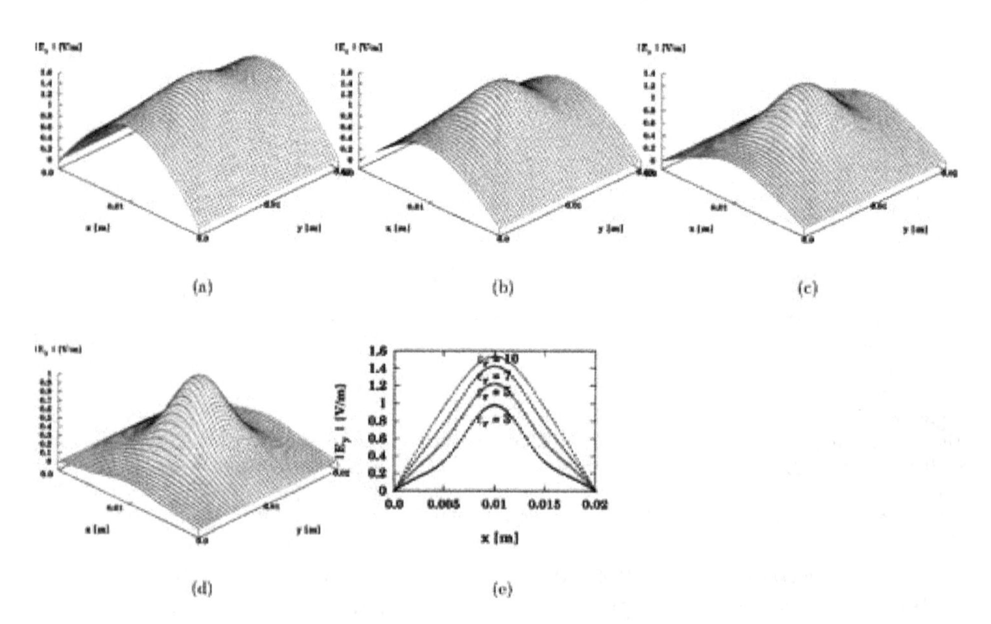

Figure 7.
*The output field as a response to a half-sine (TE_{10}) input-wave profile in the case of the circular dielectric profile (**Figure 1(c)**), where a = b = 20 mm, and the radius of the circular dielectric profile is equal to 2 mm: (a) ε_r = 3, (b) ε_r = 5, (c) ε_r = 7, and (d) ε_r = 10. The other parameters are z = 0.15 m, $k_0 = 167\ 1/m$, λ = 3.75 cm, and $\beta = 58\ 1/m$. (e) The output field for ε_r = 3, 5, 7, and 10, respectively, where y = b/2 = 10 mm.*

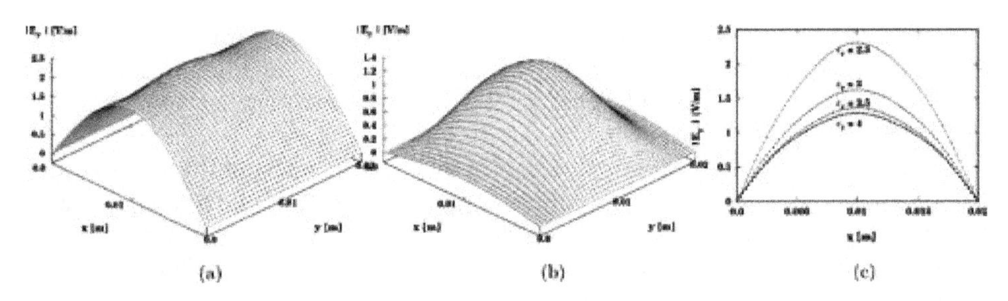

Figure 8.
*The output field as a response to a half-sine (TE_{10}) input-wave profile in the case of the hollow rectangular waveguide with one dielectric material between the hollow rectangle and the metal (**Figure 1(e)**), where $a = b = 20$ mm, $c = 14$ mm, and $d = 14$ mm, namely, $e = 3$ mm and $f = 3$ mm: (a) $\varepsilon_r = 2.5$, and (b) $\varepsilon_r = 4$. (c). The output field for $\varepsilon_r = 2.5$, 3, 3.5, and 4, respectively, where $y = b/2 = 10$ mm. The other parameters are $z = 0.15$ m, $k_0 = 167$ 1/m, $\lambda = 3.75$ cm, and $\beta = 58$ 1/m.*

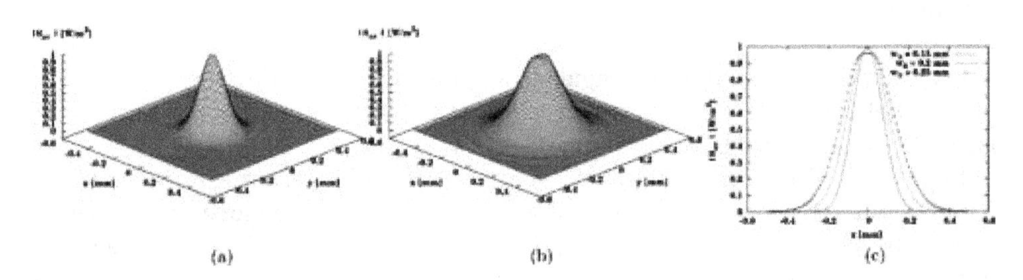

Figure 9.
*The output power density in the case of the hollow circular waveguide with one dielectric coating (**Figure 1(e)**), where $a = 0.5$ mm: (a) $w_0 = 0.15$ mm, and (b) $w_0 = 0.25$ mm. (c). The output power density of the central peak for $w_0 = 0.15$ mm, $w_0 = 0.2$ mm, and $w_0 = 0.25$ mm, respectively, where $y = b/2$. The other parameters are $z = 1$ m, $n_d = 2.2$, and $n_{(Ag)} = 13.5 - j\,75.3$.*

for $w_0 = 0.15$ mm, $w_0 = 0.2$ mm, and $w_0 = 0.25$ mm, respectively, where $y = b/2$. The other parameters are $z = 1$ m, $n_d = 2.2$, and $n_{(Ag)} = 13.5 - j\,75.3$.

Figure 1(f) and **(g)** shows two examples of discontinuous problems for circular waveguides. The practical results are demonstrated for **Figure 1(f)**.

Figure 10(a)–(c) shows also the output power density in the case of the hollow circular waveguide with one dielectric coating (**Figure 1(f)**), where a = 0.5 mm, but for other values of the spot size. **Figure 10(a)–(b)** shows the output power density for $w_0 = 0.26$ mm and $w_0 = 0.3$ mm, respectively. The output power density of the central peak is shown for $w_0 = 0.26$ mm, $w_0 = 0.28$ mm, and $w_0 = 0.3$ mm, respectively, where $y = b/2$. The other parameters are $z = 1$ m, $n_d = 2.2$, and $n_{(Ag)} = 13.5 - j\,75.3$.

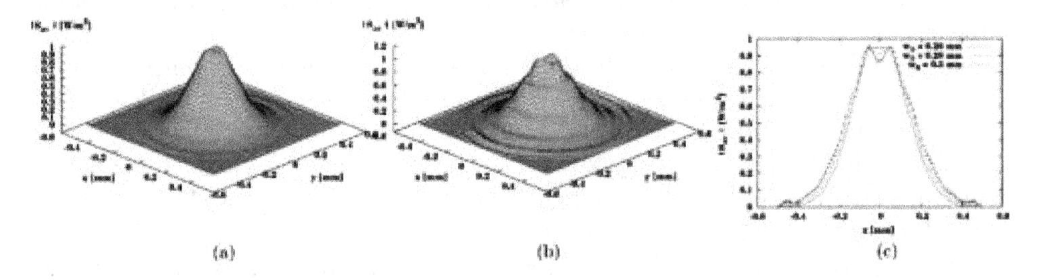

Figure 10.
*The output power density in the case of the hollow circular waveguide with one dielectric coating (**Figure 1(e)**), where $a = 0.5$ mm: (a) $w_0 = 0.26$ mm, and (b) $w_0 = 0.3$ mm. (c). The output power density of the central peak for $w_0 = 0.26$ mm, $w_0 = 0.28$ mm, and $w_0 = 0.3$ mm, respectively, where $y = b/2$. The other parameters are $z = 1$ m, $n_d = 2.2$, and $n_{(Ag)} = 13.5 - j\,75.3$.*

By changing only the values of the spot size from w_0 = 0.15 mm, w_0 = 0.2 mm, and w_0 = 0.25 mm to w_0 = 0.26 mm, w_0 = 0.28 mm, and w_0 = 0.3 mm, respectively, the results of the output power density for a = 0.5 mm are changed as shown in **Figure 10(a)–10(c)**.

The output modal profile is greatly affected by the parameters of the spot size and the dimensions of the cross section of the waveguide. **Figure 10(a)–(c)** demonstrates that in addition to the main propagation mode, several other secondary modes and symmetric output shape appear in the results of the output power density for the values of w_0 = 0.26 mm, w_0 = 0.28 mm, and w_0 = 0.3 mm, respectively.

The proposed technique in Section 4.2 is also effective to solve discontinuous problems of the straight hollow circular waveguide with three dielectric layers (**Figure 1(g)**), and some examples were demonstrated in Ref. [21].

6. Conclusions

Several examples for the rectangular and circular waveguides with the discontinuous dielectric profile in the cross section of the straight waveguide were demonstrated in this research, according to **Figure 1(a)–(g)**.

Figure 4(a)–(c) demonstrates the output field as a response to a half-sine (TE_{10}) input-wave profile in the case of the slab profile (**Figure 1(a)**), where a = b = 2 0 *mm*, c = 2 0 *mm*, and d = 2 *mm* for ε_r = 3 and 5, respectively. By increasing only the value of the dielectric profile from ε_r = 3 to ε_r = 5, the width of the output field decreased, and also the output amplitude decreased.

Figure 5(a)–(e) demonstrates the output field as a response to a half-sine (TE_{10}) input-wave profile in the case of the rectangular dielectric profile in the rectangular waveguide (**Figure 1(b)**), where a = b = 20 mm and c = d = 2 *mm*, for ε_r = 3, 5, 7, and 10, respectively. By increasing only the dielectric profile from ε_r = 3 t o ε_r = 5 , t h e width of the output field increased, and also the output amplitude increased. The output fields are strongly affected by the input-wave profile (TE_{10} mode), the location, and the dielectric profile, as shown in **Figure 4(a)–(c)** and **Figure 5(a)–(e)**.

The behavior of the output fields (**Figures 5(a)–(e)** and **6(a)–(e)**) is similar when the dimensions of the rectangular dielectric profile (**Figure 1(b)**) and the circular profile (**Figure 1(c)**) are very close. The output field (**Figure 5(a)–(e)**) is shown for c = d = 2 *mm* as regards to the dimensions a = b = 20 mm. The output field (**Figure 6(a)–(e)**) is shown where the radius of circular profile is equal to 1 *mm* (viz., the diameter 2 *mm*), as regards to the dimensions a = b = 20 mm.

Figures 6(a)–(e) and **7(a)–(e)** show the output field as a response to a half-sine (TE_{10}) input-wave profile in the case of the circular dielectric profile (**Figure 1(c)**), for ε_r = 3, 5, 7, and 10, respectively, where a = b = 20 mm, and the radius of the circular dielectric profile is equal to 1 *mm*. By changing only the value of the radius of the circular dielectric profile (**Figure 1(c)**) from 1 *mm* to 2 *mm*, as regards to the dimensions of the cross section of the waveguide (a = b = 20 *mm*), the output field of the Gaussian shape increased, and the half-sine (TE_{10}) input-wave profile decreased.

Figure 8(a)–(c) shows the output field as a response to a half-sine (TE_{10}) input-wave profile in the case of the hollow rectangular waveguide with one dielectric material between the hollow rectangle and the metal (**Figure 1(e)**), where a = b = 2 0 *mm*, c = d = 1 4 *mm*, and d = 14 *mm*, namely, e = 3 *mm* and f = 3 *mm*.

Figures 9(a)–(c) and **10(a)–(c)** show the output power density in the case of the hollow circular waveguide with one dielectric coating (**Figure 1(f)**), where a = 0.5 mm. By changing only the values of the spot size from w_0 = 0.15 mm, w_0 = 0.2 mm, and w_0 = 0.25 to w_0 = 0.26 mm, w_0 = 0.28 mm, and w_0 = 0.3 mm, respectively, the results of the output power density for a = 0.5 mm are changed as shown in **Figure 10(a)–(c)**.

The output modal profile is greatly affected by the parameters of the spot size and the dimensions of the cross section of the waveguide. **Figure 10(a)–(c)** demonstrates that in addition to the main propagation mode, several other secondary modes and symmetric output shape appear in the results of the output power density for the values of w_0 = 0.26 mm, w_0 = 0.28 mm, and w_0 = 0.3 mm, respectively.

The two important parameters that we studied were the spot size and the dimensions of the cross section of the straight hollow waveguide. The output results are affected by the parameters of the spot size and the dimensions of the cross section of the waveguide.

Author details

Zion Menachem
Department of Electrical Engineering, Sami Shamoon College of Engineering, Beer Sheva, Israel

*Address all correspondence to: zionm@post.tau.ac.il

References

[1] Chiang KS. Review of numerical and approximate methods for the modal analysis of general optical dielectric waveguides. Optical and Quantum Electronics. 1993;**26**: S113-S134

[2] Mineo M, Carlo AD, Paoloni C. Analytical design method for corrugated rectangular waveguide SWS THZ vacuum tubes. Journal of Electromagnetic Waves and Applications. 2010;**24**:2479-2494

[3] Yeap KH, Tham CY, Yassin G, Yeong KC. Attenuation in rectangular waveguides with finite conductivity walls. Radioengineering. 2011;**20**: 472-478

[4] Liu S, Li LW, Leong MS, Yeo T Sr. Rectangular conducting waveguide filled with uniaxial anisotropic media: A modal analysis and dyadic Green's function. Progress in Electromagnetics Research. 2000;**25**:111-129

[5] Lu W, Lu YY. Waveguide mode solver based on Neumann-to-Dirichlet operators and boundary integral equations. Journal of Computational Physics. 2012;**231**:1360-1371

[6] Chen TT. Wave propagation in an inhomogeneous transversely magnetized rectangular waveguide. Applied Scientific Research. 1960;**8**: 141-148

[7] Vaish A, Parthasarathy H. Analysis of a rectangular waveguide using finite element method. Progress in Electromagnetics Research C. 2008;**2**: 117-125

[8] Khalaj-Amirhosseini M. Microwave filters using waveguides filled by multi-layer dielectric. Progress in Electromagnetics Research, PIER. 2006; **66**:105-110

[9] Khalaj-Amirhosseini M. Analysis of longitudinally inhomogeneous waveguides using the method of moments. Progress in Electromagnetics Research, PIER. 2007;**74**:57-67

[10] Khalaj-Amirhosseini M. Analysis of longitudinally inhomogeneous waveguides using Taylor's series expansion. Journal of Electromagnetic Waves and Applications. 2006;**16**: 1093-1100

[11] Khalaj-Amirhosseini M. Analysis of longitudinally inhomogeneous waveguides using the Fourier series expansion. Journal of Electromagnetic Waves and Applications. 2006;**20**: 1299-1310

[12] Harrington JA, Matsuura Y. Review of hollow waveguide technology. SPIE. 1995;**2396**:4-14

[13] Harrington JA, Harris DM, Katzir A, editors. Biomedical Optoelectronic Instrumentation, 4–14; 1995

[14] Harrington JA. A review of IR transmitting, hollow waveguides. Fiber and Integrated Optics. 2000;**19**:211-228

[15] Marhic ME. Mode-coupling analysis of bending losses in IR metallic waveguides. Applied Optics. 1981;**20**: 3436-3441

[16] Menachem Z, Jerby E. Transfer matrix function (TMF) for propagation in dielectric waveguides with arbitrary transverse profiles. IEEE Transactions on Microwave Theory and Techniques. 1998;**46**:975-982

[17] Menachem Z, Tapuchi S. Influence of the spot-size and cross-section on the output fields and power density along the straight hollow waveguide. Progress in Electromagnetics Research. 2013;**48**: 151-173

[18] Vladimirov V. Equations of Mathematical Physics. New York: Marcel Dekker, Inc; 1971

[19] Miyagi M, Harada K, Kawakami S. Wave propagation and attenuation in the general class of circular hollow waveguides with uniform curvature. IEEE Transactions on Microwave Theory and Techniques. 1984;**MTT-32**: 513-521

[20] Menachem Z, Tapuchi S. Circular and periodic circular profiles in a rectangular cross section along the straight waveguide. Applied Physics Research. 2015;7:121-136

[21] Menachem Z, Tapuchi S. Straight hollow waveguide with three dielectric layers in the cross section. Journal of Electromagnetic Waves and Applications. 2016;**30**:2125-2137

2

Ferrite Materials and Applications

Tsun-Hsu Chang

Abstract

This chapter starts from a generalized permeability and aims at providing a better understanding of the ferrites behavior in the microwave fields. The formula of the generalized permeability explains why the permeability of the ferrimagnetic or even the ferromagnetic materials strongly depends on the applied magnetic bias and the polarization of the wave. Right-hand circularly polarized (RHCP) wave may synchronize with the precession of the magnetic moment, resulting in a strong resonant effect. Characterizing the ferrites' properties, including the complex permittivity, the saturation magnetization, and the resonance linewidth, will be discussed. We then utilize these properties to design and fabricate various microwave devices, such as phase shifters, circulators, and isolators. Detailed analysis and simulation to demonstrate how these ferrite devices work will be shown. The mechanism will be discussed.

Keywords: ferrite, microwave, generalized permeability, circular polarization, phase shifter, circulator, isolator

1. Introduction

Ferrimagnetism is similar to ferromagnetism in many ways [1–4]. They all have hysteresis curves as the applied magnetic field changes, resulting in the saturation magnetization ($4\pi M_s$), the coercive field (H_c), and the remnant polarization (B_r). A ferri- or ferromagnetic material can be used to fabricate a hard or a soft magnet depending on its coercivity (H_c). Hard magnets are characterized by a high H_c, indicating that their magnetization is difficult to change and will retain their magnetization in the absence of an applied field as shown in **Figure 1(a)**. On the contrary, the soft magnets have low H_c values and, normally, have very weak, remnant magnetic field (low B_r) as in **Figure 1(b)**. Hard magnetism has been extensively used to make the permanent magnets and provides a strong DC magnetic field, while the soft magnetism can be used for the AC systems.

Soft magnetic materials include electrical steels and soft ferrites [3, 4]. Unlike the ferromagnetic metals which are conductors, soft ferrites have low electric conductivity, i.e., they are dielectric materials. The electrical steels have extensive applications in low-frequency systems, such as generators, motors, and transformers, while the soft ferrites are suitable for the high-frequency applications, such as circulators, isolators, phase shifters, and high-speed switches.

This chapter will focus on the properties of the soft ferrites and their applications at high-frequency systems. The ferrites are crystals having small electric conductivity compared to ferromagnetic materials. Thus they are useful in high-frequency devices because of the absence of significant eddy current losses. Ferrites

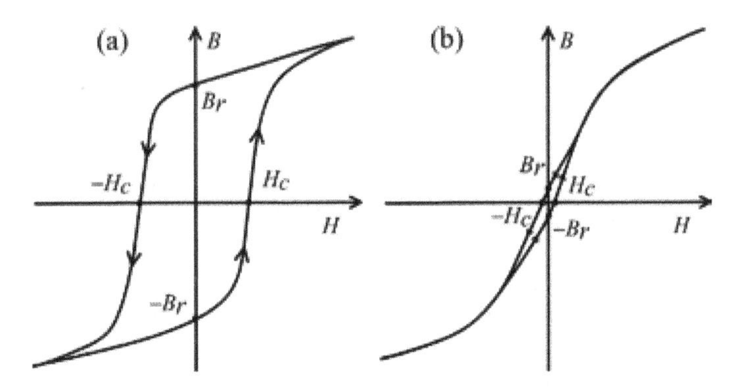

Figure 1.
The hysteresis (B-H) curves for (a) hard ferrimagnetism and (b) soft ferrimagnetism. The remnant polarization (B_r) and the coercive field (H_c) for the hard ferrites should be as large as possible. On the other hand, for the soft ferrites, the remnant polarization (B_r) and the coercive field (H_c) are very small or even close to zero.

are ceramic-like materials with specific resistivities that may be as much as 10^{14} greater than that of metals and with dielectric constants around 10 to 16 or greater. Ferrites are made by sintering a mixture of metal oxides and have the general chemical composition $MO{\cdot}Fe_2O_3$, where M is a divalent metal such as Mn, Mg, Fe, Zn, Ni, Cd, etc. Relative permeabilities of several thousands are common [5, 6]. The magnetic properties of ferrites arise mainly from the magnetic dipole moment associated with the electron spin [2].

The magnetic dipole moment precesses around the applied DC magnetic field by treating the spinning electron as a gyroscopic top, which is a classical picture of the magnetization process. This picture also explains the anisotropic magnetic properties of ferrites, where the permeability of the ferrite is not a single scalar quantity, but instead is a generally a second-rank tensor or can be represented as a matrix. The left and right circularly polarized waves have different propagation constant along the direction of the external magnetic field, resulting in the nonreciprocity of a propagating wave. Since the permeability should be treated as a tensor (matrix), not a scalar permeability, it is generally much difficult to understand and to have intuition, even for the researchers.

2. The susceptibility matrix

2.1 Magnetic dipole precesses around the applied magnetic field

The properties of ferrites are very intriguing. Without a DC bias magnetic field $\mathbf{H_0}$, the magnetic dipoles of ferrites are randomly orientated. They exhibit dielectric properties only. The dielectric loss can be high or low depending on the loss tangent of the soft ferrites. Interestingly, when the DC bias magnetic field is strong enough, and the ferrite is magnetically saturated, the magnetic dipole will precess around the DC bias field at a frequency called the Larmor frequency ($\omega_0 \propto H_0$). Waves with different circular polarizations will corotate or counter-rotate with the precession of the magnetic dipoles.

Figure 2 shows the Larmor precession with the circularly polarized fields [7]. Circular polarization may be referred to as right handed or left handed, depending on the direction in which the electric (magnetic) field vector rotates. For a

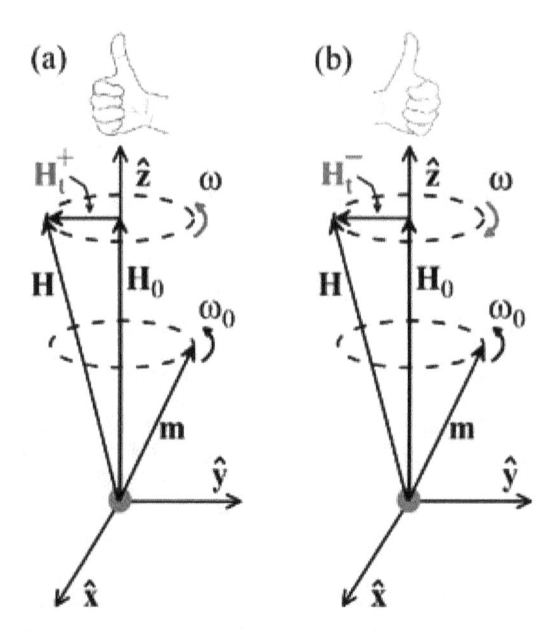

Figure 2.
Larmor precession of a magnetic moment **m** *around the applied DC bias field* $\mathbf{H}_o$ $(= H_0\hat{z})$ *with (a) a right-hand circularly polarized (RHCP) wave and (b) a left-hand circularly polarized (LHCP) wave. The frequency of the Larmor precession in both cases are the same, i.e., the Larmor frequency* ω_0 $(= \mu_0\gamma H_0)$. H_t^+ *and* H_t^- *are the transverse components of the incident waves which rotate clockwise (RHCP) and counter-clockwise (LHCP) from the source viewpoint looking in the direction of propagation. The thumb points the direction of the wave propagation, and the fingers give the rotation of the transverse components [7].*

right-hand circularly polarized (RHCP) wave, the fields rotate clockwise at a given position from the source looking in the direction of propagation. The magnetic dipole moment **m** processes around the $\mathbf{H}_0$ field vector, like a top spinning precess around the z-axis at the Larmor frequency ω_0. The spinning property depends on the applied DC bias magnetic field. **Figures 2(a)** and **(b)** shows the RHCP and LHCP waves with the gyrating frequency of ω. When the RHCP wave is propagat-ing along the direction of the DC bias field, it corotates with the precession of the magnetic dipole moments. On the other hand, the left-hand circularly polarized (LHCP) wave will counter-rotate with the precession of the dipole moments.

A linearly polarized incident wave can be decomposed into RHCP and LHCP waves of equal amplitude. The orientation of the linearly polarized wave changes after the wave propagates a certain distance because of the distinct propagation constants. The phenomenon is the famous Faraday's rotation [5, 6]. This unique property has various applications, such as phase shifters, isolators, and circulators. However, it is difficult to follow for students and even researchers in that the permeability is a tensor, not just a simple proportional constant.

Here we consider the simplest case for the pedagogic purpose—a circularly polarized plane wave is normally incident upon a semi-infinite medium. The wave characteristics such as the propagation constant k and the wave impedance Z are associated with the permeability μ, which is a tensor for the ferrite medium [5]. By finding the preferred eigenvalues, it will be shown that the properties of μ depend on the DC bias field H_0, the saturated magnetization M_s, and the operating frequency ω. By adjusting the frequency of the incident wave, i.e., ω, the permeability μ changes, especially close to the Larmor frequency (ω_0). Such an effect is called the ferrite magnetic resonance (FMR) or gyromagnetic resonance [7].

2.2 Derivation of the susceptibility matrix

The permeability $\mu_r(\omega)$ is a tensor. Note that the permittivity $\varepsilon_r(\omega)$ can be expressed in a tensor as well, but in the region of interest around 10 GHz, it can be treated as a complex proportional constant for many ceramics. Many microwave textbooks and literature have elaborated the derivation of the permeability tensor [5–7]. In this paper, a more accessible interpretation of the permeability tensor is provided. The magnetic properties of a material are due to the existence of magnetic dipole moment $\mathbf{m}$, which arise primarily from its (spin) angular momentum $\mathbf{s}$. The magnetic dipole moment and angular momentum have a simple relation, $\mathbf{m} = -\gamma\mathbf{s}$, where γ is the gyromagnetic ratio ($\gamma = e/m = 1\,759 \times 10^{11}$ C/ kg). When a DC bias magnetic field $\mathbf{B_0}$ ($= \mu_0\mathbf{H_0}$) is present, the torque $\boldsymbol{\tau}$ exerted on the magnetic dipole moment is

$$\boldsymbol{\tau} = \mathbf{m} \times \mathbf{B_0} = \mu_0 \mathbf{m} \times \mathbf{H_0}. \tag{1}$$

Since the torque is equal to the time change rate of the angular momentum, we have

$$\boldsymbol{\tau} = \frac{d\mathbf{s}}{dt} = \frac{-1}{\gamma}\frac{d\mathbf{m}}{dt}. \tag{2}$$

By comparing Eqs. (1) and (2), we obtain

$$\frac{d\mathbf{m}}{dt} = -\mu_0\gamma\mathbf{m} \times \mathbf{H_0}. \tag{3}$$

A large number of the magnetic dipole moment $\mathbf{m}$ per unit volume give rise to an average macroscopic magnetic dipole moment density $\mathbf{M}$. The torque exerting on the magnetization per unit volume $\mathbf{M}$ due to the magnetic flux $\mathbf{H}$ has the same form as Eq. (3):

$$\frac{d\mathbf{M}}{dt} = -\mu_0\gamma\mathbf{M} \times \mathbf{H}. \tag{4}$$

$\mathbf{M}$ and $\mathbf{H}$ in Eq. (4) differ slightly from $\mathbf{m}$ and $\mathbf{H_0}$ in Eq. (3) in that $\mathbf{M}$ and $\mathbf{H}$ can further be divided into two parts: the DC term and the high-frequency AC term. The DC term is, in general, much larger than the AC term. The applied DC bias magnetic field H_0 is assumed to be in the z-direction. When H_0 is strong enough, the magnetization will be saturated, denoted as M_s which aligns with the direction of H_0. If the AC term just polarizes in the transverse direction (i.e., the xy plane), the external magnetic bias field and the magnetization can be written as,

$$\mathbf{H} = H_x\hat{\mathbf{x}} + H_y\hat{\mathbf{y}} + H_0\hat{\mathbf{z}}, \tag{5}$$

$$\mathbf{M} = M_x\hat{\mathbf{x}} + M_y\hat{\mathbf{y}} + M_s\hat{\mathbf{z}}. \tag{6}$$

Since the AC terms have an $\exp(-i\omega t)$ dependence, by substituting Eq. (5) and (6) into Eq. (4) the transverse component terms read

$$(\omega_0^2 - \omega^2)M_x = \omega_0\omega_m H_x + j\omega\omega_m H_y; \tag{7}$$

$$(\omega_0^2 - \omega^2)M_y = -j\omega\omega_m H_x + \omega_0\omega_m H_y, \tag{8}$$

where $\omega_0 = \mu_0\gamma H_0$ and $\omega_m = \mu_0\gamma M_s$. In matrix representation, Eq. (7) and (8) can be rewritten as

$$\mathbf{M} = [\chi]\mathbf{H} = \begin{bmatrix} \chi_{xx} & \chi_{xy} & 0 \\ \chi_{yx} & \chi_{yy} & 0 \\ 0 & 0 & 0 \end{bmatrix} \mathbf{H}, \tag{9}$$

where $\chi_{xx} = \chi_{yy} = \omega_0\omega_m/(\omega_0^2 - \omega^2)$ and $\chi_{xy} = -\chi_{yx} = j\omega\omega_m/(\omega_0^2 - \omega^2)$. The eigenvalues of the susceptibility matrix are

$$\chi^+ = \frac{\omega_m}{\omega_0 - \omega}, \tag{10}$$

$$\chi^- = \frac{\omega_m}{\omega_0 + \omega}. \tag{11}$$

The eigenvectors corresponding to these two eigenvalues are the right-hand circularly polarized wave (RHCP, denoted as +) and the left-hand circularly polarized wave (LHCP, denoted as -), respectively. The symbols, + and -, represent positive helicity and negative helicity, respectively. The LHCP wave has a relatively mild response over the entire frequency range. On the contrary, the RHCP wave has a much more dramatic response.

The permeabilities of the RHCP and LHCP waves are

$$\mu^+ = \mu_0\left(1 + \frac{\omega_m}{\omega_0 - \omega}\right), \tag{12}$$

$$\mu^- = \mu_0\left(1 + \frac{\omega_m}{\omega_0 + \omega}\right). \tag{13}$$

Eq. (12) has a singularity when the wave frequency ω is equal to the Larmor frequency ω_0. This phenomenon is called the ferri-magnetic resonance (FMR, or called gyromagnetic resonance) [6]. On the contrary, Eq. (13) is relatively mild. Therefore, this study will mainly focus on the RHCP wave.

For a resonant cavity with a quality factor (Q), the loss effect can be introduced by using the complex resonant frequency $\omega_0(1 - i/2Q)$. By analogy with the resonant cavity the loss part can be calculated by using the complex frequency $\omega_0 \rightarrow \omega_0(1 - i\Delta H\mu_0\gamma/2\omega_0)$, where ΔH is the ferrimagnetic resonance linewidth [2, 3]. Since

$$\frac{\Delta H\mu_0\gamma}{2\omega_0} = \frac{\Delta H\mu_0\gamma}{2H_0\mu_0\gamma} = \frac{\Delta H}{2H_0}, \tag{14}$$

the permeability for the RHCP wave now reads

$$\mu_r^+ = \left(1 + \frac{\frac{\omega_m}{\omega_0}}{\left(1 - \frac{\omega}{\omega_0}\right) + i\frac{\Delta H}{2H_0}}\right). \tag{15}$$

To conduct a complete simulation of a ferrite device, we need to know its complex permittivity, the saturation magnetization, and the resonance linewidth. We will discuss how to characterize the ferrite's properties in the next section.

3. Characterization of ferrite materials

Here we will discuss the measurement of the most important properties of ferrites, including the dielectric properties ($\varepsilon_r + i\varepsilon_i$), the saturation magnetization

(M_s or $4\pi M_s$), and the resonance linewidth (ΔH). The behavior of the spin wave linewidth should be considered when the field strength of an electromagnetic wave exceeds a threshold value, that is, the high-power condition. For the general purpose, only the first three properties will be used in the ferrite simulation.

3.1 Dielectric properties

Ferrites are ceramic-like materials with relative dielectric constants around 10 to 16 or greater. The resistivities of ferrites may be as high as 10^{14} greater than that of metals. Since ferrites are dielectric materials. The dielectric properties ($\varepsilon_r + i\varepsilon_i$) always play an important role with or without the influence of the magnetic field. The perturbation method is the most commonly employed resonant technique [8, 9]. The perturbation method is very good for small-size and low-dielectric samples. When measuring the high-dielectric samples, however, the fields and the resonant frequency change drastically. The perturbation technique may lead to a reduced accuracy. Recently, the field enhancement method was proposed [10, 11]. The field enhancement method operates at a condition opposite to the perturbation method. The resonant frequency and quality factor alter significantly and depend on not only the geometry of the cavity but the sample's size and complex permittivity as well. Luckily, both the perturbation method and the field enhancement method agree well for samples with the dielectric constant below 50, which is suitable for most of the ferrites.

Figure 3 shows the ideal of the field enhancement method. **Figure 3(a)** shows the resonant frequency as functions of the dielectric constant (ε_r) using a simulation setup as in **Figure 3(b)**. The ingot-shaped sample has a diameter of 16.00 mm and thickness of 5.00 mm. The solid blue line depicts the simulation result for the field enhancement method. From the measured resonant frequency, we can derive the corresponding dielectric constant. On the other hand, the dashed black line is obtained using a sample with the same diameter but much thinner in thickness of 1.0 mm. The response of the 1-mm-thick sample quite resembles the perturbation method. The imaginary part of the permittivity (ε_i) or the loss tangent ($\tan\delta = \varepsilon_i/\varepsilon_r$) can then be determined from the measured resonant frequency and the quality factor. The field enhancement method has very wide measuring range from unity to high-κ dielectrics and from lossless to lossy materials [10, 11].

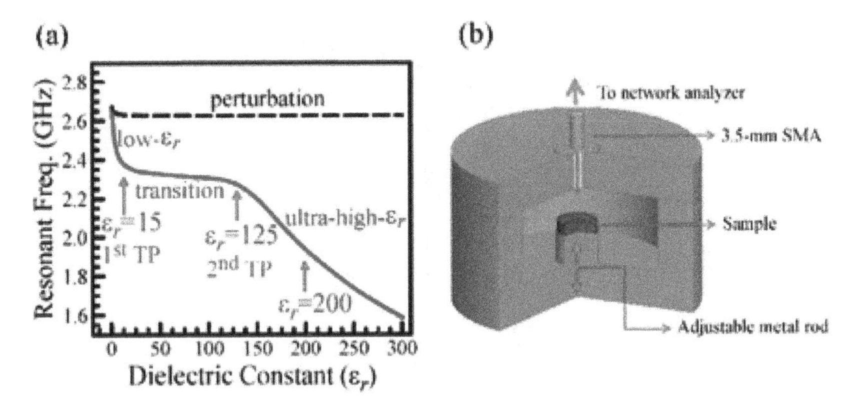

Figure 3.
(a) Resonant frequency versus dielectric constant based on full-wave simulations. The solid curve can be divided into three regions: low, transition, and ultrahigh. The dashed line is simulated with a much thinner sample of 1.00 mm in thickness, which exhibits the properties similar to those of perturbation. (b) Schematic diagram of the field enhancement method. It consists of a cylindrical resonant cavity and a metal rod. The sample is placed on the top of the metal rod. The metal rod focuses and enhances the electric field significantly. An SubMiniature version A (SMA) 3.5-mm adapter couples the wave from the top of the cavity [11].

3.2 Saturation magnetization

Ferrites have a strong response to the applied magnetic field. The magnetic properties of ferrites arise mainly from the magnetic dipole moment associated with the electron spin. Relative permeabilities of several thousands are common. The saturation magnetization (M_s or $4\pi M_s$) of a ferrite plays a key role as shown in Section 2. Researchers or engineers use the saturation magnetization as a design parameter that enters into the initial selection of a ferrimagnetic material for microwave device applications. Typical ferrimagnets exhibit values of $4\pi M_s$ between 300 gauss (G) and 5000 G. Static or low-frequency methods are generally used to measure $4\pi M_s$ [12]. From the measured hysteresis loop as shown in **Figure 4**, one can determine the saturation magnetization M_s.

Note that the saturation magnetization is denoted as M_s in the SI unit, but since the values are generally displayed in Gaussian unit (gauss, G), $4\pi M_s$ is commonly used. Also, the internal bias H_0 is different from the applied H-field (H_a). Demagnetization factor should be considered [5, 6]. The demagnetization factor allows us to calculate the H-field inside the sample denoted as H_0. In all, measurement of the saturation magnetization from the dynamic hysteresis loop characteristics can be used for the design and simulation of ferrite devices.

3.3 Resonance linewidth

The loss of ferrite material is related to the linewidth, ΔH, of the susceptibility curve near resonance. Consider the imaginary part of the susceptibility χ''_{xx} versus the bias field H_0. The linewidth ΔH is defined as the width of the curve of χ''_{xx} versus H_0, where χ''_{xx} has decreased to half of its peak value. For a fixed microwave frequency ω, resonance occurs when $\omega_0 = \mu_0\gamma H_r$, such that $\omega = \omega_0 (= \mu_0\gamma H_r)$.
The linewidth, ΔH, is defined as the width of the curve of χ''_{xx} versus H_0, where χ''_{xx} has decreased to half its peak value. This is the idea that is introduced in [5].
However, obtaining the relation of χ''_{xx} versus H_0 is not easy. Here we adopt another commonly used technique [9, 12].

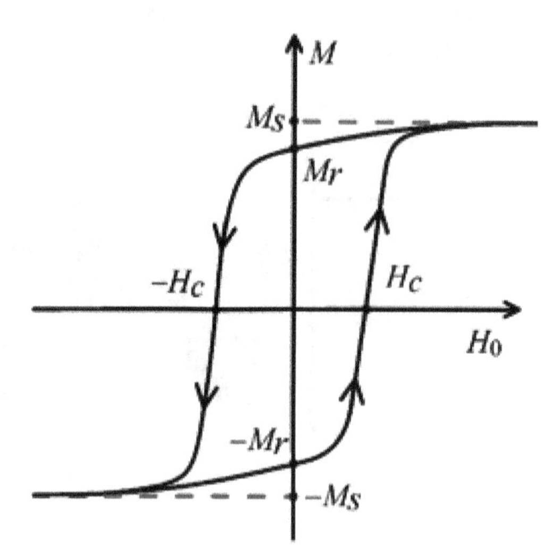

Figure 4.
The hysteresis curve regarding the magnetization M and the internal bias H_0. When the applied internal magnetic field H_0 is large enough, the magnetization will be saturated, denoted as (M_s or $4\pi M_s$). When H_0 decreases to zero, the remnant polarization is denoted as M_r. The polarization will change sign (from positive to negative) when H_0 is greater than $-H_c$ which is called the coercive field.

The idea is normally implemented using a TE_{10n} (n even) cavity in the X-band region [9, 12]. The test sample is placed at the *H*-field maximum. The sample is spherical with a diameter of approximately 0.040 inches, which is much easier to estimate than the internal bias field H_0. A cross-guide coupler is used with the coupling iris. The loaded Q (Q_L) of the empty cavity should be 2000 or greater. The sample, mounted on a fused silica or equivalent rod, is positioned away from the cavity wall at a point of minimum microwave electric field and maximum microwave magnetic field. A power meter can be used to read off the half-power points by adjusting the DC magnetic field and measuring the difference in H_0-field directly [12]. **Figure 5** shows how the resonant linewidth is determined.

The three key parameters are obtained in three experimental setups under different sizes and geometries of the samples. If the samples' properties are slightly different or the machining error is not negligible, the error will be large or even unacceptable. The ultimate goal is to integrate the measurements and to extract the parameters using one experimental setup. The three key parameters will be used in the design of the microwave ferrite devices in the next session.

4. Applications of ferrite materials

The ferrites are crystals having small electric conductivity compared to ferromagnetic materials. Thus they are useful in high-frequency situations because of the absence of significant eddy current losses. Three commonly used ferrite devices are discussed below. These are phase shifters, circulators, and isolators [13–16].

4.1 Phase shifter

The phase shifters are important applications of ferrite materials, which are two-port components that provide variable phase shift by changing the bias magnetic field. Phase shifters find application in test and measurement systems, but the most significant use is in phase array antenna where the antenna beam can be steered in space by electronically controlled phase shifters. Because of the demand, many different types of phase shifters have been provided. One of the most useful designs is the latching nonreciprocal phase shifter using a ferrite toroid in the rectangular waveguide. We can analyze this geometry with a reasonable degree of approximation using the double ferrite slab geometry.

Figure 6 shows the full-wave simulation for a two-port phase shifter using high-frequency structure simulator (HFSS, ANSYS). A standard waveguide WR-90 is employed with a width of 22.86 mm and height of 10.16 mm. The field patterns are

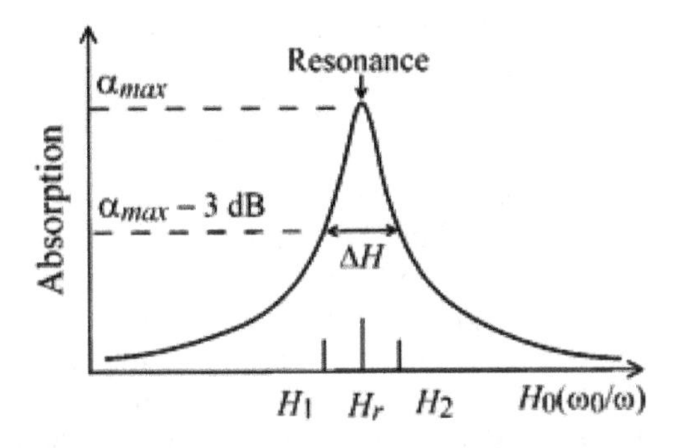

Figure 5.
The idea is to place the sample at the maximum of the H-field. It exhibits resonant absorption when the internal bias field is changed to H_r. By changing the magnetic field, we will obtain the absorption. H_1 and H_2 are associated with the 3-dB absorption. The difference between these two values is the resonance linewidth ΔH.

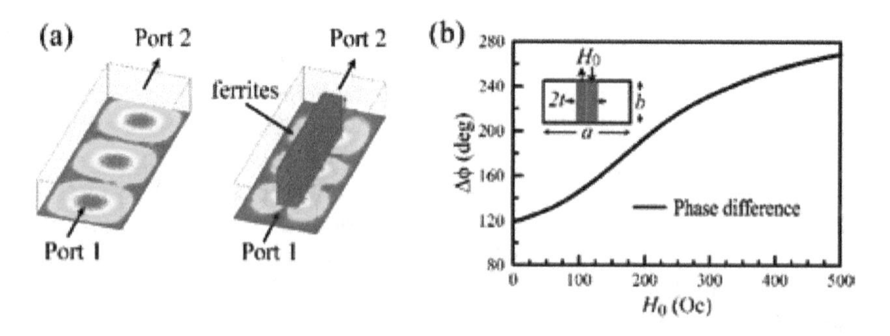

Figure 6.
Simulation results. (a) The field pattern to the left is for the empty waveguide. The right figure shows field strength with ferrites. (b) The phase difference $\Delta\phi$ as a function of the internal bias field H_0.

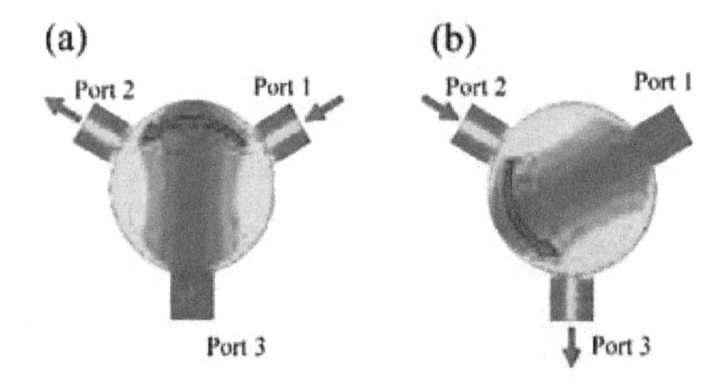

Figure 7.
Schematic diagrams of the operation of a stripline circulator. (a) The incident wave is injected from Port 1 and the transmitted wave ideally goes to Port 2. (b) The incident wave from Port 2 will go to Port 3. The color spectrum is the electric field pattern inside the ferrite disks. It is a three-port, nonreciprocal device. A full-wave solver, high-frequency structure simulator (HFSS), is used [18].

displayed in **Figure 6(a)** for an empty waveguide and a waveguide with two ferrite slabs. The phase difference $\Delta\phi$, shown in **Figure 6(b)**, is calculated under two conditions: with and without ferrites. The simulation parameters are the same as example 9.4 of Pozar's textbook. The saturation magnetization ($4\pi M_s$) is 1786 G, the dielectric constantϵ' is 13.0, and the resonance linewidth ΔH is 20 Oe. The length and thickness of the ferrite are L = 37.50 mm and t = 2.74 mm.

4.2 Circulator

Circulator, a nonreciprocal device, has been widely used in various microwave systems. **Figure** 7 schematically shows the function of a stripline circulator. The circulator is, in general, a three-port device. If the incident wave is injected from Port 1, then the wave will ideally go to Port 2, while Port 3 will be isolated as shown in **Figure 7(a)**. On the other hand, if the wave is injected from Port 2, it will go to Port 3 and isolate from Port 1 as shown in **Figure 7(b)**. There are three figures of merit for a circulator: transmission, reflection, and isolation. The transmission from Port 1 to Port 2 should be as high as possible, i.e., the insertion loss should be as small as possible. The reflection received at Port 1 due to the incident wave of Port 1 (S_{11}) and the isolation from Port 1 to Port 3 (S_{13}) should be as small as possible. The nonreciprocity of the circulator can be used to protect the oscillators from the damage of the reflected power in plasma or material processing systems. It can also be used to separate the transmitted and the received waves in radar or communication systems [13–17].

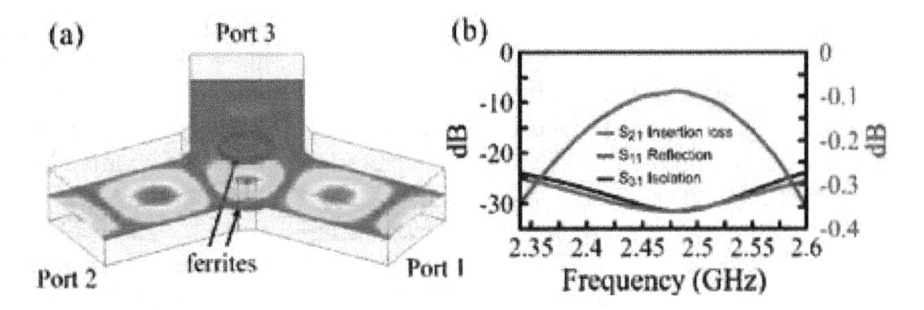

Figure 8.
(a) Schematic diagrams of the operation of a waveguide circulator. A full-wave solver, HFSS, is used with the saturation magnetization ($4\pi M_s$) is 1600 G, the dielectric constantε' is 13.0, and the resonance linewidth ΔH is 10 Oe. The radius and thickness of the ferrite disks in rust red are R = 21.0 mm and t = 5 mm, respectively. The waveguide is a standard WR 340 with 86.36 × 43.18 mm^2. The electric field pattern is displayed in color. (b) Simulation results of the waveguide circulator like the one in Part (a). The solid red curve is the transmission or insertion loss; the blue curve represents the reflection loss, and the black is the isolation.

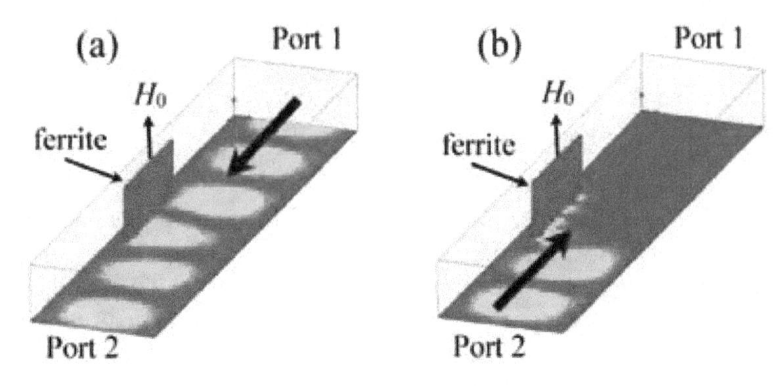

Figure 9.
The simulated field strength for a two-port isolator using the full-wave solver. (a) The high forward transmission (S_{21}) and (b) the low reverse transmission (S_{12}). The saturation magnetization ($4\pi M_s$) is 1700 G, the dielectric constant ε' is 13.0, and the resonance linewidth ΔH is 200 Oe. The length and thickness of the ferrite are L = 24.0 mm and t = 0.5 mm.

In addition to the stripline circulator, there are other types such as the microstrip circulator and the waveguide circulator. The microstrip circulator is similar to the stripline circulator in many ways. Here we show a waveguide circulator which is capable of high-power operation. **Figure 8(a)** shows the structure of the nonreciprocal device and the simulated electric field strength. The simulation parameters are described in the caption. The circulator is, in general, a three-port device. If the incident wave is injected from Port 1, then the wave will ideally go to Port 2, while Port 3 will be isolated as shown in **Figure 8(b)**. On the other hand, if the wave is injected from Port 2, it will go to Port 3 and be isolated from Port 1 as shown in **Figure 8(b)**.

4.3 Isolator

The isolator is one of the useful microwave ferrite components. As shown in **Figure 9**, the isolator is generally a two-port device having unidirectional transmission characteristics (nonreciprocity). From Port 1 to Port 2 (S_{21}), the forward transmission is high (i.e., low insertion loss in **Figure 9(a)**). However, from Port 2 to Port 1 (S_{12}), the reverse transmission is low (i.e., high isolation in **Figure 9(b)**). Besides, the reflection (S_{11} and S_{22}) should be as low as possible. The simulation parameters in **Figure 9** are the same as Ex. 9.2 of Pozar's textbook [5]. The simulation parameters and the sample's geometry are described in the caption.

An isolator is commonly used to prevent the high reflected power from damaging the precious and expansive microwave source. For example, the impedance of a plasma system changes a lot when the plasma is ignited. The radical change of the impedance will result in impedance mismatch and cause serious reflection which might kill the source instantly. An isolator can be used in place of a matching or tuning network. However, it should be realized that the reflected power will be absorbed by the ferrite of the isolator, as shown in **Figure 9(b)**. When the ferrite absorbs the reflected energy, the temperature will rise and the performance will be compromised. Therefore, a simple isolator can be implemented by using a circulator with one port well terminated [19]. For example, if Port 3 in **Figure 8(a)** is matched with water load, the power injected from Port 2 will go to Port 3 and will be isolated from Port 1. The power handling capability can be improved.

Circulator and isolator can be implemented using the self-bias [20–22], just like the latch phase shifter ($H_0 = 0$) shown in **Figure 6** using the remnant field B_r or M_r only. The self-biased ferrite devices will simplify the design and fabrication, but the overall performance is still not good enough. Further theoretical and experimental study is needed.

5. Conclusions

Ferrimagnetism and ferromagnetism share many magnetic properties in common, such as hard and soft magnets, but the conductivity differentiates these two materials. Ferrites are ceramic materials and suitable for the high-frequency operation. The electromagnetic properties of ferrite materials are difficult to understand in that the magnetic susceptibility is a tensor and depends on the saturated magnetization M_s, the internal bias field H_0, and the resonance linewidth ΔH. The magnetic susceptibility also depends on the frequency of the microwave ω as well as the polarization of the wave. The first two sessions explain the basic properties.

The complex permittivity $\varepsilon_r + i\varepsilon_i$, the saturation magnetization M_s, and the resonance linewidth ΔH are the most important electromagnetic properties of ferrites. How to measure the ferrite's properties are discussed in Section 3. The full-wave simulation is conducted to demonstrate how the phase shifter, circulator, and isolator work, which are shown in Section 4. Although the examples are discussed in many textbooks, Section 4 offers in-depth simulation results for the first time.

At high-power operation, the ferrite devices will be heated. The spin wave linewidth may be taken into account. Besides, the ferrites will become paramagnetism when the temperature exceeded the Curie temperature. These two factors are important for high-power operation, which are not considered in this chapter.

Acknowledgements

This chapter was supported in part by the Ministry of Science and Technology of Taiwan and in part by China Steel Company/HIMAG Magnetic Corporation, Taiwan. The author is grateful to the Taiwan Branch of ANSYS Inc. for technical assistance and to Dr. Hsein-Wen Chao and Mr. Wei-Chien Kao for their assistance in the full-wave simulation. Dr. Hsin-Yu Yao and Mr. Shih-Chieh Su are appreciated for the discussion of the ferrites' characterization.

Author details

Tsun-Hsu Chang
Department of Physics, National Tsing Hua University, Hsinchu, Taiwan

*Address all correspondence to: thschang@phys.nthu.edu.tw

References

[1] Okamoto A. The invention of ferrites and their contribution to the miniaturization of radios. IEEE Globecom Workshops. 2009:1-42. DOI: 10.1109/GLOCOMW.2009.5360693. ISBN 978-1-4244-5626-0

[2] Néel L. Magnetic Properties of Ferrites: Ferrimagnetism and Antiferromagnetism. Annales de Physique. 1948;**3**:137-198

[3] Chen CW. Magnetism and Metallurgy of Soft Magnetic Materials. Elsevier; 1977. 15–60 p. DOI: 10.1016/B978-0-7204-0706-8.X5001-1. ISBN: 9780444601193

[4] McGrayne SB, Suckling EE, Kashy E, Robinson FNH, Bleaney B. Magnetism Physics; 2018. Available from: https://www.britannica.com/science/magnetism/ [Accessed: 2018-12-20]

[5] Pozar DM. Chap. 9. In: Microwave Engineering. 4th ed. Hoboken, New Jersey: John Wiley & Sons, Inc.; 2011

[6] Collin RE. Chap. 6. In: Foundations for Microwave Engineering. 2nd ed. New York: McGraw Hill; 1992

[7] Chang TH. Gyromagnetically-induced transparency for ferrites. American Journal of Physics. 2016;**84**:279-283

[8] Cohn SB, Kelly KC. Microwave measurement of high-dielectric-constant materials. IEEE Transactions on Microwave Theory and Techniques. 1966;**14**:406-410

[9] Chen LF, Ong CK, Neo CP, Varadan VV, Varadan VK. Microwave Electronics: Measurement and Materials Characterization. Hoboken, New Jersey: John Wiley & Sons Inc.; 2004

[10] Chao HW, Wong WS, Chang TH. Characterizing the complex permittivity of high- κ dielectrics using enhanced field method. The Review of Scientific Instruments. 2015;**86**:114701

[11] Chao HW, Chang TH. Wide-range permittivity measurement with a parametric-dependent cavity. IEEE Transactions on Microwave Theory and Techniques. 2018;**66**:4641-4648

[12] Skyworks. Test for Line Width and Gyromagnetic Ratio [Internet]. 1999. Available from: http://www.skyworksinc.com/uploads/documents/Test_for_Line_Width_Gyromagnetic_Ratio_202837B.pdf [Accessed: 2018-12-20]

[13] Fuller AJB. Ferrites at Microwave Frequencies. London: Peter Peregrinus; 1987

[14] Pardavi-Horvath M. Microwave applications of soft ferrites. Journal of Magnetism and Magnetic Materials. 2000;**215**:171-183

[15] Schloemann E. Advances in ferrite microwave materials and devices. Journal of Magnetism and Magnetic Materials. 2000;**209**:15-20

[16] Harris VG, Geiler A, Chen Y, Yoon SD, Wu M, Yang A, et al. Recent advances in processing and applications of microwave ferrites. Journal of Magnetism and Magnetic Materials. 2009;**321**:2035-2047

[17] Helszajn J. The Stripline Circulators: Theory and Practice. Hoboken, New Jersey: John Wiley & Sons; 2008

[18] Chao HW, Wu SY, Chang TH. Bandwidth broadening for stripline circulator. The Review of Scientific Instruments. 2017;**88**:024706

[19] Linkhart DK. Microwave Circulator Design. 2nd ed. Norwood, MA: Artech House Microwave Library; 2014

[20] Wang J, Yang A, Chen Y, Chen Z, Geiler A, Gillette SM, et al. Self-biased Y-junction circulator at K_u-band. IEEE Microwave and Wireless Components Letters. 2011;**21**:292-294

[21] Peng B, Xu H, Li H, Zhang W, Wang Zhang YW. Self-biased microstrip junction circulator based on barium ferrite thin films for monolithic microwave integrated circuits. IEEE Transactions on Magnetics. 2011;**47**: 1674-1677

[22] Zuo X, How H, Somu S, Vittoria C. Self-biased circulator/isolator at millimeter wavelengths using magnetically oriented polycrystalline strontium M-type hexaferrite. IEEE Transactions on Magnetics. 2003;**39**: 3160-3162

3

Terahertz Sources, Detectors and Transceivers in Silicon Technologies

Li Zhuang, Cao Rui, Tao Xiaohui, Jiang Lihui and Rong Dawei

Abstract

With active devices lingering on the brink of activity and every passive device and interconnection on chip acting as potential radiator, a paradigm shift from "top-down" to "bottom-up" approach in silicon terahertz (THz) circuit design is clearly evident as we witness orders-of-magnitude improvements of silicon THz circuits in terms of output power, phase noise, and sensitivity since their inception around 2010. That is, the once clear boundary between devices, circuits, and function blocks is getting blurrier as we push the devices toward their limits. And when all else fails to meet the system requirements, which is often the case, a logical step forward is to scale these THz circuits to arrays. This makes a lot of sense in the terahertz region considering the relatively efficient on-chip THz antennas and the reduced size of arrays with half-wavelength pitch. This chapter begins with the derivation of conditions for maximizing power gain of active devices. Discussions of circuit topologies for THz sources, detectors, and transceivers with emphasis on their efficacy and scalability ensue, and this chapter concludes with a brief survey of interface options for channeling THz energy out of the chip.

Keywords: terahertz, detector, source, transceiver, silicon

1. Introduction

The significance of terahertz electronics is self-evident for readers of this book. The general consensus among silicon THz circuit designers (!) is that silicon will be the dominant technology for the lower end of the THz spectrum (300 GHz to around 1 THz) in light of recent breakthroughs of silicon circuits in terms of effective isotropic radiated power (EIRP), phase noise, and receiver sensitivity. For many applications, silicon circuits are on par or even superior to III/V compound technologies and optical-based techniques in this frequency range now. This chapter aims to introduce the reader to the fascinating world of silicon THz circuit design through a step-by-step approach: We examine conditions for extracting the most power gain out of a given active device. Popular topologies for silicon sources, detectors, and transceivers are discussed next, and this chapter concludes with a brief survey of THz interface options for efficient energy transfer between circuits and the outside world.

2. How to have THz power and radiate it too

Due to the excessive loss and scarcity of power gain for silicon devices in the THz region, one should strive to extract the most power out of a given device during the whole design phase. This involves making sure that the device is working under the optimum condition (i.e., the device is embedded in the right impedance environment for maximum power gain), the topology of the circuit is optimum for the intended application, and the power is transferred from the circuit through the most efficient interface. This section gives an overview of these areas.

2.1 Power gain maximization for a given active device

The active devices in THz circuits are connected to the rest of the circuits through passive elements, such as capacitors, inductors, and transmission lines. The overall circuit performance is decided both by the active device and these passive elements. Thus, to maximize the circuit performance, a "divide-and-conquer" approach is the logical choice. That is, we first find the "best" active device in a given technology under certain constraints such as power consumption or noise performance. We then decide the best passive network into which the device should be embedded. The problem is there is no such thing as "pure" active device; passive elements are always present in a given active device. Mason [1] has thus defined a figure of merit for active devices:

$$U = \frac{|Y_{21} - Y_{12}|^2}{4(G_{11}G_{22} - G_{12}G_{21})} \tag{1}$$

G_{ij} is the real part of Y_{ij} in Eq. (1). The above FOM is called Mason's invariant U, since it is invariant to passive embedding environments that are linear, lossless, and reciprocal [1, 2].

A device is active if its U is larger than one, which means this device is capable of providing real power. The maximum oscillation frequency (f_{max}) of a device is defined as the frequency where its U equals one, i.e., beyond which frequency it is no longer active. The maximum power gain of this device embedded in the two-port also drops to unity at the maximum oscillation frequency (f_{max}).

U is also the maximum power gain of the device after unilateralization, that is, when Y_{12} is made to zero. Generally speaking, higher U means higher power gain at a given frequency.

For a given two-port shown in **Figure 1(b)**, the power gain is defined as

$$G_p = \frac{P_L}{P_{IN}} = \frac{-0.5 \times \mathrm{Re}(V_2 \times I_2^*)}{0.5 \times \mathrm{Re}(V_1 \times I_1^*)} = \frac{\mathrm{Re}(V_2 \times V_2^* \times Y_L^*)}{\mathrm{Re}(V_1 \times V_1^* \times Y_{IN}^*)} = |A_V|^2 \frac{\mathrm{Re}(Y_L^*)}{\mathrm{Re}(Y_{IN}^*)} \tag{2}$$

P_L and P_{IN} are the real power delivered to the load and to the two-port. V_1, I_1, and V_2, I_2 in **Figure 1** are the voltage and current at port 1 and port 2, respectively. A_V is the voltage gain of the two-port. Y_s and Y_L represent the source and load admittance presented to the device. I_s and Y_s form the Norton equivalent circuit of the signal source.

For an unconditionally stable two-port (it does not oscillate for any passive load and source admittance) at a given frequency, the power gain could be maximized by biconjugate matching at the input and output. Conjugate matching is achieved when the load admittance is equal to the conjugate of the source admittance at a given node; biconjugate matching means that this condition is satisfied both at the input and the output port. For a given two-port, its maximum power gain is

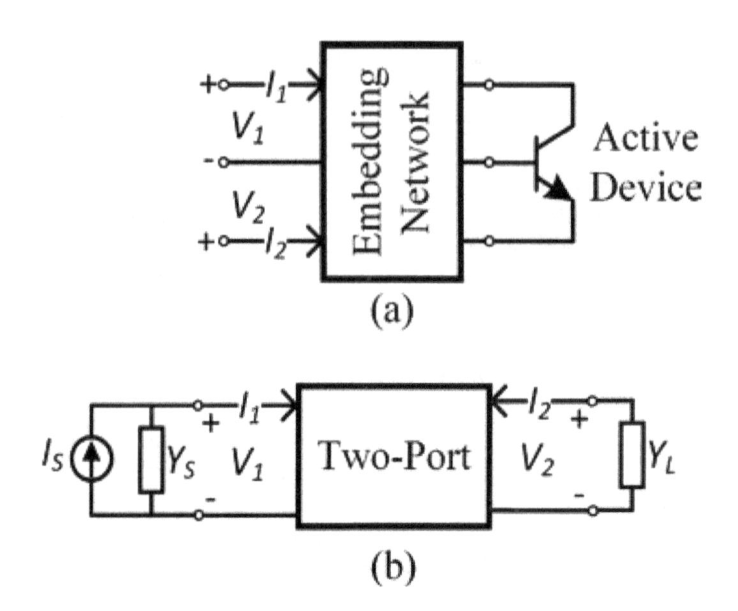

Figure 1.
The two-port representation of an embedded active device. (a) An active device embedded in a linear, lossless, and reciprocal passive network resulting in a two-port and (b) the two-port interfacing to signal source and load.

$$G_{\max} = \left|\frac{Y_{21}}{Y_{12}}\right| \times \left(k - \sqrt{k^2 - 1}\right) \tag{3}$$

where K is the stability factor defined as

$$\frac{2 \times \mathrm{Re}(Y_{11}) \times \mathrm{Re}(Y_{22}) - \mathrm{Re}(Y_{12}Y_{21})}{|Y_{12}Y_{21}|} \tag{4}$$

Biconjugate matching is possible when K is equal or greater than 1. For a given two-port, it is unconditionally stable when the following conditions are simultaneously satisfied:

$$g_{11} \geq 0 \tag{5}$$

$$g_{22} \geq 0 \tag{6}$$

$$K \geq 1 \tag{7}$$

Keep in mind that unlike U, which is invariant to the embedding network, G_{max} is sensitive to its environment. We can modify the embedding environment to make G_{max} larger. It can be shown that the maximum G_{max} for a given device is $\left(\sqrt{U} + \sqrt{U-1}\right)^2$. For a detailed discussion about U and G_{max}, please refer to [3–6] which present ways of designing the embedding network for maximizing G_{max} while maintaining stability under process variations. The basic idea is to utilize feedback to generate negative resistance, such as adding capacitive degeneration to CE (common emitter) amplifiers or adding inductance to the base node of CB (common base) amplifiers.

Another important THz circuit is oscillators. Here, we need to make a distinction between amplifiers and oscillators. The former is one kind of driven circuit, the output of which is controlled by its input. Oscillators belong to the group of autonomous circuit, which generates time-varying signals without time-varying stimulus.

By definition, amplifiers operate below f_{max} to provide power gain. For oscillators, the situation is more complicated: The power gain of active devices within the oscillator should be greater than unity to start the oscillation. As the oscillation amplitude grows, the power gain of the two-port (including device parasitic and loading) gradually compresses to unity when the circuit reaches steady-state oscillation. Thus, it could be argued that devices within oscillators operate at the f_{max} of its embedded two-port in steady-state oscillation. To take the gain compression into account, large-signal parameters should be used for the analysis.

Unlike analog circuit designers who deal exclusively with voltage and current gains, microwave circuit designers are more comfortable with power gain. Momeni [7] has thus shown a refreshing view about the optimum voltage gain and phase shift for a given two-port to oscillate at f_{max}:

$$A_V = \frac{V_2}{V_1} = A_{OPT} = \sqrt{\frac{G_{11}}{G_{22}}} \tag{8}$$

$$\phi = \angle\frac{V_2}{V_1} = \phi_{OPT} = (2k+1)\pi - \angle(Y_{12} + Y_{21}^*) \tag{9}$$

A_{OPT} and Φ_{OPT} are the optimum voltage gain and phase shift for the two-port at f_{max}. Equations (8) and (9) are derived assuming no clipping to the power rails occurs inside the circuit. If clipping happens, another set of equations apply for A_{OPT}; Φ_{OPT} remains the same [8, 9].

It can be shown that biconjugate matching automatically satisfies Eqs. (8) and (9) at f_{max}.

Under biconjugate matching, G_p reaches unity at f_{max}. Equation (2) can be rewritten as

$$|A_V| = |A_{OPT}| = \sqrt{\frac{\mathrm{Re}(Y_{IN}^*)}{\mathrm{Re}(Y_L^*)}} = \sqrt{\frac{\mathrm{Re}(Y_{S,OPT})}{\mathrm{Re}(Y_{L,OPT}^*)}} = \sqrt{\frac{\mathrm{Re}(Y_{S,OPT})}{\mathrm{Re}(Y_{L,OPT})}} \tag{10}$$

$Y_{S,OPT}$ and $Y_{L,OPT}$ are the optimal source and load admittance for biconjugate matching:

$$Y_{S,OPT} = \frac{|Y_{12}Y_{21}| \times \sqrt{K^2-1} + j \times (\mathrm{Im}(Y_{12}Y_{21}) - 2 \times \mathrm{Re}(Y_{22}) \times \mathrm{Im}(Y_{11}))}{2\mathrm{Re}(Y_{22})} \tag{11}$$

$$Y_{L,OPT} = \frac{|Y_{12}Y_{21}| \times \sqrt{K^2-1} + j \times (\mathrm{Im}(Y_{12}Y_{21}) - 2 \times \mathrm{Re}(Y_{11}) \times \mathrm{Im}(Y_{22}))}{2\mathrm{Re}(Y_{11})} \tag{12}$$

By substituting Eqs. (11) and (12) into (10), we have

$$|A_{OPT}| = \sqrt{\frac{G_{11}}{G_{22}}} \tag{13}$$

We are now in the position to derive Φ_{OPT}. First, we have to define the net power flowing into the two-port shown in **Figure 1(b)**:

$$P = V_1 \times I_1^* + V_2 \times I_2^* \tag{14}$$

At f_{max}, the power gain of the two-port drops to unity, which means the real power consumed by the two-port equals the real power generated. Thus, the real part of Eq. (14) equals zero:

$$P_R = |V_1|^2 \times \left[G_{11} + G_{22} \times |A_V|^2 + \mathrm{Re}\left(Y_{12}^* A_V^* + Y_{21}^* A_V\right)\right] = 0 \tag{15}$$

Substituting Eq. (13) into (15), we have

$$2\sqrt{G_{11}G_{22}} + \mathrm{Re}\left[\left(Y_{12} + Y_{21}^*\right)e^{j\phi}\right] = 0 \tag{16}$$

Since the Mason's invariant equals unity at f_{max}, Eq. (1) could be rearranged as

$$\frac{\left|Y_{12} + Y_{21}^*\right|^2}{4G_{11}G_{22}} = 1 \tag{17}$$

Equation (16) thus equals

$$2\sqrt{G_{11}G_{22}} \times \left[1 + \cos\left(\phi + \angle\left(Y_{12} + Y_{21}^*\right)\right)\right] = 0 \tag{18}$$

We have

$$\phi_{OPT} = (2k+1)\pi - \angle\left(Y_{12} + Y_{21}^*\right) \tag{19}$$

Since the above derivation is restricted to f_{max}, it would be interesting to observe the possible deviations of Eqs. (8) and (9) with respect to the two-port's voltage gain and phase shift under biconjugate matching when operating at frequency below f_{max}. A SiGe HBT transistor is used as an example. The emitter width and length of the transistor is 0.12 and 2.5 μm, and the emitter current density is biased for peak f_{max}. The source and load admittance are adjusted for biconjugate matching under each frequency evaluated (**Figures 2** and **3**).

It is clear that Eqs. (8) and (9) are only strictly valid at f_{max}, but the optimum phase shift calculated with Eq. (9) tracks reasonably well with the results obtained with biconjugate matching over a wide frequency range.

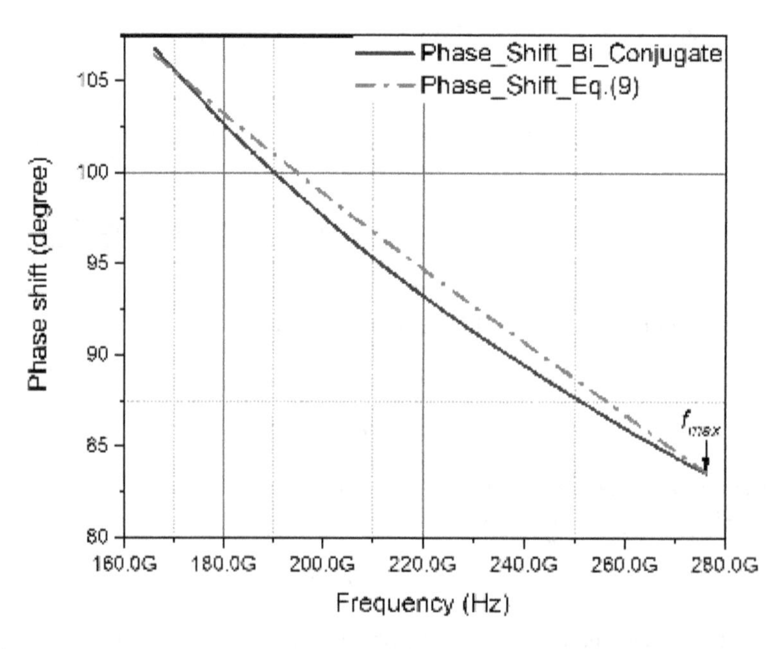

Figure 2.
Comparison of the phase shift of a SiGe HBT transistor under biconjugate matching and the optimum phase shift calculated with Eq. (9).

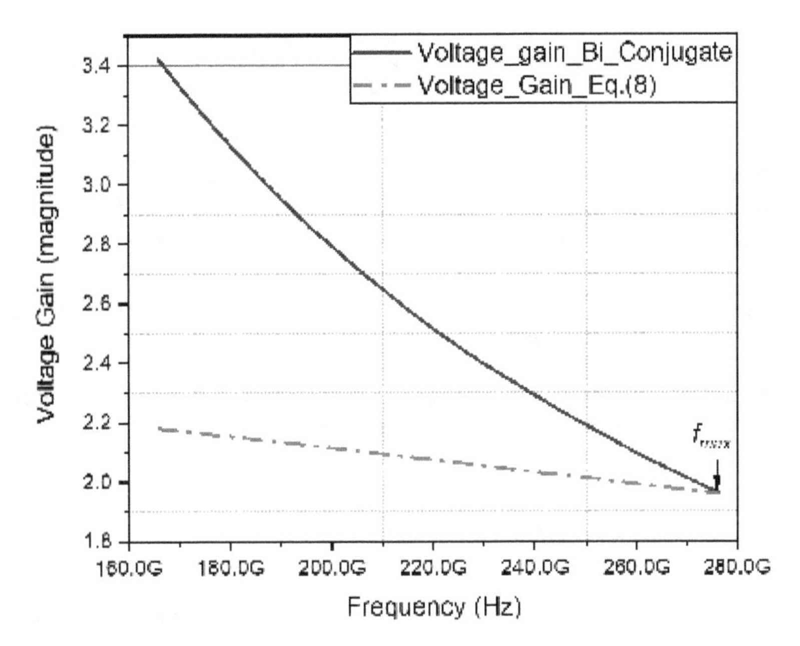

Figure 3.
Comparison of the voltage gain of a SiGe HBT transistor under biconjugate matching and the optimum voltage gain calculated with Eq. (8).

2.2 Circuit topology for THz sources, detectors, and transceivers

Among the many potential benefits offered by THz application, the large bandwidth available is the most obvious one. However, a lot of design issues need to be addressed in order to truly harness this bandwidth potential. We discuss this problem in terms of *SNR* at the receiver:

$$SNR \times B = \frac{P_r}{kTBF} = \left(\frac{\lambda}{4\pi d}\right)^2 \times \frac{P_t G_t G_r}{kTFL} \tag{20}$$

B is the receiver bandwidth. For communications, we would like *B* to scale with frequency. For imaging applications, sometimes *B* is not that important once it reaches certain value as only the range resolution scales with $1/B$. The cross-range resolution scales inversely with wavelength λ. Thus, we lump SNR and B together for trade-offs. P_t and P_r are the power transmitted and received by the transmitter (Tx) and receiver (Rx). G_t and G_r are the gain of the transmitting and receiving antenna. *K* is the Boltzmann constant. *T* is the ambient temperature. *d* is the distance between the transmitter and receiver. *F* is the noise factor of the receiver. *L* represents the loss in the Tx and Rx system, and we assume it scales with $f^{0.5}$ to f here.

Assuming constant drive power, P_t for terahertz transmitters approximately follows a $P_t \propto 1/f^2$ relationship since the maximum unilateral gain *U* follows a -20dB/decade slope above $f_{max}/2$. *F* generally scales with *f*. $SNR \times B$ then scales with $1/f^{5.5}$ to $1/f^6$, which is a really disheartening result. This partly explains why the current silicon THz links are usually demonstrated with link distances ranging from centimeters to meters.

Before leaving this chapter in despair, we can try to manipulate Eq. (20) a little bit further:

$$SNR \times B = \frac{1}{16\pi^2 k} \times \frac{\lambda^2}{Td^2 L} \times \frac{P_t G_t G_r}{F} \tag{21}$$

The first term is by all means beyond our control, and we do not want to change the second term for now. So, what can we do about the last term? It happens that if we were to keep the two-antenna size constants while increasing the frequency, G_t and G_r each come with a nice λ^2 on the denominator. Equation (21) thus equals

$$SNR \times B = \frac{1}{16\pi^2 k} \times \frac{\lambda^2}{Td^2 L} \times \frac{4\pi A_{tp} \varepsilon_t P_t}{\lambda^2} \times \frac{G_r}{F} = \frac{A_{tp} \varepsilon_t P_t}{4\pi k T d^2 L} \times \frac{G_r}{F} \tag{22}$$

A_{tp} and ε_t are the physical area and the aperture efficiency of the transmitting antenna; ε_t is between zero and unity. For active phased arrays, Eq. (22) could be rewritten as

$$SNR \times B = \frac{A_{tp} \varepsilon_t N_t P_{te}}{4\pi k T d^2 L} \times \frac{N_r G_{re}}{F_e} \tag{23}$$

where N_t and N_r are the numbers of transmitters and receivers. P_{te} is the output power for each transmitter. G_{re} is the gain of the receiving antenna for each receiver. F_e is the noise factor of each receiver. For active phased arrays, antenna elements with their corresponding transmitters and receivers are evenly distributed with a pitch of about $\lambda/2$. Thus, N_t and N_r are proportional to $1/\lambda^2$. It is proven that for phased array with lossless combing, the G_r/F term in Eq. (22) scales with N_r [10]. Assuming constant ε_t and G_{re} with respect to f, we see that $SNR \times B$ scales with $f^{-0.5}$ to f^0!

The six orders of magnitude difference of $SNR \times B$ deduced from Eqs. (20) and (23) give us a hint of the size of the design space for silicon THz systems.

2.2.1 THz Sources

When talking about silicon THz sources, a plethora of options is available that varies in functionality, complexity, and performance. For incoherent imaging applications, the most important metrics are output power and efficiency, whereas for spectroscopy, the bandwidth is the most important specification. Perhaps the most demanding application is for THz communications, for which output power, power efficiency, tuning range, phase noise, harmonics, and spurious suppression are all important parameters. This subsection aims to give a brief and incomplete introduction to what has been done in this area.

THz signal can be generated either by frequency multipliers or by on-chip oscillators.

In multipliers, the MOS or bipolar transistor is driven heavily to generate highly nonlinear current. The intended frequency component is then extracted with other components filtered. If efficiency is important, the active device should be conjugate matched for the fundamental and the intended harmonic. The impedance presented to the device at other harmonics is usually short or open circuit to maximize energy transfer between the fundamental and the intended harmonic. But we should not be overzealous about this goal; usually taking care of the first two or three harmonics is enough since the higher harmonics are insignificant. The transistor also has to be biased correctly for maximum harmonic generation. For MOS transistor, the conduction angle is specified. Like power amplifiers, efficient

MOS multiplier works in the class AB, B, or C region depending on the frequency, multiplication factor, and input power. For bipolar transistor, this efficiency is a function of V_{be} (or collector current density) [11].

Relationship of phase noise between the harmonic and the fundamental for multipliers is [11]:

$$S_{out}(\Delta\omega) = N^2 S_{fund}(\Delta\omega) + S_{harm}(\Delta\omega) + S_{amp}(\Delta\omega) \tag{24}$$

where $S_{out}(\Delta\omega)$ and $S_{fund}(\Delta\omega)$ represent the spectral density of phase fluctuations for the harmonic and fundamental signal with $\Delta\omega$ radians offset from the carrier. It is in dBc/Hz form. N is the frequency multiplication ratio. The last two terms in Eq. (24) represent the added noise from the harmonic-generating device and the ensuing amplifiers (if any). The combined value is usually less than 3 dB for reasonably designed circuits.

Multipliers are usually compact and broadband, but they are not as efficient as (well designed) oscillators. A 90–300 GHz transmitter based on distributed quadrupler is designed for spectroscopy and imaging [12]. It resembles the distributed amplifier (DA) in that the input and output capacitance of active device are absorbed in the input and output transmission line. Differential quadrature signal is used to drive two groups of quadrupler diff-pairs, the current of which is then combined to cancel the second harmonic. As another example, quadrupler is used in an 8-element 400 GHz transmitter phased array to replace power amplifier [13]. This also simplifies phase shifter design since the fundamental signal only needs to be shifted within 90 degrees as the phase shift is multiplied by four.

In oscillators, the transistor is made unstable by intentionally introducing positive feedback around it. Steady-state oscillation occurs at the frequency where the open-loop transfer function equals −1 (Barkhausen's criteria). Since the f_{max} of most silicon devices is below 300 GHz, harmonic generation is employed. The fundamental oscillation frequency is usually around 100 GHz for better phase noise, since larger oscillation amplitude and hence better SNR are easier to obtain at lower frequencies.

A high efficiency and scalable 4 × 4320 GHz oscillator array is built in SiGe BiCMOS technology [9]. The oscillator shown in **Figure 4(a)** oscillates at 160 GHz and is optimized for optimum transistor voltage gain and phase shift as discussed in Section 2.1. Y_1 and Y_2 represent the source and load admittance for the transistor. A transmission line with impedance Z_0 and electrical length θ_{TL} is used to introduce

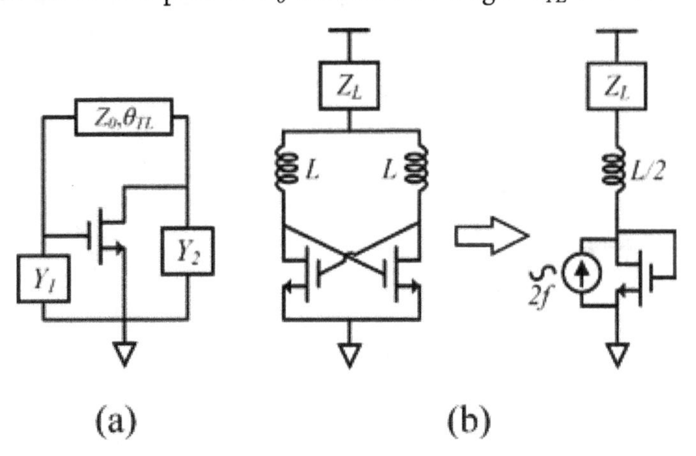

Figure 4.
Topology of (a) self-sustained oscillator and (b) conventional cross-coupled oscillator and its equivalent circuit at second harmonic.

feedback. The transmission line spans approximately a quarter wavelength at the second harmonic to transform the relatively small impedance of the gate node to high values at the drain node. Since harmonic signals are in the current form, boosting the output impedance at the harmonic frequency substantially improves output harmonic power. The DC-to-THz radiation efficiency is 0.54%. Early reports of THz oscillators based on push-push oscillators are less efficient than this partly due to the insufficient output impedance at second harmonic. As is shown in **Figure 4(b)**, the cross-coupled pair in a push-push oscillator is effectively a diode-connected transistor at second harmonic.

Frequency tuning of oscillators is usually done by varying the capacitance of varactor in the oscillation tank. Higher oscillation frequency translates to smaller capacitance, which is problematic for small varactors as its parasitic capacitance would swamp the variable capacitance. This would severely constrain the oscillator's tuning range.

Shown in **Figure 5(a)** and **(b)** are the cross section of a NMOS varactor and its small signal model with its source AC grounded. To increase its Q factor, the minimum channel length is usually employed, which further increases the overlap capacitance (C_{ov}) between the two terminals.

A straightforward way to increase the tuning ratio of the varactor is to place an inductor to partially absorb the parasitic capacitance. A 300 GHz differential Clapp push-push VCO with 8.5% tuning range and phase noise of −85 dBc at 1 MHz offset is reported in [14]. Its simplified schematic and equivalent small signal circuit for calculation of the input impedance seen at the base is shown in **Figure 6**. Note that the base resistance, the depletion capacitance between base and collector, and the output resistance of the transistor are ignored.

The input impedance seen from the base is

$$Z_B = -\frac{1}{j\omega C_{pi}}\left(1+\frac{g_m + j\omega C_{pi}}{1/R_{var} + j\omega C_{eff}}\right) \tag{25}$$

where C_{eff} is

$$C_{eff} = C_{\text{var}} \times \left(1 - \frac{1}{\omega^2 L_{\text{deg}} C_{\text{var}}}\right) = C_{\text{var}} \times \left(1 - \frac{\omega_0^2}{\omega_{\text{OSC}}^2}\right) \tag{26}$$

R_{var} represents the loss in varactor; its conductance is $1/Q$ of the varactor at the evaluation frequency of the quality factor. The resonant frequency ω_O for C_{var} and L_{DEG} is set below the main oscillation frequency.

The equivalent series resistance and capacitance for Z_B is

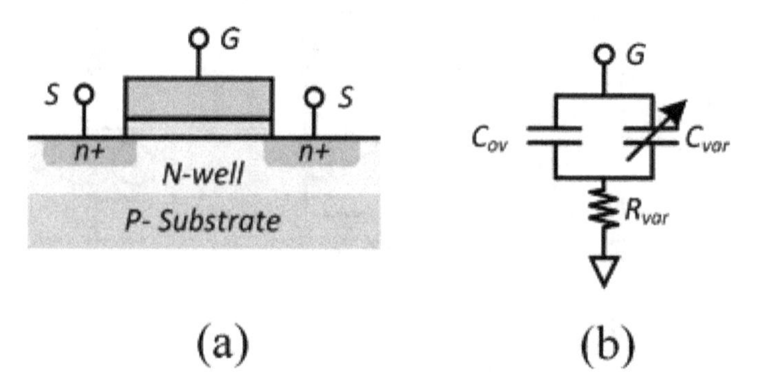

Figure 5.
Cross section of NMOS varactor (a) and equivalent circuit with grounded source (b).

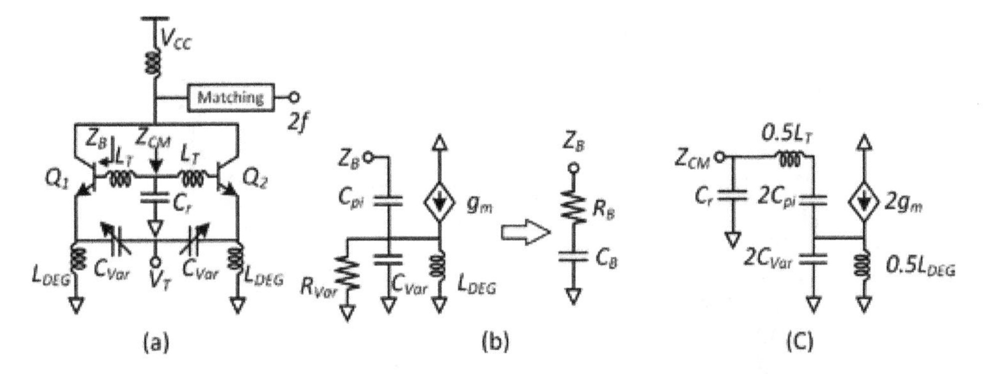

Figure 6.
Push-push VCO with common-mode resonance: (a) schematic, (b) base input impedance for differential mode, and (c) input impedance for common mode.

$$R_B = \frac{1/R_{\text{var}} - g_m C_{eff}/C_{pi}}{1/R_{\text{var}}^2 + \omega^2 C_{eff}^2} \approx -\frac{g_m}{\omega^2 C_{eff} C_{pi}} \tag{27}$$

$$C_B = \left(1 + \frac{g_m/R_{\text{var}} + \omega^2 C_{pi} C_{eff}}{1/R_{\text{var}}^2 + \omega^2 C_{eff}^2}\right) \times \frac{1}{C_{pi}} \approx \frac{C_{eff} \times C_{pi}}{C_{eff} + C_{pi}} \tag{28}$$

Another interesting property of the circuit is that the oscillation frequency for common mode is intentionally set to the second harmonic. C_r in **Figure 6(a)** and **(c)** forms a series resonator with the tank, raising the second harmonic voltage seen at the base significantly. This boosts the second harmonic generation. The output power for the two versions of the VCO is 0.6 dBm and 0.2 dBm, respectively.

As is evident from Eq. (27), the negative resistance seen at the base of the capacitively degenerated transistor could be used to mitigate the loss of the varactor [15]. This property is used in [16] to build a 300 GHz triple-push VCO. The tuning range is 8%, and the phase noise is −101.9 dBc/Hz at 1 MHz offset for the 100 GHz main loop. That translates to a phase noise of −80.28 dBc/Hz at 1 MHz offset for 300 GHz assuming noiseless multiplication.

An interesting observation is that inductors actually have better quality factor than varactors at higher frequency. A carefully designed inductor has a Q of 15–20 at 100 GHz, whereas the quality factor for varactor is around 2–5 at that frequency. It would be nice if we can replace the varactor with a high-quality metal-insulator-metal (MIM) or metal-oxide-metal (MOM) capacitor; we then need to figure out how to tune the inductance of these nice inductors. **Figure 7** shows one such circuit [17] which is also based on Clapp oscillators.

We know that inductance at the base generates a negative resistance and a positive inductance seen from the emitter, as LNA designers can attest to [18]. Careful derivation of Z_E leads to the following results [14]:

$$Z_E \approx \frac{1}{g_m} - \frac{\omega^2 L_{BB}}{\omega_T} + \frac{\omega^2 R_{BB}}{\omega_T^2} + j\omega\left(\frac{R_{BB}}{\omega_T} - \frac{1}{g_m \omega_T} + \frac{\omega^2 L_{BB}}{\omega_T^2}\right) \tag{29}$$

Thus, we can tune the inductance by varying the g_m of the transistor. For bipolar transistors, g_m equals the emitter current divided by the thermal voltage V_T (26 mV in room temperature).

The impedance at the emitter is mapped to the resonant tank through the transformer formed by L_E and L_T in **Figure 7(b)**. The currents of the two active inductors are controlled by tail current source Q_5. The second harmonic current is extracted through L_C, and the degeneration resistor R_{EE} is used to improve

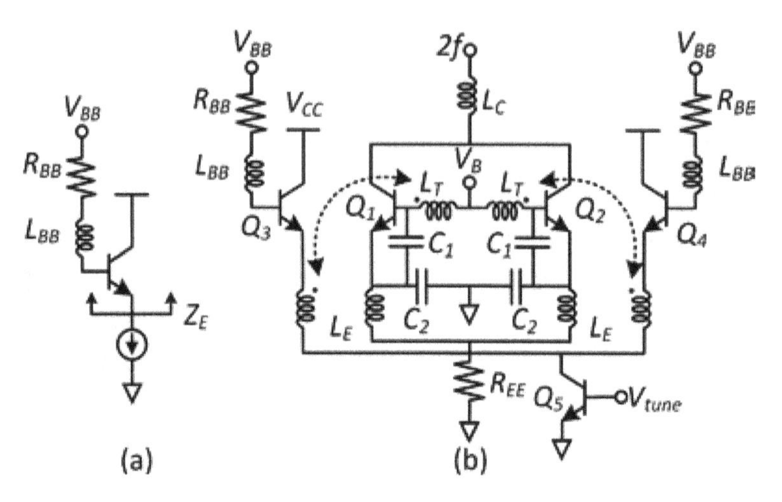

Figure 7.
Schematics for (a) tunable active inductor and (b) complete VCO.

common-mode rejection. Two versions of this VCO with different oscillation frequency are built; the tuning range are 3.5 and 2.8%, respectively. The harmonic power suffers due to the inclusion of R_{EE}, the output power at 201.5 and 212 GHz are −7.2 and−7.1 dBm, respectively. But the phase noise performance is very good, with −87 and −92 dBc at 1 MHz offset for the two VCOs. This translates to −83.5 and −89 dBc at 1 MHz offset if they oscillate at 300 GHz, which is comparable to the circuit shown in **Figure 6** that generates 0 dBm at 300 GHz.

This raises interesting question as smaller output power usually means inferior phase noise performance. One possible explanation is that the noise current at the second harmonic in Q1 and Q2 of **Figure 6** generates large noise voltage at the emitter since these nodes are open circuit due to resonance, thus amplifying the noise current at second harmonic. It should be noted that the phase noise could be improved substantially by breaking the noise current path at second harmonic [19].

A salient feature of Clapp oscillator is the inherent isolation of the load from the tank. This helps to preserve the quality factor of the tank, leading to a better phase noise and less load pulling (variation of oscillation amplitude and frequency caused by load variation). The problem of low output impedance at second harmonic is also mitigated in this topology as the base is isolated from the drain.

One starts to wonder if there is a way to further improve the phase noise performance. We know there is a trade-off between noise performance and power consumption, but we are kind of stuck here: We need larger transistor to burn larger power, but the larger capacitance of the device means smaller inductors in the tank, which complicates the design and ultimately degrades the *Q*. We need a larger design space here.

One way to do that is to build an array of *N* oscillators and lock them together. Theoretically the phase noise would drop by *N* as the *SNR* increases by *N*, and the output power would also increase by *N*. Better still, if we distribute them evenly with a certain pitch (normally λ/2) and radiate the power out collectively, the energy would focus in certain direction, improving the EIRP by N^2. Tousi et al. shows one such design [20]:

Each individual oscillator shown in **Figure 8** is a cross-coupled push-push oscillator. It is designed for optimum fourth harmonic generation by making sure that the gate is isolated from the drain at the fourth harmonic. The second harmonic is rejected by the narrow band on-chip antenna. Each oscillator is coupled to other oscillators as shown in **Figure 8(a)** through active phase shifters. **Figure 8(a)** forms

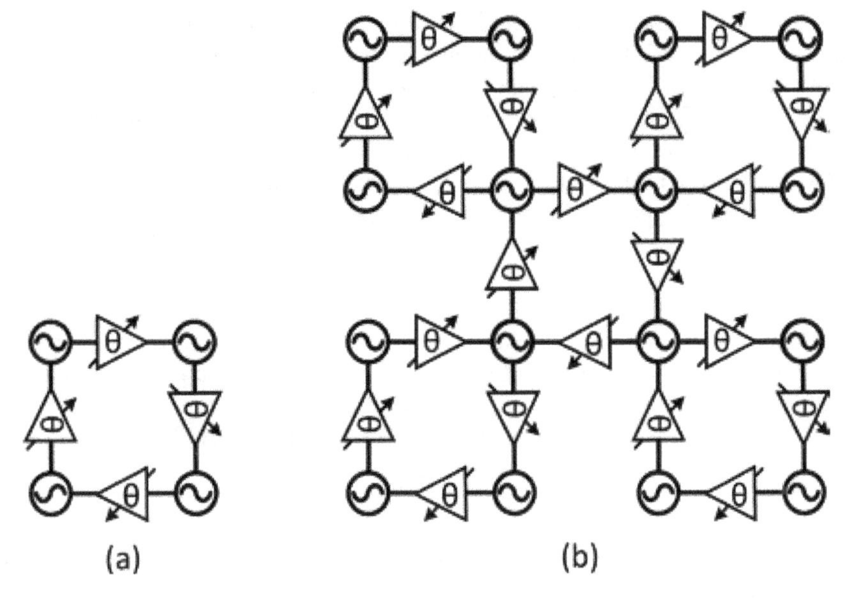

Figure 8.
Block diagram for (a) a unit cell and (b) 2-D oscillator lattice.

a unit cell through which a 2-D oscillator lattice could be formed as shown in **Figure 8(b)**. One nice feature of this design is the use of phase shifter as coupling elements, as this provides us with a new way of tuning the frequency of these injection-locked oscillators [21]:

$$\frac{\Delta f_0}{f_0} = k \sin\left(\Delta\phi(t)\right) \tag{30}$$

Equation (30) is derived from the Adler's equation under locked conditions. Δf_O is the frequency difference of the injected signal and the free-running frequency f_O of the slave oscillator. $\Delta\phi(t)$ is the phase difference. K is a factor relating to the quality factor of the oscillation tank, amplitude of the injected signal, and the free-running oscillation amplitude of the tank.

Since the total phase shift through the loop in **Figure 8(a)** is $2k\pi$, the phase shift through the oscillator plus that of the phase shifter is constrained to a set of fixed values. The oscillation frequency of the whole array is changed if we apply a common shift to all the phase shifter in **Figure 8(b)**. Interesting thing happens when we tune the phase shifter connected with an individual oscillator. If we apply a common shift to the four shifters, this disrupts the local loop momentarily as the instantaneous phase of the affected oscillator jumps to a new value to accommodate the change. This ability of changing the phase of radiating elements individually turns this design into a phased array. Measurements show that the beam could be steered $\pm$50 degree in the E plane and $\pm$ 45 degree in the H plane. The frequency tuning range is 2.1%. The peak EIRP is 17.1 dBm, and the phase noise is 93 dBc – at 1 MHz offset for this 338 GHz array, which show substantial improvements over single oscillators.

A brief comparison between multipliers and harmonic injection-locked VCOs is also in place: The former is generally compact and wideband. They add negligible phase noise if designed properly as dictated by Eq. (22). The biggest issue is the harmonics, which leads to annoying LO spurs that leads to spurious emission and corrupts received signals. The latter comes with higher efficiency and much higher harmonic rejection ratio, but the bandwidth is limited. The close-in phase noise is dominated by the source just like the multiplier, but the far-out phase noise is dominated by the VCO. For mm-wave frequency synthesizers, the two options generally achieve comparable phase noise performances [11]. Since we have to use multipliers to get from mm-wave to THz anyway, this same conclusion holds for THz. For communication applications, it is advisable to use PLL-locked or injection-locked VCOs to generate relatively high LO frequency and use multipliers to boost it to THz frequency (N-push VCOs does this in one place).

2.2.2 THz detectors

THz detectors utilize the nonlinearity of active devices to directly rectify THz signal to DC. A lot of devices could be used, like diode-connected NMOS transistor [22], CE (common emitter) [23] or CB (common base) connected [24, 25] SiGe HBT, CMOS-compatible Schottky diode [26], or P+/n-well diode [27]. THz reception with detectors is incoherent, that is, only the amplitude information is recovered at the receiving side, which limits THz detectors almost exclusively to incoherent THz imaging applications. The strength of THz detectors lies within their simplicity: They do not need LO (local oscillator) signal to do the THz down-conversion. The received THz signal self-mix themselves to DC through even-order nonlinearity of the device. This makes scaling extremely easy, as only low-frequency routing is needed, whereas LO-driven mixer needs cumbersome and power-hungry LO tree which quickly becomes unmanageable when the array gets large. A 1024 pixel NMOS detector array [22] in 65-nm CMOS process showcases the impressive scalability of THz detectors.

However, this flexibility comes with a price: The gain and noise performance of detectors is quite limited, and the specification "responsivity" and "noise equivalent power (NEP)" are used in place of conversion gain and noise figure. The responsivity is defined as the voltage output divided by received power, and NEP is defined as the output noise voltage density divided by responsivity.

The bandwidth of imaging application is usually below 1 MHz; thus, technologies with lower 1/f noise corner frequencies like SiGe HBT or P+/n-Well diode are preferred. The 1/f noise corner frequencies for SiGe HBT and P+/n-well diode are below 1 kHz and 10 kHz, respectively, whereas for NMOS transistor or Schottky diodes, the numbers are well above 1 MHz. For SiGe HBTs, it is shown that CB-connected topology has higher responsivity than CE-connected topology when operating above f_{max} [24].

The principle of THz imaging with detectors largely follow their optical counterpart: They use THz lenses to do the focusing. The problem is that THz wavelength is 2–3 orders of magnitude larger than visible lights, thus large and bulky THz optics are required for reasonable imaging resolutions. They require a lot of effort to set up the imaging setup with the invisible THz radiations [25]. This is the innate deficiency with incoherent imaging. Coherent imaging with THz transceivers could get rid of those optics.

2.2.3 THz transceivers

For transceivers working at lower frequencies, the transmitter and receiver are usually integrated on one chip, and they share the common RF port through switches or duplexers (bandpass filters tuned for simultaneous transmit and receive on different bands). Up to now, neither option is satisfactory for fully integrated silicon THz circuits.

One solution for integrated THz transceiver is to share the antenna and figure out ways to isolate the Tx (transmitter) and Rx (receiver). Park et al. have shown a fully integrated 260 GHz transceiver based on shared leaky-wave antenna [28]. The leaky-wave antenna resembles a lossy transmission line (TL); thus, the Tx and Rx ports could be placed on either end of the antenna. When the transmitter is working, the receiver is turned off and terminates the TL on its side. The same holds true for the receiving mode. The problem with the leaky-wave antenna is that they are relatively long (1.2 mm or 2.5 λ in this design). Statnikov et al. [29] have shown a fully integrated 240 GHz frequency-modulated continuous wave (FMCW) radar transceiver based on shared dual-polarization antenna. A quadrature hybrid coupler is used as a polarizer for the dual-polarization antenna and duplexer for the Tx and Rx. Isolation of the Tx and Rx depends on the orthogonality of left hand circular polarized (LHCP) and right hand circular polarized (RHCP) waves. The Tx and Rx interface with two orthogonal port of the branch-line coupler and are isolated from each other. In Tx mode, the branch-line coupler excites the LHCP mode of the antenna. When the transmitted wave hits a target and bounces back, it changes to RHCP and is subsequently routed to the receiver through the coupler. This scheme is not directly applicable for point-to-point communications, just like frequency-division duplexing (FDD)-based transceivers could not communicate directly with each other.

Another solution is to use two antennas. For FMCW radars, the leakage from the TX to the Rx results in strong interferences around DC [30]. This raises the noise floor in the range spectrum. With area permitting, the Tx and Rx antenna should be separated further apart for better isolation. The measured crosstalk between the two antennas with a separation of about 1.8 mm in a 160 GHz FMCW radar transceiver is below 31 dB [31]. This isolation might be adequate for FMCW radar applications, but it is still wanting for communications.

Transceiver-based THz imaging makes coherent imaging possible, as both the magnitude and phase information of the signal from targets are retained. With both information available, it is possible to get rid of bulky THz optics by sampling the THz field directly and do the focusing digitally. The THz field is usually sampled on a 2-D plane with different THz frequencies; this is fulfilled by raster scanning a FMCW transceiver (or the sample). For a given point in space, the round-trip phase delay from the transceiver to that point is a function of its position and sampling frequency. By raster scanning the transceiver or the sample under different frequencies, its phase delay variation is orthogonal to every other point in the sampling space. This forms the basis for the 3-D imaging through the back-projection algorithm. 3-D imaging based on SiGe FMCW transceivers is reported by several groups [32, 33], showcasing the great potential for low-cost THz imaging applications.

For communication applications, the modulation scheme plays a major role in deciding the transceiver architecture. Low-complexity modulation schemes like on-off keying (OOK) and binary phase shift keying (BPSK) lead to robust and power-efficient design, but the spectrum efficiencies are relatively low. Modulation schemes like 32 QAM and 128 QAM lead to much higher spectrum efficiency, but they are quite demanding on linearity and phase noise performance, and they require image-rejection architectures as the spectra of QAM are asymmetric around the carrier. The upper sideband (USB) and lower sideband (LSB) of the spectra become each other's own image when converted to baseband, and image-rejection is needed to avoid signal corruption. Image-rejection modulation/demodulation is difficult in THz range as I/Q mixers are required. It is very difficult to guarantee phase and amplitude matching for the I/Q LO signal for adequate image-rejection at THz frequency.

A 210 GHz fundamental transceiver chipset with OOK modulation is demonstrated in a 32-nm SOI CMOS process [34]. Ideally speaking, power amplifier (PA)-based fundamental operation is more power-efficient than frequency multipliers.

This helps to boost efficiency of the whole system as PAs are usually the most power-hungry circuits in transceivers. Perhaps the most difficult part of this design is controlling the oscillator pulling effect. Since the PA works at the same frequency as the on-chip VCO, significant coupling could occur between PA and VCO. The injection-locking effect would impact the phase noise performance heavily. The on-chip antenna used in this design only makes things more difficult. To improve the VCO performance, a stacked cross-coupled VCO topology is used to boost oscillation amplitude, improving its robustness in response to interferences.

A 240-GHz direction-conversion transceiver in SiGe BiCMOS technology is demonstrated with BPSK capability. BPSK is a constant-envelope modulation, which means the PA could be driven to saturation for better power efficiency. The spectra of BPSK modulation are symmetric around its carrier (symmetrically modulated), making direct conversion easier to implement as no image-rejection is needed. A 30 GHz LO signal is supplied to this transceiver, and on-chip ×8 multipliers are used for the 240 GHz LO generation. This helps to alleviate the detrimental effect of LO spurs caused by multipliers since they are separated by twice the baseband bandwidth (15 GHz). An on-chip antenna with 1-dB bandwidth of 33 GHz is achieved partly due to the local back etching (LBE) technology used. The silicon substrate below the antenna is removed, resulting in a low-loss air cavity below the antenna. The transceiver link is tested with 15 cm separation, and an impressive 6-dB bandwidth of 35 GHz is obtained. A 25 Gbps wireless link is demonstrated by this transceiver with no equalization. One problem with direct conversion using no I/Q demodulation is that the demodulated signal's SNR is dependent on the phase difference between the Tx LO and Rx LO. A phase shifter is used in this test in case manual tuning is required to boost the SNR.

A 300 GHz QPSK transmitter for dielectric waveguide communication is demonstrated in a 65-nm CMOS process [35]. Again, off-chip LO signal is used to drive on-chip frequency multipliers. The targeted data rate is 30 Gbps, which translates to around 20 GHz baseband bandwidth for QPSK assuming a roll-off factor of 0.3. Thus, the off-chip LO signal frequency is set to 45 GHz. An on-chip quadrature modulator is used to modulate the baseband data to an IF frequency of 135 GHz. It is further shifted by a double-balanced mixer to 315 GHz. Such a high IF alleviates the need for image-rejection mixers since the image frequency falls completely out of band. A 30 Gbps QPSK is demonstrated with on-chip probing.

A 230 GHz direct-conversion 16-QAM 100-Gbps wireless link is demonstrated with a communication distance of 1 meter [36]. The I/Q mixer directly interfaces with on-chip antenna to avoid bandwidth limitation introduced by LNA. On-chip LO multiplier chain is used to convert the external 13.75–16 GHz LO to 220–256 GHz. The baseband bandwidth is around 14 GHz; this poses challenge as the spacing of LO spurs is comparable to this bandwidth. This leads to spurious modulation that overlaps with desired signal. Nevertheless, 100 Gbps with an EVM of 17% is demonstrated.

A 300 GHz 32-QAM and 128-QAM transmitter with 105-Gbps data rate is demonstrated in a 40 nm CMOS process [37]. As there is no PA available, an array of eight square mixers (i.e., mixing through the second-order nonlinearity) is power combined at the output stage. A heterodyne topology is used, and the LO frequencies for the two up-conversion stage are both set at 135 GHz. The IF frequency for the first stage is around 10 GHz, and high-pass filtering is used to suppress the LSB by approximately 10 dB. Single-balanced mixer is used in the first stage to intentionally leak LO signal to the second stage. The second-order nonlinearity of NMOS transistor is used to mix the (IF+LO) signal with LO leakage to obtain the desired intermodulation signal (IF+2LO). Unwanted second harmonics of LO and IF signal is canceled at the output rat-race balun. On-chip probing validates the operation;

the 32-QAM modulation with an EVM of 8.9% is achieved with 105 Gbps. No on-chip antenna is used as this chip is intended to drive high-power THz devices like traveling-wave tubes.

2.3 THz interface for efficient power transfer

The THz interface serves as a gateway between the circuit and the outside world. The efficiency of this interface greatly impacts the performance of the overall system. A simple derivation of the transceiver's link budget would highlight the importance of this interface (**Figure 9**):

$$P_R = P_T - IL_2 - IL_0 - IL_1 \tag{31}$$

where P_R is power received at the receiver, P_T is the output power of the transmitter, and IL_2 and IL_1 are the loss for THz interface at the transmitter and the receiver. IL_0 is the loss associated with the propagation medium. If the medium is free space, which is often the case, IL_0 equals

$$IL_0 = 20 \times \log_{10}\left(\frac{\lambda_0}{4\pi d}\right) + G_{TX} + G_{RX} \tag{32}$$

λ_0 is the free-space THz wavelength, d is the propagation distance, and G_{TX} and G_{RX} are the gain of antenna employed at the transmitter and receiver. Note that Eqs. (31) and (32) are in dB form. Also keep in mind that Eq. (32) is only valid under far-field conditions, that is, the radiation field seen by the receiver is not reactive and could be approximated by plane waves. Common criterion for far-field condition is that the receiver is separated from the transmitter by at least $2D^2/\lambda$, where D is the maximum overall dimension of the transmitting antenna.

To maximize P_R, we have to minimize the loss at the interface and increase the gain of the antennas. This section covers both areas.

For THz silicon chips, grounded coplanar waveguide (GCPW) is the most prevalent medium for on-chip THz routing. It is a combination of coplanar waveguide with microstrip line. This configuration have several merits: First, the ground plane of the microstrip line shields the signal from the electrically thick silicon substrate, which reduces loss and prevents the signal from leaking into substrate modes; the coplanar waveguide makes interface with outside world easier, be it through flip-chip bonding or on-chip probing since the ground conductor lies in proximity with the signal trace.

A 90–300 GHz transmitter and a 115–325 GHz receiver are flip-chip bonded to a liquid-crystal polymer (LCP) substrate [12]. This connects the chip to the 100–280 GHz Vivaldi antenna on the LCP substrate. Such wideband antenna is extremely difficult to realize on-chip. As another example, a CMOS 300 GHz transmitter chip is flip-chip bonded to a GCPW-to-WG transition module [38] implemented on a

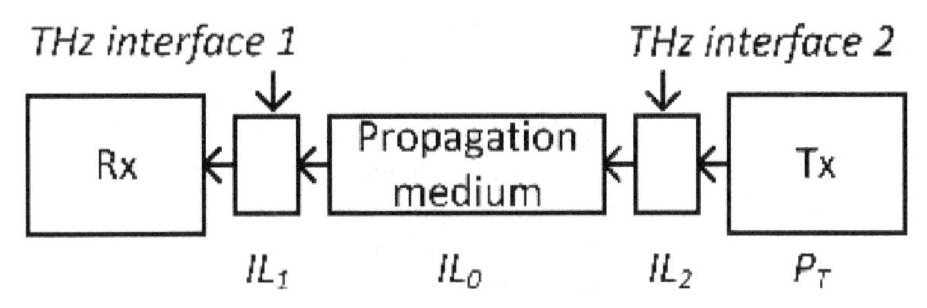

Figure 9.
Link budget analysis of a THz transceiver system.

low-cost glass epoxy PCB. Once transitioned to waveguide interfaces, the chip could be interfaced to a plethora of THz components like horn antennas and high-power amplifier modules. The packaging loss is 8 dB, which includes the transition loss (flip-chip bonding and the GCPW-to-WG), impedance mismatch, and loss in the epoxy material. It should be noted that gold stud bumping is used in both cases, which is compatible with conventional wire bonders and is quite convenient for R&D labs.

An effective way to lower transition loss is to radiate the THz signal directly from chip. Since the high permittivity silicon substrate readily traps the THz radiation, there are basically two lines of thoughts regarding on-chip antenna design. One is to accept this coupling and take this into account while designing antennas; the other is to eliminate the substrate mode altogether.

The first approach tries to make use of the electrically thick antenna to improve the antenna bandwidth. Metal reflector are placed underneath the substrate to reflect energy back, which is often the top layer of the PCB under the chip. To make sure the reflected power adds constructively with the surface radiation, the chip should be $(2k + 1)/4\lambda$ thick [34, 39, 40]. The phase delay of the reflected wave increases linearly with frequency, which limits the bandwidth. Artificial magnetic conductor (AMC) reflector in place of the solid metal reflector on the PCB is used to compensate this phase shift [41], which extends the on-chip antenna bandwidth substantially. The antenna efficiency is within 0.5dB of the peak efficiency between 200 and 300 GHz. Note that the chip should be $1/2\lambda$ thick in this case as AMC introduces zero phase shift at the center frequency.

Although promising, allowing reflections within the substrate increases couplings between antenna elements and on-chip passive components. This makes circuit performance sensitive to antenna location and substrate dimensions (both lateral dimension and chip thickness), which is undesirable for designing arrays.

There are two widely adopted approach to eliminate the substrate mode. The first approach involves attaching a high-k dielectric lens on the back side of the chip, and the antenna radiates through the lens [22, 42–44]. This approach offers high directivity and improves efficiency. The most obvious drawback is the need for nonstandard packaging, which could be quite costly. An intuitive way of understanding why this structure can eliminate the substrate mode can be found in [8].

The second approach is more straightforward: The substrate is shielded from the radiating element using one or a few low-level metal layers. This eliminates the possibility of coupling with substrate modes, but it limits the bandwidth and radiation efficiency severely since the radiating element is only a few microns away from the ground due to the restriction of silicon backend process. The radiating element thus forms a high Q tank with the ground, and the bandwidth of the antenna is on the order of a few percent [13, 20, 25]. The obvious way to increase bandwidth and radiation efficiency is to lower the Q, which could be accomplished by increasing the volume of the resonant tank. One solution is to add a dielectric superstrate above the radiating element [45, 46], which diverts some of the electric fields from the dielectric layer below the radiating element, increasing the resonance volume considerably. The radiation efficiency increases with the superstrate till the onset of the $TE1$ mode, which limits the thickness of the superstrate to $\lambda_0/4\sqrt{\varepsilon_r - 1}$ where ε_r is the dielectric constant of the superstate. Dielectric resonator antenna (DRA) is another option [47, 48]. This antenna composes of a dielectic resonator on top of the chip and a feed element in the top metal layer of the chip. The feed element is used to excite resonance inside the resonator, through which the THz signal radiates. The size of the resonator and relative position of feed could be adjusted for intended oscillation frequency and resonance mode. For a 270 GHz microstrip antenna, the gain is enhanced by 4 dB, and the 3-dB gain bandwidth is extended by 100% when an dielectric resonator is used [49].

For phased array applications, the patch antenna with ground shield is the most straightforward approach.

3. Summary

Silicon THz circuit design is an active research area open to innovations on multiple levels. We need better passive components, better circuits, and the most important thing is we need to come up with better ways of building THz arrays. Scaling is the key for significantly boosting the performance of silicon THz systems as we venture into this last untapped spectrum [50].

Author details

Li Zhuang*, Cao Rui, Tao Xiaohui, Jiang Lihui and Rong Dawei
Key Lab of Aperture Array and Space Applications, He Fei, China

*Address all correspondence to: lizhuang@mail.ustc.edu.cn

References

[1] Mason SJ. Power gain in feedback amplifier. Transactions of the IRE Professional Group on Circuit Theory. 1954;**1**(2):20-25

[2] Gupta MS. Power gain in feedback amplifiers, a classic revisited. IEEE Transactions on Microwave Theory and Techniques. 1992;**40**(5):864-879

[3] Spence R. Linear Active Networks. London, UK: Wiley; 1970

[4] Khatibi H, Khiyabani S, Afshari E. A 173 GHz amplifier with a 18.5 dB power gain in a 130 nm SiGe process: A systematic Design of High-Gain Amplifiers above $f_{max}/2$. IEEE Transactions on Microwave Theory and Techniques. 2018;**66**(1):201-214

[5] Khatibi H, Khiyabani S, Afshari E. A 183 GHz desensitized unbalanced Cascode amplifier with 9.5-dB power gain and 10-GHz band width and −2 dBm saturation power. IEEE Solid-State Circuits Letters. 2018;**1**(3):58-61

[6] Wang Z, Heydari P. A study of operating condition and design methods to achieve the upper limit of power gain in amplifiers at near-fmax frequencies. IEEE Transactions on Circuits and Systems I: Regular Papers. 2017;**64**(2): 261-271

[7] Momeni O, Afshari E. High power terahertz and millimeter-wave oscillator design: A systematic approach. IEEE Journal of Solid-State Circuits. 2011; **46**(3):583-597

[8] Han R, Afshari E. A CMOS high-power broadband 260-GHz radiator array for spectroscopy. IEEE Journal of Solid-State Circuits. 2013;**48**(12): 3090-3104

[9] Han R et al. A SiGe terahertz heterodyne imaging transmitter with 3.3 mW radiated power and fully-integrated phase-locked loop. IEEE Journal of Solid-State Circuits. 2015; **50**(12):2935-2947

[10] Lee JJ. G/T and noise figure of active array antennas. IEEE Transactions on Antennas and Propagation. 1993;**41**(2): 241-244

[11] Wang CC, Chen Z, Heydari P. W-band silicon-based frequency synthesizers using injection-locked and harmonic triplers. IEEE Transactions on Microwave Theory and Techniques. 2012;**60**(5):1307-1320

[12] Chi T, Huang MY, Li S, Wang H. A packaged 90-to-300GHz transmitter and 115-to-325GHz coherent receiver in CMOS for full-band continuous-wave mm-wave hyperspectral imaging. In: International Solid-State Circuits Conference. 2017. pp. 304-305

[13] Yang Y, Gurbuz OD, Rebeiz GM. An eight-element 370–410-GHz phased-array transmitter in 45-nm CMOS SOI with peak EIRP of 8–8.5 dBm. IEEE Transactions on Microwave Theory and Techniques. 2016;**64**(12):4241-4249

[14] Ahmed F, Furqan M, Heinemann B, Stelzer A. 0.3-THz SiGe-based high-efficiency push–push VCOs with >1-mW peak output power employing common-mode impedance enhancement. IEEE Transactions on Microwave Theory and Techniques. 2018;**66**(3):1384-1398

[15] Heydari P. Millimeter-wave frequency generation and synthesis in silicon. In: 2018 IEEE Custom Integrated Circuits Conference (CICC). 2018. pp. 1-49

[16] Chiang PY, Wang Z, Momeni O, Heydari P. A silicon-based 0.3 THz frequency synthesizer with wide locking range. IEEE Journal of Solid-State Circuits. 2014;**49**(12):2951-2963

[17] Chiang P-Y, Momeni O, Heydari P. A 200-GHz inductively tuned VCO With-7-dBm output power in 130-nm SiGe BiCMOS. IEEE Transactions on Microwave Theory and Techniques. 2013;**61**(10):3666-3673

[18] Schmid RL, Coen CT, Shankar S, Cressler JD. Best practices to ensure the stability of sige HBT cascode low noise amplifiers. In: Proceedings of the IEEE Bipolar/BiCMOS Circuits and Technology Meeting. 2012. pp. 2-5

[19] Murphy D, Darabi H, Wu H. Implicit common-mode resonance in LC oscillators. IEEE Journal of Solid-State Circuits. 2017;**52**(3):812-821

[20] Tousi Y, Afshari E. A high-power and scalable 2-D phased array for terahertz CMOS integrated systems. IEEE Journal of Solid-State Circuits. 2015;**50**(2):597-609

[21] Bhansali P, Roychowdhury J. Gen-Adler: The generalized Adler's equation for injection locking analysis in oscillators. In: 2009 Asia and South Pacific Design Automation Conference. 2009. pp. 522-527

[22] Al Hadi R et al. A 1 k-pixel video camera for 0.7-1.1 terahertz imaging applications in 65-nm CMOS. IEEE Journal of Solid-State Circuits. 2012; **47**(12):2999-3012

[23] Uzunkol M, Gurbuz OD, Golcuk F, Rebeiz GM. A 0.32 THz SiGe 4×4 imaging array using high-efficiency on-chip antennas. IEEE Journal of Solid-State Circuits. 2013;**48**(9):2056-2066

[24] Al Hadi R, Grzyb J, Heinemann B, Pfeiffer UR. A terahertz detector array in a SiGe HBT technology. IEEE Journal of Solid-State Circuits. 2013;**48**(9): 2002-2010

[25] Li Z, Qi B, Zhang X, Zeinolabedinzadeh S, Sang L, Cressler JD. A 0.32-THz SiGe imaging array with polarization diversity. IEEE Transactions on Terahertz Science and Technology. 2018;**8**(2):215-223

[26] Han YR, Zhang Y, Kim D, Kim Y, Shichijo H, Kenneth O. Terahertz image sensors using CMOS Schottky barrier diodes International SoC Design Conference, pp. 254–257, 2012

[27] Ahmad Z, Kenneth O. THz detection using P+/n-well diodes fabricated in 45-nm CMOS. IEEE Electron Device Letters. 2016;**37**(7): 823-826

[28] Park J, Kang S, Thyagarajan SV, Alon E, Niknejad AM. A 260 GHz fully integrated CMOS transceiver for wireless chip-to-chip communication. In: 2012 Symposium on VLSI Circuits (VLSIC). 2012. pp. 48-49

[29] Statnikov K, Sarmah N, Grzyb J, Malz S, Heinemann B, Pfeiffer UR. A 240 GHz circular polarized FMCW radar based on a SiGe transceiver with a lens-integrated on-chip antenna. In: Proceedings of European Radar Conference (EuRAD). 2014. pp. 447-450

[30] Park J, Kang S, Niknejad AM. A 0.38 THz fully integrated transceiver utilizing a quadrature push-push harmonic circuitry in SiGe BiCMOS. IEEE Journal of Solid-State Circuits. 2012;**47**(10):2344-2354

[31] Hitzler M et al. Ultracompact 160-GHz FMCW radar MMIC with fully integrated offset synthesizer. IEEE Transactions on Microwave Theory and Techniques. 2017;**65**(5):1682-1691

[32] Jaeschke T, Bredendiek C, Pohl N. 3-D FMCW SAR imaging based on a 240 GHz SiGe transceiver chip with integrated antennas. In: The German Microwave Conference (GeMIC). 2014. pp. 1-4

[33] Hamidipour A, Feger R, Poltschak S, Stelzer A. A 160-GHz system in package for short-range mm-wave applications. International Journal of Microwave and Wireless Technologies. 2014;**6**(3–4): 361-369

[34] Wang Z, Chiang PY, Nazari P, Wang CC, Chen Z, Heydari P. A CMOS 210-GHz fundamental transceiver with OOK modulation. IEEE Journal of Solid-State Circuits. 2014;**49**(3):564-580

[35] Zhong Q, Chen Z, Sharma N, Kshattry S, Choi W, Kenneth KO. 300-GHz CMOS QPSK transmitter for 30-Gbps dielectric waveguide communication. In: 2018 IEEE Custom Integrated Circuits Conference (CICC). 2018. pp. 1-4

[36] Rodríguez-Vázquez P, Grzyb J, Heinemann B, Pfeiffer UR. A 16-QAM 100-Gb/s 1-M wireless link with an EVM of 17% at 230 GHz in an SiGe technology. IEEE Microwave and Wireless Components Letters. 2019; **29**(4):297-299

[37] Takano K et al. 17.9 A 105Gb/s 300GHz CMOS transmitter. In: 2017 IEEE International Solid-State Circuits Conference (ISSCC). 2017. pp. 308-309

[38] Takano K et al. 300-GHz CMOS transmitter module with built-in waveguide transition on a multilayered glass epoxy PCB. In: 2018 IEEE Radio and Wireless Symposium (RWS). 2018. pp. 154-156

[39] Kang S, Thyagarajan SV, Niknejad AM. A 240 GHz fully integrated wideband QPSK transmitter in 65 nm CMOS. IEEE Journal of Solid-State Circuits. 2015;**50**(10):2256-2267

[40] Thyagarajan SV, Kang S, Niknejad AM. A 240 GHz fully integrated wideband QPSK receiver in 65 nm CMOS. IEEE Journal of Solid-State Circuits. 2015;**50**(10):2268-2280

[41] Zhong Q, Choi W, Miller C, Henderson R, Kenneth KO. A 210-to-305GHz CMOS receiver for rotational spectroscopy. Digest of Technical Papers - IEEE International Solid-State Circuits Conference. 2016;**59**:426-427

[42] Filipovic DF, Gearhart SS, Rebeiz GM. Double-slot antennas on extended hemispherical and elliptical silicon dielectric lenses. IEEE Transactions on Microwave Theory and Techniques. 1993;**41**(10):1738-1749

[43] Statnikov K, Ojefors E, Grzyb J, Chevalier P, Pfeiffer UR. A 0.32 THz FMCW radar system based on low-cost lens-integrated SiGe HBT front-ends. In: European Solid-State Circuits Conference. 2013. pp. 81-84

[44] Pfeiffer UR et al. A 0.53 THz reconfigurable source module with up to 1 mW radiated power for diffuse illumination in terahertz imaging applications. IEEE Journal of Solid-State Circuits. 2014;**49**(12):2938-2950

[45] Edwards JM, Rebeiz GM. High-efficiency elliptical slot antennas with quartz superstrates for silicon RFICs. IEEE Transactions on Antennas and Propagation. 2012;**60**(11):5010-5020

[46] Ou YC, Rebeiz GM. Differential microstrip and slot-ring antennas for millimeter-wave silicon systems. IEEE Transactions on Antennas and Propagation. 2012;**60**(6):2611-2619

[47] Mongia RK, Bhartia P. Dielectric resonator antennas—A review and general design relations for resonant frequency and bandwidth. International Journal of Microwave and Millimeter-Wave Computer-Aided Engineering. 1994;**4**(3):230-247

[48] Li C, Chiu T. 340-GHz low-cost and high-gain on-Chip higher order mode dielectric resonator antenna for THz applications. IEEE Transactions on

Terahertz Science and Technology. 2017;**7**(3):284-294

[49] Hou D, Chen J, Yan P, Hong W. A 270 GHz × 9 multiplier chain MMIC with on-chip dielectric-resonator antenna. IEEE Transactions on Terahertz Science and Technology. 2018;**8**(2):224-230

[50] Afshari E. "The Last Untapped Spectrum: Should Industry Care About Terahertz?" 2016. [Online]. Available from: https://sscs.ieee.org/event-record ing-and-slides-now-available-the-last-untapped-spectrum-should-ind ustry-care-about-terahertz-presented-by-ehsan-afshari

4

Dielectric Losses of Microwave Ceramics Based on Crystal Structure

Hitoshi Ohsato, Jobin Varghese and Heli Jantunen

Abstract

So far, many microwave dielectric materials have been investigated for a range of telecommunication applications. In dielectrics, the three main dielectric properties are quality factor (*Q*), dielectric constant and temperature coefficient of resonant frequency. Among these, the most essential dielectric property is *Q*. More specifi-cally, *Q* is the inverse of the dielectric loss ($\tan\delta$); thus $Q = 1/\tan\delta$. There are two kinds of losses: those depending on crystal structure and losses due to external factors. The former is intrinsic losses such as ordering, symmetry, and phonon vibration. The latter is extrinsic losses due to factors such as grain size, defects, inclusions and distortion. In this chapter, the authors present the origin of dielectric losses based on the crystal structure. An ideal and well-proportional crystal structure constitutes a low loss material. Most dielectric materials are paraelectrics with inversion symmetry *i* and high symmetry. In general, it is believed that ordering gives rise to a high *Q*, on which many researchers are casting doubt. In the case of complex perovskites, the symmetry changes from cubic to trigonal. Ordering and symmetry should be compared with the structure. In this chapter, three essential conditions for the origin of high *Q* such as high symmetry, compositional ordering and compositional density are presented.

Keywords: microwave dielectrics, *Q*-factor, ordering, symmetry, indialite/cordierite, pseudo tungsten-bronze, complex perovskite

1. Introduction

Microwave and millimetre-wave dielectric materials [1–6] have been investigated for a wide range of telecommunication applications, such as mobile and smartphones, wireless local area network (LAN) modules and intelligent transport system (ITS). Millimetre-wave dielectric materials with high quality factor *Q* and low dielectric constant ε_r are required for the next 5G telecommunication applications used for noncondensed high data transfer on LAN/ personal area networks (PAN) and the higher frequency radar on autonomous cars.

In microwave dielectrics, there are three fundamental dielectric properties: quality factor (*Q*), dielectric constant (ε_r) and temperature coefficient of resonant frequency (TCf/τ_f) [1, 2, 6]. Microwave dielectrics have been used as the criti-cal constituents of wireless communications [7–10], such as resonators, filters and temperature-stable capacitors with a near zero ppm/°C $TC\varepsilon_r$ (temperature coefficient of the dielectric constant). Among the dielectric properties, the most essential property is Q the inversion of the dielectric loss ($\tan\delta$); thus $Q = 1/\tan\delta$. The dielectric losses of microwave dielectrics should be small. So, most of the microwave dielectrics are paraelectrics with inversion symmetry i, while most of the electronic materials are ferroelectrics with spontaneous polarity showing substantial dielectric losses [11–13]. The microwave dielectrics attract attention as a high potential material, which have an over-well-proportional rigid crystal structure with symmetry. That is, the structure should be without electric defects, nondistortion and without strain.

Under the influence of an electric field, four types of polarisation mechanisms can occur in dielectric ceramics, that is, interfacial, dipolar, ionic and electronic. In general, the microwave dielectric properties such as ε_r and Q are mostly influenced by ionic or electronic polarisation. The dielectric polarisation generates the dielectric losses in the presence of an electromagnetic wave. When the frequency is increased to millimetre-wave values, the dielectric losses may be increased or decreased depending on the polarisation mechanism. There are two kinds of losses: those depending on crystal structure and losses due to external factors. It was believed that the intrinsic losses are due to the ordering/disordering, symmetry and phonon vibration, while extrinsic losses are due to factors such as grain size, defects, inclusions, density and distortion from stress.

In this chapter, the origins of high Q are discussed based on the intrinsic factors related to the crystal structure, such as symmetry, compositional ordering and compositional density. Although it has previously been believed that ordering based on the order-disorder phase transition brings high Q [14], the authors propose that it is primarily a high symmetry that leads to high Q [15]. The following focused studies relate to specific examples; indialite with high symmetry showing higher Q than cordierite with an ordered structure [16–18]; pseudo tungsten-bronze solid solutions without phase transition showing high Q based on the compositional ordering [19–21]; complex perovskite compounds with order-disorder transitions depending on density and grain size [22, 23] and complex perovskites with composition deviated from the stoichiometric depending on the compositional density showing a high Q [24–29].

2. Focused studies

2.1 Indialite/cordierite glass ceramics

2.1.1 Indialite Q-factor improved by Ni-substitution

Cordierite ($Mg_2Al_4Si_5O_{18}$) has two polymorphs: cordierite and indialite, as shown in **Figure 1(a)** and **(c)**, respectively [30, 31]. Cordierite is of low symmetry form: orthorhombic crystal system *Cccm* (No. 66), which has $Si_4Al_2O_{18}$ six-membered tetrahedron rings with ordered SiO_4 and AlO_4 tetrahedra as shown in **Figure 1(b)**. On the other hand, indialite is of high symmetry form: hexagonal crystal system *P6/mcc* (No. 192), which has disordered $Si_4Al_2O_{18}$ equilateral hexagonal rings as shown in **Figure 1(c)**.

Cordierite shows a lower ε_r of 6.19 which depends on the silicates and a near-zero *TCf* of −24 ppm/°C [32] as compared to other silicates as shown in **Figure 2(a)**. Based on these properties, Terada et al. carried out initiative research on these micro-wave dielectrics [16]. They reported an excellent *Qf* by substituting Ni for Mg as shown in **Figure 2(b)**. The *Qf* was improved from 40 × 10^3 GHz to 100 × 10^3 GHz by Ni substitution of $x = 0.1$ in $(Mg_{1-x}Ni_x)_2Al_4Si_5O_{18}$. The Ni substitution did not change the ε_r value

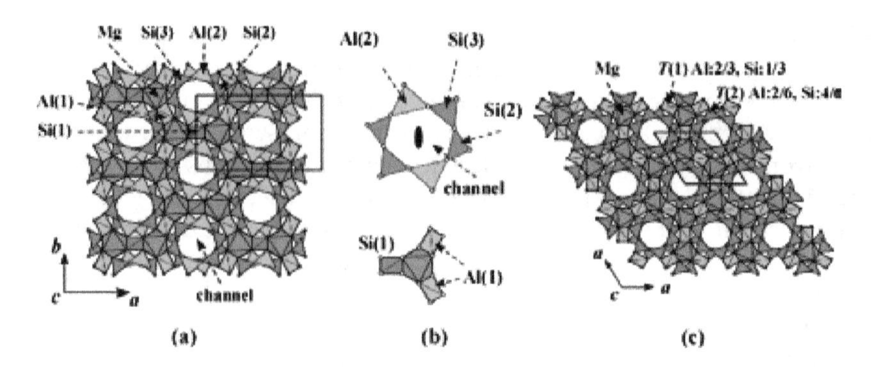

Figure 1.
Schematic representation of cordierite (a), six-membered tetrahedron ring with ordered SiO_4 and AlO_4 (b) and indialite (c).

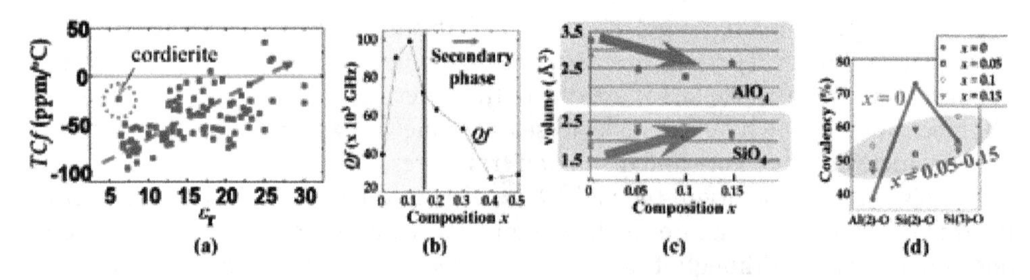

Figure 2.
Cordierite with near zero ppm/°C deviated from other compounds (a). Ni-substituted cordierite Qf (b), volume of AlO_4 and SiO_4 (c) and covalencies of Si-O and Al-O as a function of composition x (d).

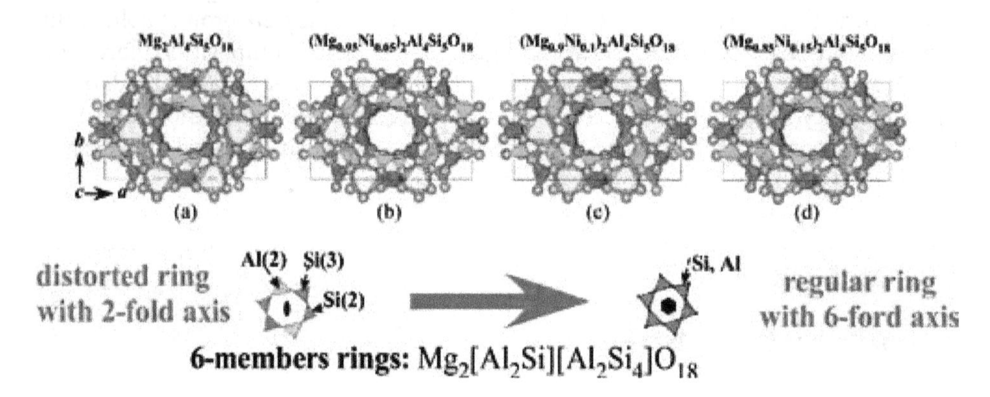

Figure 3.
Crystal structure of Ni-substituted cordierite: $(Mg_{1-x}Ni_x)_2Al_4Si_5O_{18}$ with composition x = 0 (a), 0.05 (b), 0.1 (c) and 0.15 (d).

considerably, but the *TCf* was degraded from −24 to −30 ppm/°C [16]. For $x > 0.1$, the properties were affected by the formation of the secondary phase of $NiAl_2O_4$.

Terada et al. also analysed the crystal structure by the Rietveld method [33] to clarify the origin of the improved *Qf* value. The X-ray powder diffraction (XRPD) pattern was obtained by a multi-detector system (MDS) [34] in the synchrotron radiation "Photon Factory" of the National Laboratory for High Energy Physics in Tsukuba, Japan. **Figure 3(a)–(d)** shows the crystal structures of Ni-substituted cordierite $(Mg_{1-x}Ni_x)_2Al_4Si_5O_{18}$ with $x = 0$, 0.05, 0.1 and 0.15. The crystal structure showed a tendency to deform to indialite with high symmetry on the hexagonal ring composed of corner-sharing of (Si, Al)O_4 tetrahedra in the *a-b* plane. Ni-substituted cordierite $(Mg_{1-x}Ni_x)_2Al_4Si_5O_{18}$ with composition $x = 0.1$

(**Figure 3(c)**) was obviously closer to equilateral hexagonal rings compared to $(Mg_{0.95}Ni_{0.05})_2Al_4Si_5O_{18}$ (**Figure 3(b)**) and $Mg_2Al_4Si_5O_{18}$ (**Figure 3(a)**).

The transformation from cordierite to indialite, represented by the ratio of disordering between the SiO_4 and AlO_4 tetrahedra, is based on the volumes and covalencies of the SiO_4 and AlO_4 tetrahedra [35]. The volume was calculated using atomic coordi-nates obtained by Rietveld crystal structural analysis as shown above. The covalency (f_c) of the cation-oxygen bond was estimated from the following equation [36].

$$f_c = as^M \quad (1)$$

The empirical constants a and M depending on the inner-shell electron number 10 are 0.54 v.u. and 1.64, respectively [37], where s is the bond length obtaining from the following equation:

$$s = (R/R1)^{-N} \quad (2)$$

where, R is defined as the bond length, and $R1$ and N are the measured parameter reliant on the cation site and each cation-anion pair, respectively.

Figure 2(c) and **(d)** depicts the calculated volume and covalency of SiO_4 and AlO_4 octahedra, respectively. These figures show the phase changing from cordierite to indialite as substitution of Ni in the Mg site. In the cordierite $Mg_2Al_4Si_5O_{18}$ (**Figure 1(a)**), Si/Al ions in the tetrahedra are ordered. Therefore, the volume and covalency of tetrahedra are different values, but the values are becoming similar to the substitution of Ni in the Mg site. This is due to the disordering of Si/Al ion phase transition in the cordierite (**Figure 1(a)**) to indialite (**Figure 1(c)**). In the indialite, the disordered $Si_4Al_2O_{18}$ equilateral hexagonal rings with 6-ford axis are the main framework as analysed by the Rietveld method as shown in **Figure 3(d)**. The improvement of *Qf* as shown in **Figure 2(b)** should be based on the disordering due to high symmetry instead of an ordering of SiO_4 and AlO_4 tetrahedra by order-disorder transition. It is one example of high symmetry bringing a higher Q than ordering by the order-disorder transition [18].

2.1.2 Indialite glass ceramics with high Q

As described in the previous section, the *Qf* value of indialite derived by substituting Ni for Mg was improved to three times that of cordierite. Based on the new knowledge, Ohsato et al. proposed the synthesis of indialite with superior microwave dielectric properties [17]. The indialite, being a high-temperature form, could not be synthesised by the solid-state reaction because the order-disorder phase transition is hindered by the incongruent melting to form mullite and liquid. On the other hand, indialite is an intermediate phase during the crystallisation process from glass with a cordierite composition to cordierite, as shown in **Figure 4**. Therefore, fabrication of indialite glass ceramics has been attempted [17, 39]. Although the indialite is a metastable phase transforming to cordierite at higher temperatures, it is a relatively stable phase which occurs in nature formed by the crystallisation of natu-ral glass. As this occurrence is in India, the mineral was named indialite. Another phase of μ-cordierite precipitating in the early stage of the crystallisation of cordierite glass is β-quartz solid solutions. The naming of μ-cordierite is not correct because of the differ-ent crystal structure, so the name that should be used is β-quartz solid solutions [38].

The cordierite composition was melted at 1550°C and was cast into a cylindrical rod with the diameter φ = 10 mm and l = 30 mm in a graphite mould. In order to avoid fracture due to internal strain, the cast glass rod was annealed at 760°C below

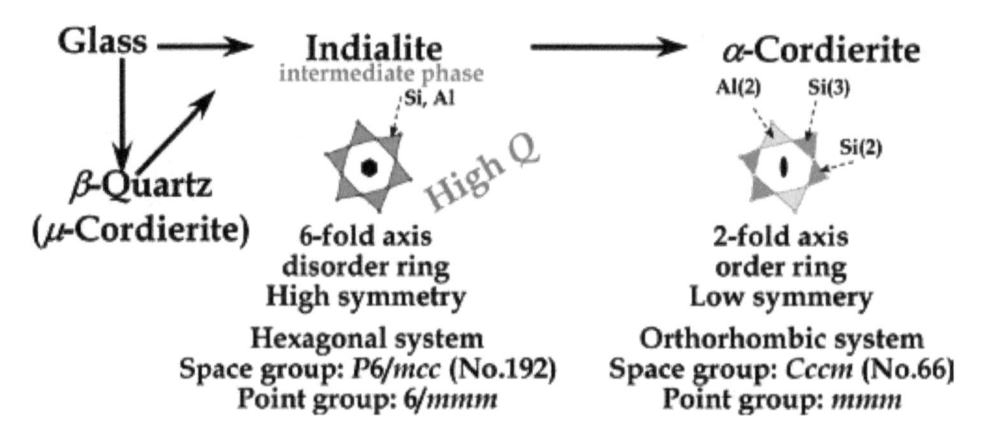

Figure 4.
Polymorphism of cordierite: indialite is the high temperature/high symmetry form, and cordierite is the low temperature/low symmetry form. In addition, indialite is an intermediate phase during crystallisation from glass to cordierite.

the glass transition point of 778°C [38]. The 10 mm diameter glass rod was cut to form a resonator with a height of 6 mm. The glass pellets were crystallised at temperatures in the range 1200–1470°C/10 and 20 h. The crystallised pellets had two problems: deformation by the formation of glass phase and cracking by anisotropic crystal growth from the surface (**Figure 5(a)**) [39]. **Figure 5(b)** and **(c)** shows photographs taken by a polarising microscope of a thin section of the crystallised samples. The needle-like crystals grown from the surface had an orientation with *c*-axis elongation. The microwave dielectric properties of the sample with cracking had a wide scattering range of the data [17, 39].

Figure 6(a) shows the volume of indialite/cordierite examined by the Rietveld method [40], which is estimated with two phases such as indialite and cordierite. Hereabout, the residual % is compared to that of cordierite. At 1200°C, the precipitated phase of indialite was about 96.7%. The volume of indialite reduced as the temperature and to 17.1% (82.9% for cordierite) at 1400°C. **Figure 6(b)** and **(c)** shows the microwave dielectric properties of indialite/cordierite glass ceramics and remarkably high *Qf* value of more than 200 × 10^3 GHz at 1300°C/20 h [17]. This is much better than the highest *Qf* value of 100 × 10^3 GHz obtained by substitution with Ni using the conventional solid-state reaction as previously described (**Figure 2(b)**) and is feasible for millimetre-wave dielectrics. The *Qf* values decreased as crystallisation temperature. In comparison with the amount of indialite as shown in **Figure 6(a)**

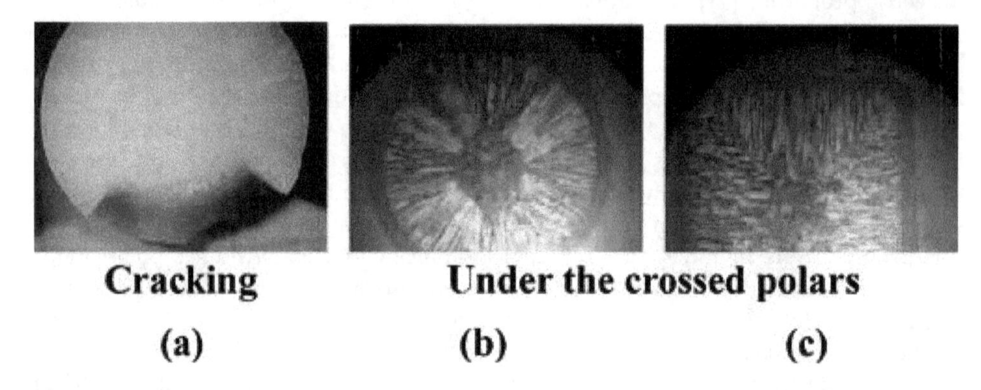

Figure 5.
Cracking of crystallised pellets (a) and anisotropic crystal growth of the pellets under the crossed polars with a sensitive test plate (b) and (c).

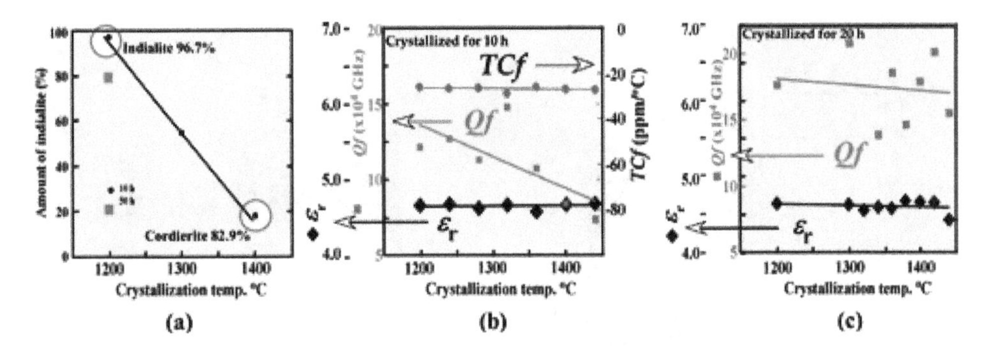

Figure 6.
Amount of indialite (a) and microwave dielectric properties of crystallised indialite at 1200–1440°C for 10 (b) and 20 (c) hours.

[39] and its Qf values as shown in **Figure 6(b)** and **(c)** [17], it is clear that the indialite glass ceramics present a higher Qf than that of cordierite. The ε_r was the lowest among the silicates, about 4.7 as shown in **Figure 6(b)** and **(c)**, and the TCf was −27 ppm/°C as shown in **Figure 6(b)**. Therefore, from these figures, indialite shows a higher Qf than cordierite. This TCf value of −27 ppm/°C is better than that of other silicates having a low TCf of approximately −60 ppm/°C [39].

2.1.3 Conclusions for indialite/cordierite glass ceramics

- Indialite/cordierite glass ceramics are one of the examples of high symmetry bringing a higher Q than ordering by order-disorder transition. Indialite glass ceramics with disordered high symmetry have higher Qf properties than cordierite with ordered low symmetry.

- Cordierite with substituted Ni for Mg synthesised by solid-state reaction exhibited an improved Qf from 40×10^3 to 100×10^3 GHz (**Figure 2(b)**). Rietveld crystal structure analysis showed that the cordierite was transformed to indialite [16].

- A novel idea from glass ceramics suggested the fabrication of indialite as an intermediate phase. Glass ceramics crystallised at 1200°C were almost completely indialite at 96.7% with a high Qf of 150×10^3 GHz, and those crystallised at 1400°C were cordierite at 82.9% with a lower Qf of 80×10^3 GHz. (**Figure 6**) [17, 39].

- Indialite/cordierite crystallised from cordierite glass at 1300°C/20 h showed good microwave dielectric properties of $\varepsilon_r = 4.7$, $Qf > 200 \times 10^3$ GHz and $TCf = -27$ ppm/°C (**Figure 6**) [17, 39].

2.2 Pseudo tungsten-bronze solid solutions: compositional ordering bringing high Q

2.2.1 Introduction

The pseudo tungsten-bronze solid solutions $Ba_{6-3x}R_{8+2x}Ti_{18}O_{54}$ (R = rare earth) located on the tie-line of $BaTiO_3$-$R_2Ti_3O_9$ are shown in **Figure 7(a)** and have been utilised in mobile phones because of their high dielectric constant of 80–90 [20, 21]. This solid solution was first reported by Varfolomeev et al. [41], based on Nd and Sm systems. The composition ranges $0.0 < < x < < 0.7$ for R = Nd and $0.3 < < x < < 0.7$ for

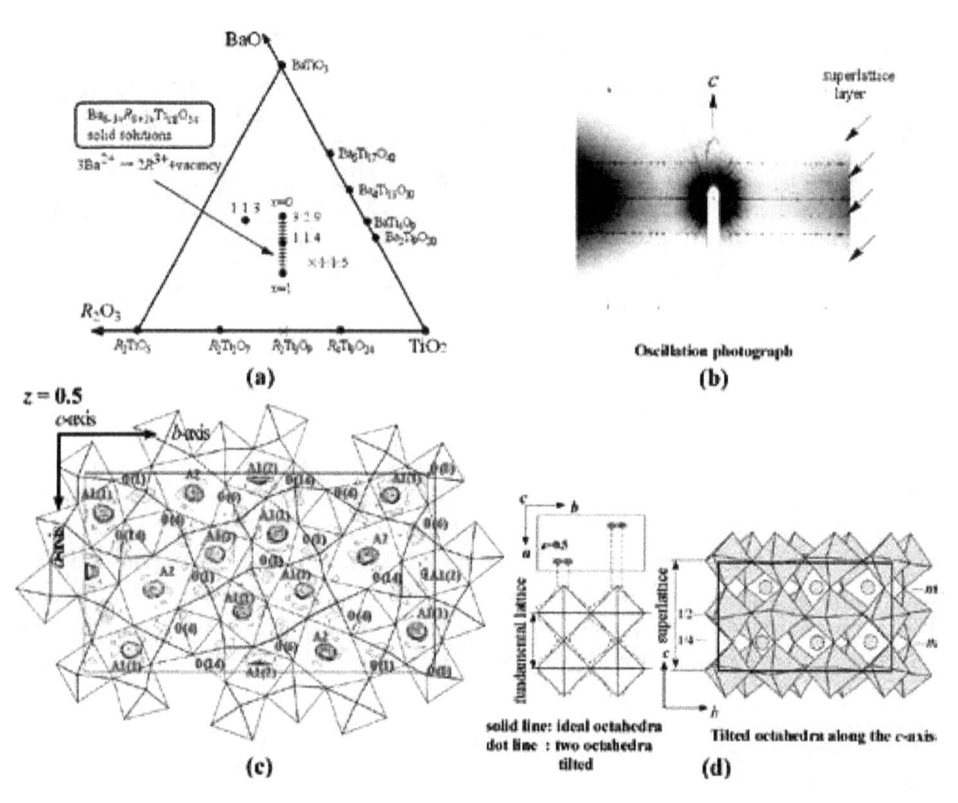

Figure 7.
A part of the BaO-R_2O_3-TiO_2 ternary phase diagram with pseudo tungsten-bronze type solid solutions (a). Oscillation photograph along c-axis of pseudo tungsten-bronze type solid solutions (b). Electron density map (Fourier map) of the fundamental structure superimposed on a superstructure framework (c) and TiO_6 tilting octahedra along the c-axis on the super-lattice (d) deduced from the splitting of oxygen in the fundamental structure (c), and the splitting of oxygen atoms based on the tilting of octahedra as shown in left side figure of the fundamental lattice (d). Right side schematic figure: super structure produced by tilting octahedral (d).

Sm [42] were reported by Ohsato et al. [19] and Negas et al. [43]. The composition range of the solid solutions becomes narrower with the decrease in the ionic radius of the *R*-ion, and Ga and Eu form only $BaO{\cdot}R_2O_3{\cdot}4TiO_2$ composition [44].

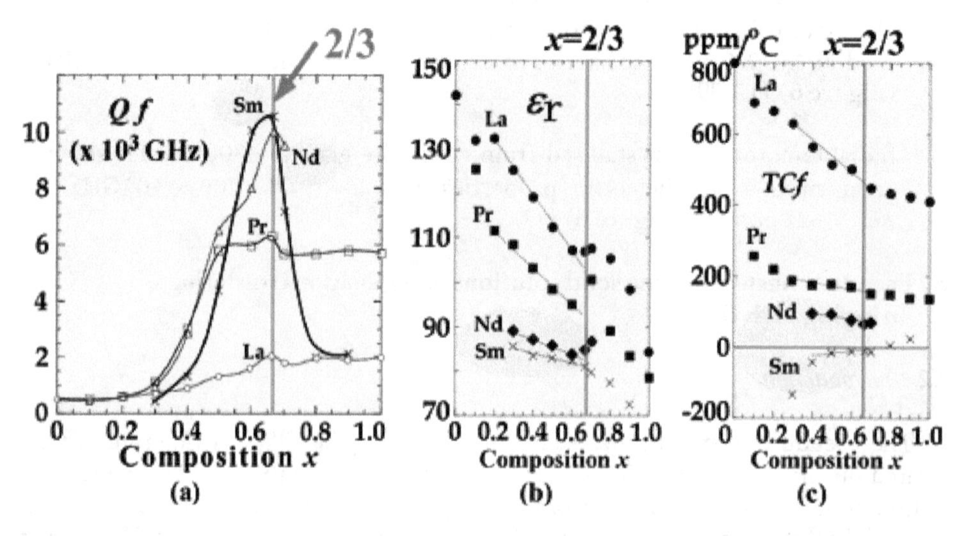

Figure 8.
Qf (a), ε_r (b) and TCf (c) of Sm, Nd, Pr and La system as a function of composition x.

Ohsato et al. and Negas et al. reported the microwave dielectric properties for the Sm, Nd, Pr and La systems as a function of composition x as shown in the **Figure 8(a)** [20, 43, 45] and Fukuda et al. reported the Pr system [46]. On the solid solutions, the composition with $x = 2/3$ was found by Ohsato et al. [42], at which the *Qf* value becomes the highest due to the ordering in the rhombic and pentagonal sites. The dielectric constants ε_r and *TCf* (**Figure 8(b)** and **(c)**) are decreased as a function of the composition x and are affected by volume and tilting angle of the TiO_6 octahedra and the polarizabilities of *R* and Ba ions [20]. The Clausius-Mosotti equation deter-mined the temperature coefficient of the dielectric constant $TC\varepsilon_r$ as a function of the ratio of the mean radii (r_a/r_b) of *A*- and *B*-site ions by Valant et al. [47]. Hither, ra/rb is connected to the tilting of the TiO_6 octahedra. In this study, on the system without order-disorder phase transition that is without symmetry change, it is discussed that the ordering especially compositional ordering brings high *Qf*.

2.2.2 Crystal structure of pseudo tungsten-bronze solid solutions

2.2.2.1 Structure

The crystal structure of the pseudo tungsten-bronze $Ba_{6-3x}R_{8+2x}Ti_{18}O_{54}$ (*R* = rare earth) solid solutions [48–51] includes the perovskite blocks of 2 × 2 unit cells with rhombic (*A*1) sites and pentagonal (*A*2) sites, as shown in **Figure 9**, which are named after the tetragonal tungsten-bronze structure with 1 × 1 perovskite blocks and pentagonal sites [20, 48, 50]. On this compound, two large sites including Ba- and *R*-ions are placed such as *A*1 and *A*2. The Ba-ions engaged on the pentagonal *A*2-sites and *R*-ions *A*1-sites on the perovskite blocks. Two more sites, *B* and *C* are positioned on the tungsten-bronze crystal structure. The *B*-site is the same as the TiO_6 octahedral place in the perovskite, and the *C*-site is a triangular site which is usually empty. This crystal structure of this compound has a special relation-ship with the perovskite structure. If the two ions are the same size, the structure will change to perovskite with only an *A*1-site owing to the combination of the

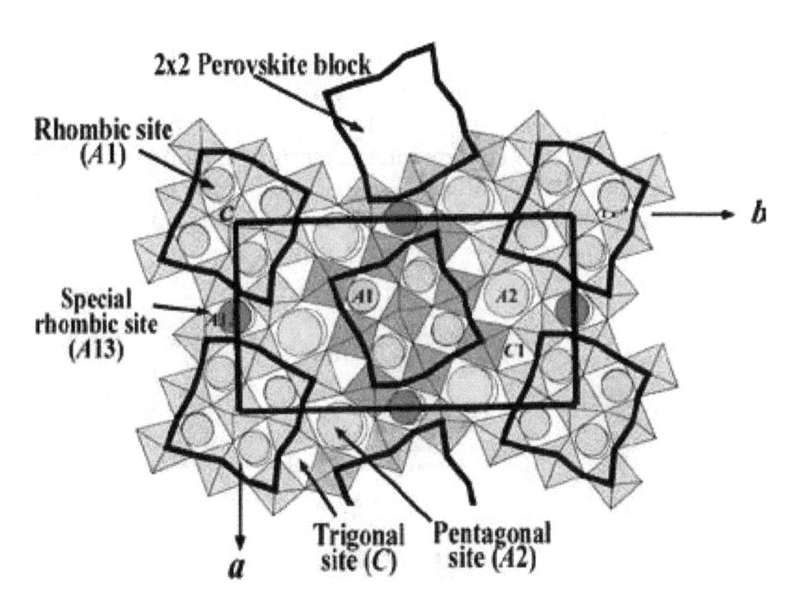

Figure 9.
Crystal structure of the pseudo tungsten-bronze solid solutions. Rhombic (A1) sites located in 2 × 2 unit cells of perovskite blocks, and pentagonal (A2) and trigonal sites (C).

pentagonal *A*2-site and the trigonal *C*-sites [20, 52]. The crystal data are as follows: orthorhombic crystal system of space group *Pbnm* (No. 62), point group *mmm* and lattice parameters a = 12.13, b = 22.27, c = 7.64 Å, Z = 2 and Dx = 5.91 g/cm^3.

2.2.2.2 Tilting

The structure has a super lattice along the *c*-axis with a double lattice of perovskite as shown in **Figure 7(b)** of an oscillation photograph with super diffraction lines [53, 54]. The crystal data of the fundamental lattice are as follows: orthorhombic crystal system of space group *Pbam* (No. 55), point group *mmm* and lattice parameters a = 12.13, b = 22.27, c = 3.82 Å, Z = 1 and Dx = 5.91 g/cm^3. The super lattice is depending on the tilting of TiO_6 octahedra as shown in **Figure 7(d)**. The tilting was endowed in the density map (**Figure 7(c)**) which is of the fundamental lattice superimposed on a superstructure framework. The top oxygen ions (O(1), O(4), O(6), O(8) and O(14)) of octahedra are separated into two along the *c*-axis. The left figure of **Figure 7(d)** shows the reason for splitting of the top oxygen [20]. However, this super lattice is not depending on the order-disorder phase transition as complex perovskite as explained at 2.3 section. The tilting of octahedra might be depending on the size of *A*-ion in the perovskite block.

2.2.2.3 Compositional ordering

The chemical formula of the solid solutions is $Ba_{6-3x}R_{8+2x}Ti_{18}O_{54}$, and the structural formula is $[Ba_4]_{A2}[Ba_{2-3x}R_{8+2x}]_{A1}Ti_{18}O_{54}$. Here, the amount of Ba in the *A*1-sites becomes zero ($2 - 3x = 0$), that is, $x = 2/3$. This composition is special as the fol-lowing sentence: the structural formula is $[Ba_4]_{A2}[R_{8+4/3}]_{A1}Ti_{18}O_{54}$ which is occupied separately by Ba in *A*2 and by *R* in *A*1 as shown in **Figure 10(b)**. This special composition is called "compositional ordering" [20, 21, 42].

2.2.3 Microwave dielectric properties of pseudo tungsten-bronze solid solutions

Figure 8 shows the microwave dielectric properties of the solid solutions as a function of composition x of $Ba_{6-3x}R_{8+2x}Ti_{18}O_{54}$ [20, 21, 42]. The quality factor (*Qf*) changes nonlinearly and has the highest value at particular point $x = 2/3$ with compositional ordering specified above [55]. The highest *Qf* value might be depending on the internal strain. **Figure 11(a)** confers internal strain η obtained from the slope of equation $\beta\cos\theta = r/t + 2\eta\sin\theta$. The internal strain η of the special point $x = 2/3$ is the lowest with the compositional ordering as a function of composition x as shown

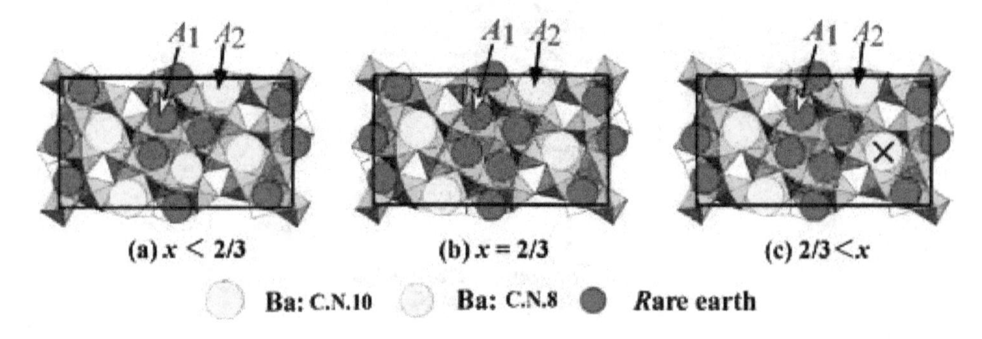

Figure 10.
Structure of disordering (a), compositional ordering (b) and defects in A2-sites (c), depending on the x values of $Ba_{6-3x}R_{8+2x}Ti_{18}O_{54}$.

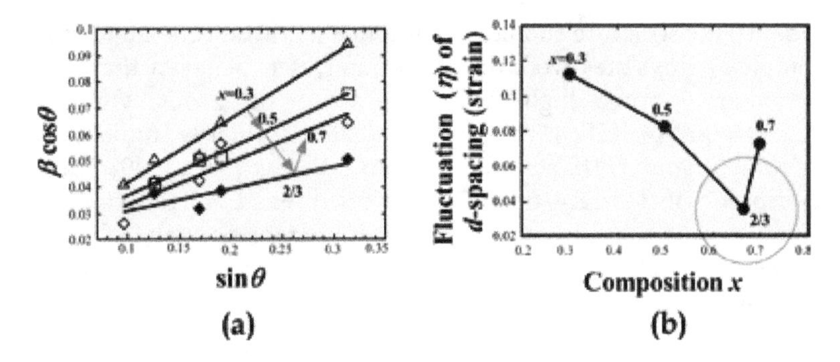

Figure 11.
Internal strain η values obtained from the slope of equation βcosθ = r/t + 2ηsinθ as a function of sinθ for x = 0.3, 0.5, 2/3 and 0.7 (a) and strain η (d-spacing) as a function of composition x (b).

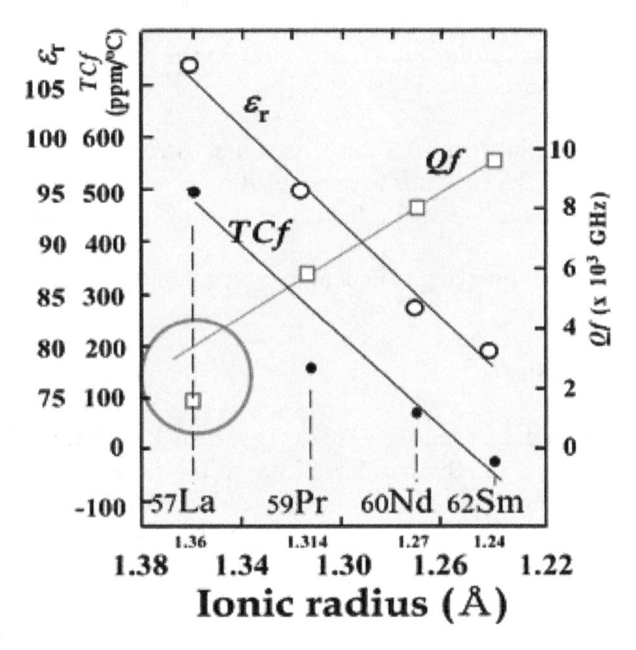

Figure 12.
Microwave dielectric properties as a function of ionic radius of R ion.

in **Figure 11(b)**. The internal strain comes from the fluctuation of *d*-spacing of the lattice broadening the full width at half maximum (FWHM) [20, 21, 56].

The *Qf* value at the special point $x = 2/3$ shows the highest of 10.5×10^3 GHz in the Sm system, 10.0×10^3 GHz in the Nd system and 2.0×10^3 GHz in the La system as depicted in **Figure 8(a)** [20, 21, 56]. The *Qf* values reducing in the order of Sm, Nd, Pr and La are depending on the ionic radius relating size difference between Ba and *R* [57], and that of La is deviating from the *Qf* line through the Sm, Nd and Pr as shown in **Figure 12**. If the sizes are similar, the crystal structure should become perovskite structure. In the case of Sm, the difference is maximum which introduces the stability of the crystal structure. The size of La ion is similar to Ba, so the structure might be unstable to be low *Qf*.

2.2.4 Symmetry and ordering for Q

On the microwave dielectrics, high Q has been brought by a high potential material, which has an over-well-proportional rigid crystal structure with symmetry

[11–13]. That is, the structure should be without electric defects, nondistortion and strain. Complex perovskites were described later, it is believed that ordering by long time sintering brings high Q, but we are pointing out symmetry is the predominant factor [14, 15]. In the case of indialite/cordierite, indialite with high symmetry shows higher Q than cordierite with ordering [17, 18, 39]. This case has an order-disorder phase transition. On the other hand, in the case of pseudo tungsten-bronze solid solutions which has no phase transition, one of ordering that is the composi-tional ordering brings high Q [20, 21]. In the case of no symmetry change, ordering is predominant.

2.2.5 Conclusions for pseudo tungsten-bronze

- The pseudo tungsten-bronze solid solutions have been used for mobile phones for miniaturisation based on their high Qf and high ε_r.

- The compound has a unique point of $x = 2/3$ on the $Ba_{6-3x}R_{8+2x}Ti_{18}O_{54}$ chemical formula which shows the highest Qf value.

- The special point of $x = 2/3$ on the structural formula of $[Ba_4]_{A2}[Ba_{2-3x}R_{8+2x}]_{A1}Ti_{18}O_{54}$ is the composition at which Ba-ions disappear on the $A1$-sites because $2 - 3x = 0$. That is the point of compositional ordering.

- The compositional ordering brings high Q by maintaining the stability of the crystal structure.

2.3 Complex perovskites

There are many kinds of complex perovskites such as 1:1, 1:2 and 1:3 type in B-site and 1:1 type in A-sites [21]. In this chapter, 1:2 type complex perovskite compounds $A^{2+}(B^{2+}_{1/3}B^{5+}_{2/3})O_3$ are presented such as $Ba(Zn_{1/3}Ta_{2/3})O_3$ (BZT), $Ba(Mg_{1/3}Ta_{2/3})O_3$ (BMT) and $Ba(Zn_{1/3}Nb_{2/3})O_3$ (BZN). These complex perovskite compounds have order-disorder phase transitions (**Figure 13(a)** and **(b)**) [58]. The ordered phase that appears at low temperatures is a trigonal (rhombohedral) structure of space group $P\bar{3}m1$ (No. 164), and the disordered phase appearing at high temperatures is a high symmetry cubic structure of $Pm\bar{3}m$ (No. 221), as shown

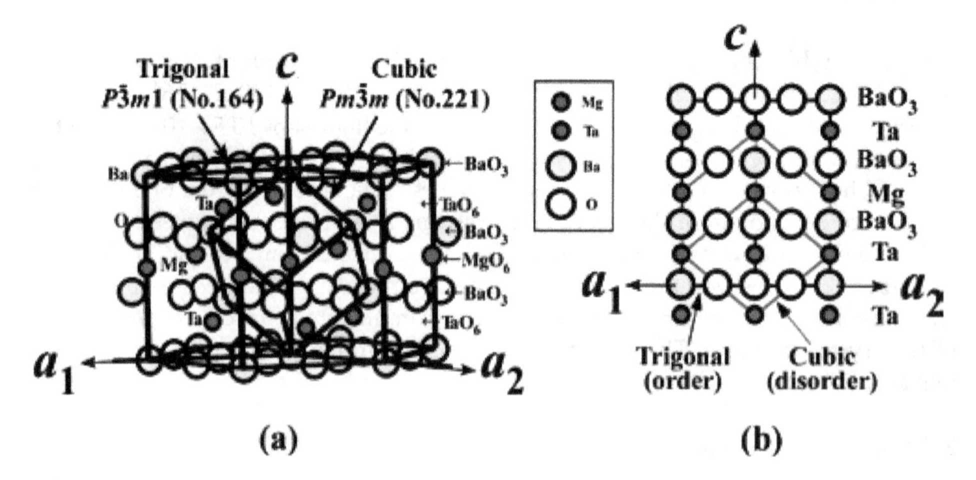

Figure 13.
Complex perovskite crystal structure composed by Mg/TaO$_6$ octahedra located between BaO$_3$ closed packing layer, showing relationship between cubic and trigonal crystal lattice. Perspective figure (a) and (110) plane (b).

in **Figure 13** [21]. In the ordered form of BMT, Mg^{2+} and Ta^{5+}-ions located among the adjacent packing layers of BaO_3 are ordering as -Mg-Ta-Ta-Mg-Ta-Ta-Mg-, as shown in **Figure 13**. On the other hand, in the disordered form, Mg^{2+} and Ta^{5+}-ions occupy *B*-sites statistically.

2.3.1 Introduction

Kawashima et al. [14] presented that ordering brings a high *Q* based on BMT with long duration sintering, which showed high *Qf* and ordering. This has previously been believed to be the case because long duration sintering samples generally show high *Qf* and ordering. However, some examples have arisen that contradict this relation, such as BMT-$Ba(Co_{1/3}Ta_{2/3})O_3$ [59], BMT-$BaZrO_3$ [60], BMT-$BaSnO_3$ [61] and BZT-$(SrBa)(Ga_{1/2}Ta_{1/2})O_3$ [62]. Koga et al. [23–26] presented the relation-ship between high *Qf* and the ordering ratio as determined by the Rietveld method, the high *Qf* samples with disordered structure synthesised by spark plasma sin-tering (SPS) [63] and the effects of annealing of disordered BZN with an order-disorder transition point of 1350°C [26]. HRTEM and Rietveld studies confirmed the ordering and disordering of BZN samples [64]. Partial ternary phase diagrams such as BaO-ZnO-Ta_2O_5, BaO-MgO-Ta_2O_5 and BaO-ZnO-Nb_2O_5 were studied on the composition with high *Qf* that deviated from the stoichiometric composition of BZT/BMT/BZN by Kugimiya et al. [22, 27], Koga et al. [24, 26] and Kolodiazhnyi [29]. Kugimiya pointed out that the solid solutions with high density and high *Qf* located on the tie-line BMT-$Ba_5Ta_4O_{15}$, which have completed the ideal chemical formula without oxygen defects. It is one of the conditions for high Q that the high compositional density brings high *Qf*.

2.3.2 Origin of high Q for microwave complex perovskite

In this section, it is explained that ordering has no relation with *Qf* based on the following three sets presented by Koga et al. [23, 25, 26, 63].

2.3.2.1 Ordering ratio and Qf

The ordering of BZT was observed on the samples with high *Qf* sintered at 1350° C [23] over 80 h. **Figure 14** presents the XRPD patterns (**a**) with super lat-tice lines (asterisked), and the high angle diffraction patterns (**b**) which depicts splitting of 420 cubic diffraction peak into two peaks, namely 226 and 422 in the trigonal system. These data are consistent with the report by Kawashima et al. [14].

Koga et al. investigated the amount of BZT ceramic as ordering ratio by the Rietveld method [23], which is shown in **Figure 15(a)**. The ordering ratio saturates at about 80%, but the *Qf* values increase up to 100 × 10^3 GHz. This shows that the effect of ordering on the *Qf* is not so significant. However, the *Qf* values are affected by density and grain size as shown in **Figure 15(b)** and **(c)**, respectively [15, 23].

2.3.2.2 BZN with a clear order-disorder transition

Many complex perovskites such as BMT and BZT have the order-disorder phase transition at high temperature, and the order-disorder transition is not so clear. On the other hand, BZN shows clearly the phase transition at lower temperature 1350° C [26]. **Figure 16(a)** shows *Qf* as a function of sintering temperature. Under the transition temperature such as 1200 and 1300°C, the sintered samples show order with under 50 × 10^3 GHz of *Qf*. Moreover, at 1400°C, higher than the transition

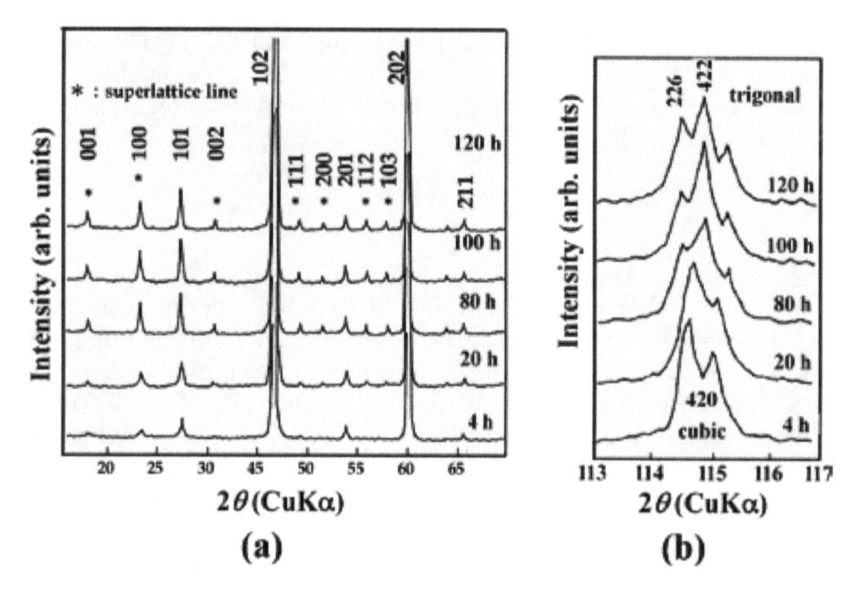

Figure 14.
XRPD patterns of BZT ceramics with different sintering times at 1350°C (a), here, asterisks are super lattice diffractions, and Magnified XRPD patterns around 2θ = 115° in which 420 diffraction peak split to 226 and 422(b).

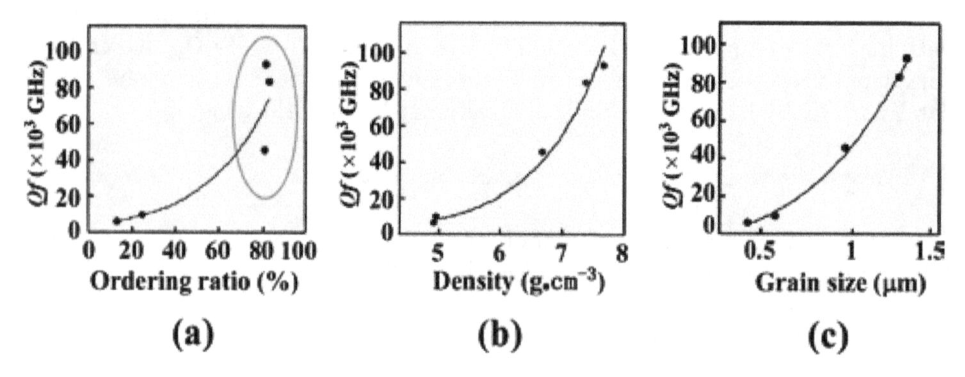

Figure 15.
The Qf as functions of ordering ratio (a), density (b) and grain size (c) of BZT ceramics.

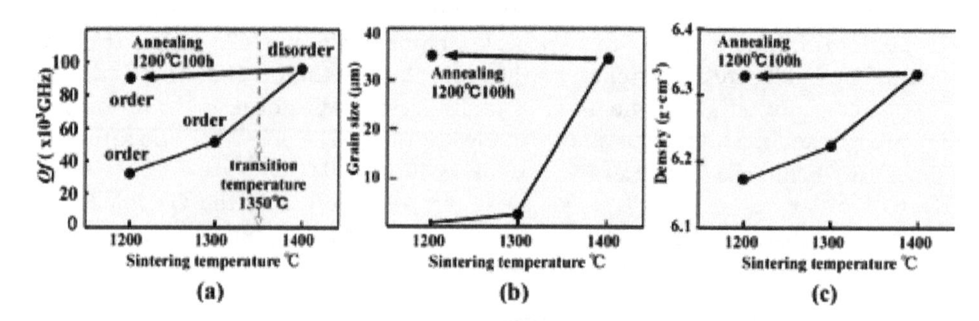

Figure 16.
Qf (a), grain size (b) and density (c) as a function of sintering temperature of BZN ceramics.

temperature, the *Qf* values increased to 90 × 10^3 GHz with disordering structure. This shows that the high symmetry form with disorder performs higher *Qf* than ordering form. Moreover, the sample annealed at 1200°C/100 h transformed to order form, but the *Qf* value did not improve and slightly decreased. Grain size and densities as shown in **Figure 16(b)** and **(c)** also increased as the sintering temperature from 1200 to 1400°C [15, 26]. As if the sample sintered at 1400°C annealed at 1200°C/100 h, the grain size and densities were not changed. Because of annealing, the slight decrease in *Qf* might be a result of the low symmetry that accompanies order. On the contrary, Wu et al. [65] presented annealing of BZN at 1300°C brings high *Qf* with ordering. The annealing temperature is high enough for sintering, so sintering was proceeded with ordering the same as Kawashima's results [14].

The BZN samples A and B are also studied by XRPD and HRTEM, which sintered at 1400°C/100 h above the order-disorder phase transition point and subsequently annealed at 1200°C/100 h below the transition point, respectively [26, 64]. The two samples were identified by conventional XRPD as shown in **Figure 17(a)**. As the super lattice lines are not clear, the high angle XRPD patterns around 2θ~115° were measured (**Figure 17(b)**). On the XRPD pattern, the sample A shows a single peak of the 420 diffraction, so it was confirmed as disorder phase. On the other hand, the sample B shows the peak splitting of 422 and 226 depending ordering. These results are comparable with Koga's data [23]. These two samples were analysed by the Rietveld method.

HRTEM figures as shown in **Figure 18** for most area of sample A (**Figure 18(a)**) and B (**Figure 18(c)**) are disordered and ordered area along the [111]c direction, respectively. A fast Fourier transform (FFT) image is inserted in **Figure 18(a)** of a disordered area without further reflections along the [111]c direction and in **Figure 18(c)** of a ordered area with additional two reflection points for super lattice. In the both sample A and B, mixed area of disordered and ordered area existed in **Figure 18(b)**, and in the sample B, ordered area showing twin-related anti-phase domain boundary also existed as shown in **Figure 18(d)**. The FFT image of twin area shows superimposed of ordered diffrac-tions with four additional points.

Figure 19 depicts the high-resolution XRPD pattern of sample A and B using syn-chrotron radiation [64]. The super lattice diffraction 100 t peaks (reciprocal lattice plane 100 in the trigonal crystal system) are observed in both samples. The diffrac-tion intensity of sample A is lower than that of sample B. These super lattice diffrac-tion intensity peaks are comparable with the ordering ratios, that is the sample A and B have the value of 27.6 and 54.2%, respectively, obtained by the Rietveld method. Although the degree of ordering of sample B is large compared to that of sample A, it was assumed about 80% ordering for a whole sample, as in the case of BZT [23].

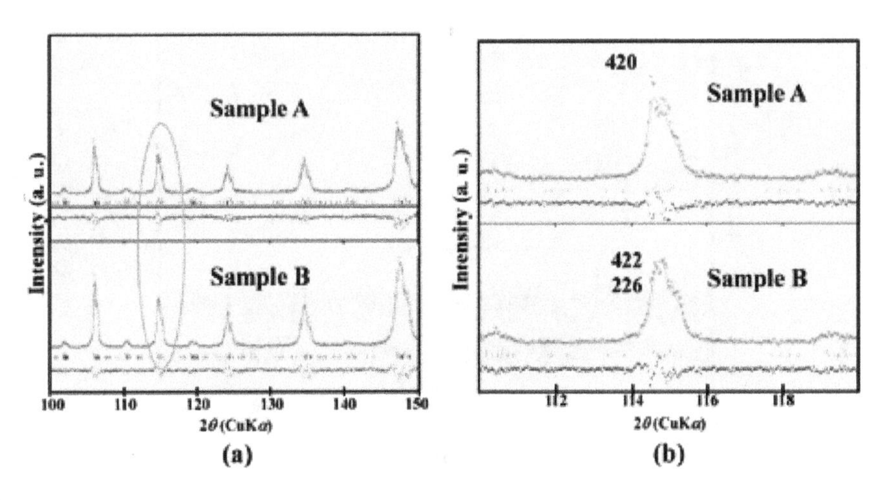

Figure 17.
XRPD patterns for BZN ceramics sintered at 1400°C (sample A) and annealed at 1200°C (sample B) (a) and magnified high angle XRPD patterns around 2θ~115° (b).

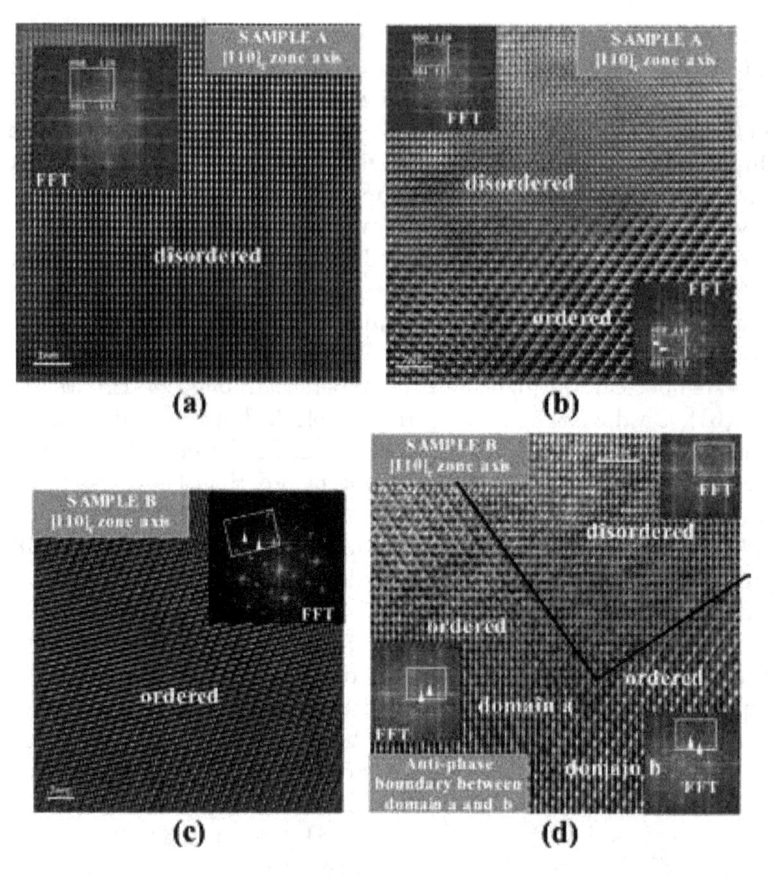

Figure 18.
HRTEM images of sample A and B with FFT image along the [111]c direction: disordered area in sample A (a), mixed area of disordered and ordered area in sample A (b), ordered area in sample B (c) and twin related anti-phase domain boundary in sample B (d).

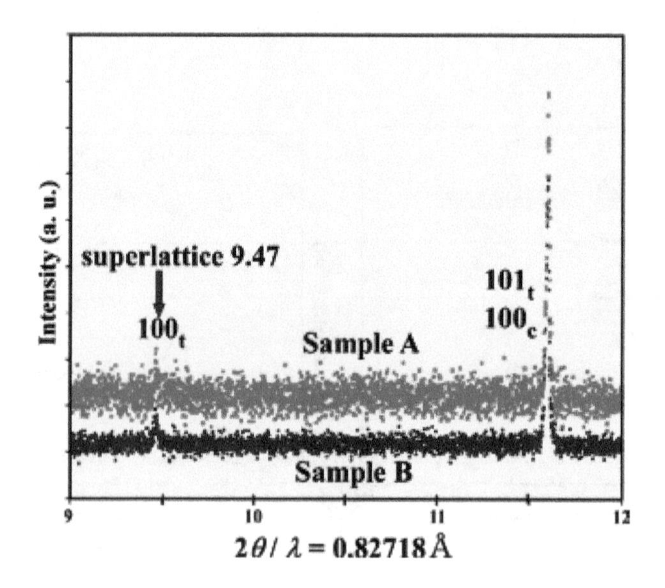

Figure 19.
High-resolution synchrotron XRPD patterns (λ = 0.82718 Å) for sample A and B with super lattice peak 100_t. Here, subscript t is trigonal, and c is cubic.

It is revealed that the degree of ordering increased from 27.6 to 54.2% due to the annealing. However, the *Qf* values, grain size and the density have no influence on the degree of ordering (**Figure 16**). While the disordered area of sample A (sintered above the transitional temperature) changes to the low-temperature phase with ordering by the annealing, the *Qf* values were expected to be increased. However, the *Qf* values changed only somewhat from 95.7×10^3 GHz to 95.0×10^3 GHz [64]. The effect of ordering is not acceptable to change the *Qf* value considerably.

2.3.2.3 BZT with disordering leaded high Qf by SPS

The ordered and disordered BZT ceramics can be achieved by varying the sinter-ing duration in the conventional solid-state reaction (SSR). A high density and high Q ceramics of ordered BZT were obtained by SSR with a long sintering time of over 80 h, while the disordered BZT was not possible to fabricate by using SSR. Koga et al. [63] reported the high density disordered BZT ceramics for a short sintering time of 5 mins by using spark plasma sintering (SPS). **Figure 20(a)** presents the *Qf* as a function of the densities of BZT fabricated using SSR and SPS [15, 63]. The fabricated SPS samples were shown to be disordered cubic type of perovskite as depicted in the XRPD pattern (**Figure 20(b)**) with a peak of 420 reflection in com-pared with the ordered trigonal type with peaks separations of 422 and 226 when sintered using SSR (1400°C 100 h). The ceramics were sintered at the temperature between 1150 and 1300°C under 30 Mpa pressure [63].

This may result in the disordered BZT with a high density of 7.62 g/cm^3, which is approximately 20% higher than that of low-density samples of 5.0–6.0 g/cm^3 syn-thesized by conventional SSR. The full width at half maximum (FWHM) of the 420 peak became narrower with an increase in the temperature from 1100 to 1300°C (**Figure 20(b)**) indicates that the degree of crystallisation of the disordered cubic phase is improved without the need for conversion to the ordered trigonal phase. Regardless of the method of synthesis, *Qf* is strongly dependent on density, and *Qf* values were improved with increasing density. The dense disordered BZT ceramics synthesized by SPS showed a significantly high *Qf* (= 53.4×10^3 GHz) comparable to that of the ordered BZT with the same density (= ca. 7.5 g/cm^3) synthesized by SSR. The crystallisation with densification of BZT ceramics should play a more critical role in the improvement of the Q factor in the BZT system rather than the structural ordering.

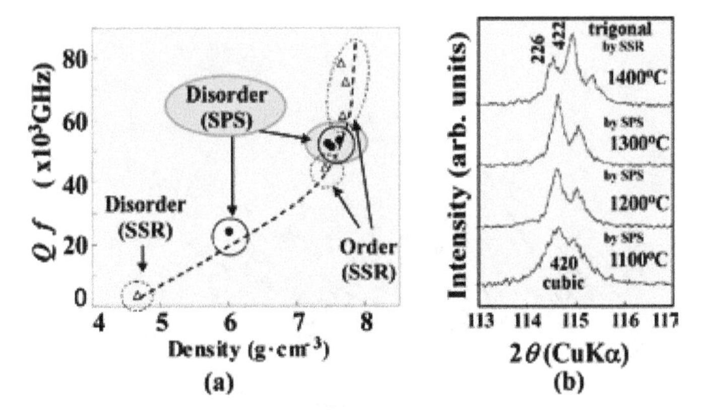

Figure 20.
Qf of BZT by solid-state reaction (SSR) and spark plasma sintering (SPS) as a function of density, Disordered BZT by SPS shows high Qf (a). Nonsplitting XRPD patterns around 420 diffraction of BZT sintering by SPS with different sintering temperatures compared with ordered sample by SSR with splitting pattern (b).

2.3.3 *Deviated compositions with high Qf from stoichiometric complex perovskite composition*

In a BaO-Mg/ZnO-Ta_2O_5 partial ternary ceramic (BMT/BZT system), complex perovskite such as BMT and BZT are forming solid solutions, and the *Qf* values varied intrinsically based on the crystal structure in the solid solutions depending on the density and defects. In this section, the crystal structure and properties on the varied compositions from the stoichiometric complex perovskite composition are reviewed for high *Qf* research.

2.3.3.1 *The highest Qf composition with intrinsic compositional density by Kugimiya's research*

Kugimiya [22, 27] presented the highest *Qf* composition with intrinsic compositional density on the Ta and Ba rich side near the BMT-$Ba_5Ta_4O_{15}$ tie-line in a BaO-MgO-$TaO_{5/2}$ partial system (BMT system), as shown in **Figure 21.** He presented three areas divided by the following two lines as shown in **Table 1** and **Figure 21.**

$$\alpha = 5\gamma/4 \tag{3}$$

$$\alpha = \gamma/2 \tag{4}$$

Here, α and γ are as written in the formula $\alpha BaO{\cdot}\gamma TaO_{5/2}$. On the $\alpha = 5\gamma/4$ line, $Ba_{1+\alpha}(Mg_{1/3}Ta_{2/3} + {}_{4\alpha/5}V_{\alpha/5})O_{3+3\alpha}$ solid solutions are formed as the ideal compositions without vacancies in the *A*- and O-sites. In the *B*-site, the vacancy is neutralized and without charge.

In **Figure 21**, the composition with intrinsic compositional high density shows the highest Q of 50.0 × 10^3 on the tie-line between BMT and $Ba_5Ta_4O_{15}$ ($\alpha\alpha\alpha\alpha = 5\gamma\gamma\gamma\gamma/4$). The contour lines in **Figure 21** show Q values from 2.0 × 10^3 in the outer area to 25.0 × 10^3 in the centre. The contour is elongated parallel to the Q max line as drawn in **Figure 21,** and it changes steeply on the perpendicular to this line. So, the compositions without oxygen vacancy and with neutralised charge vacancies

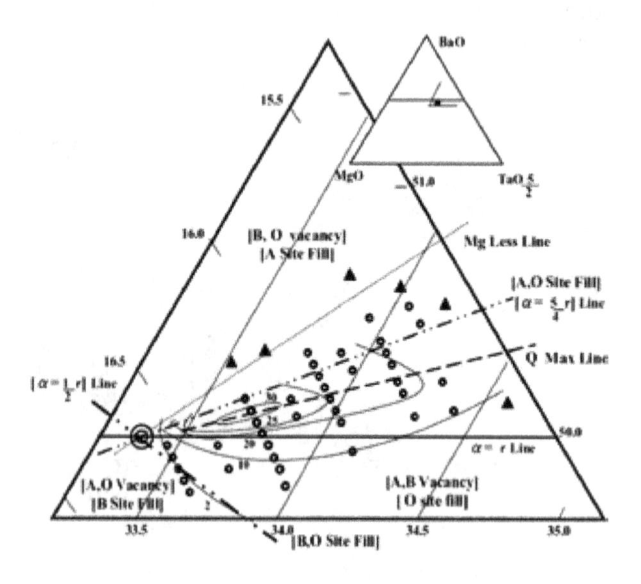

Figure 21.
BaO-MgO-$TaO_{5/2}$ partial system (BMT system).

α	Chemical formula	Vacancy
$\alpha > 5\gamma/4$	$Ba_{1+\alpha}(Mg_{1/3}Ta_{2/3+\gamma}V_{\alpha-\gamma})O_{3+\alpha+5\gamma/2}V_{2\alpha-5\gamma/2}$	*A*: fill, *B*, O: vacancy
$\alpha = 5\gamma/4$	$Ba_{1+\alpha}(Mg_{1/3}Ta_{2/3+4\alpha/5}V_{\alpha/5})O_{3+3\alpha}$	*A*, O: fill, *B*: vacancy
$5\gamma/4 > \alpha > \gamma/2$	$Ba_{1+\alpha}V_{5\gamma/6-2\alpha/3}(Mg_{1/3}Ta_{2/3+\gamma}V_{\alpha/3-\gamma/6})O_{3+\alpha+5\gamma/2}$	*A*, *B*: vacancy, O: fill
$\alpha = \gamma/2$	$Ba_{1+\alpha}V_{\alpha}(Mg_{1/3}Ta_{2/3+\gamma})O_{3+6\alpha}$	*A*: vacancy, *B*, O: fill
$\alpha < \gamma/2$	$Ba_{1+\alpha}V_{\gamma-\alpha}(Mg_{1/3}Ta_{2/3+\gamma})O_{3+\alpha+5\gamma/2}V_{\gamma/2-\alpha}$	*A*, O: vacancy, *B*: fill

Table 1.
*The chemical formula for three areas divided by two lines: $\alpha = 5\gamma/4$ and $\alpha = \gamma/2$, here, α and γ are in $Ba_{\alpha}Ta_{\gamma}O_{\alpha+5\gamma/2}$ and vacancies on the **A**-, **B**- and O-sites [22].*

are ideal for microwave dielectrics, and the density is high due to the partial substitution of Ta in the site of Mg, which is denoted as intrinsic compositional density [28]. Other regions have some defects degrading the *Qf* values, which were explained on the references [21, 22, 27, 28].

2.3.3.2 Phase conditions in the vicinity of BZT by Koga's research

Koga et al. [24, 25] showed the highest *Qf* composition shifted from the stoichiometric BZT composition. The ordering ratio of the deviated composition was not higher than that of the stoichiometric composition, which was calculated by the Rietveld method. These results were presented by the study of the phase relations in the vicinity of BZT in the BaO-ZnO-Ta_2O_5 ternary system, as shown in **Figure 22** [24, 25]. These samples were sintered at 1400°C/100 h as reported in Koga's paper. These diffrac-tion patterns fit the Rietveld method well [23, 24]. Ordering ratios obtained are shown in **Figure 23(a)**. Three areas in the vicinity of BZT are presented as shown in **Figure 22**. 1st one (I) is ordering area with BZT single phase, the 2nd one (II) is ordering area with secondary phase and 3rd one (III) is disordering area with BZT single phase.

The first area (I) is characterised as a BZT single phase with an ordered structure and a high *Qf*. The varied compositions E and K have high *Qf* values about 50% higher than that of the stoichiometric BZT composition A. The ordering ratios at E and K are lower than that of stoichiometric BZT at A, but the density at E is the same as that of A [25]. The second (II) is composed by an ordered BZT

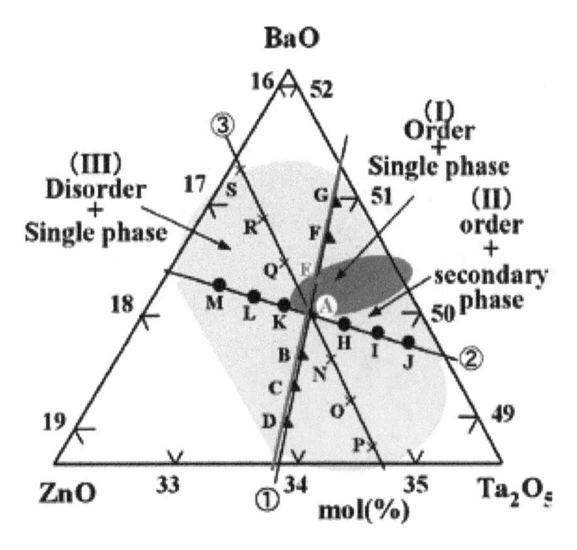

Figure 22.
Phase relations in the vicinity of BZT in the BaO-ZnO-Ta_2O_5 ternary system.

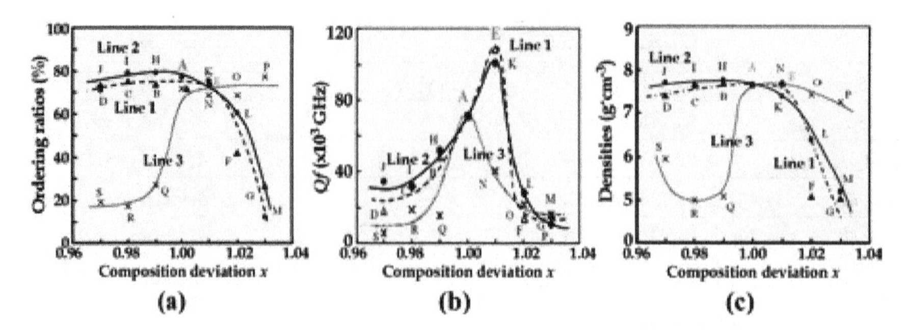

Figure 23.
Ordering ratio (a), Qf (b) and density (c) as a function of composition deviation from stoichiometric BZT.

accompanied by a secondary phase $BaTa_2O_6$ with a specific amount of Zn determined by X-ray microanalyser (XMA). The ordering ratio in this area is high at about 70–80% (**Figure 23(a)**). Although the structure is ordered, the *Qf* values decrease in the order of A-N-O-P from stoichiometric BZT (**Figure 23(b)**). The ordered BZT with the secondary phase is located on the Ta_2O_5 rich side as a eutectic phase diagram system. The third (III) with a disordered single phase shows low *Qf* and low density (**Figure 23(c)**). The low density comes from the numerous pores.

2.3.3.3 Phase conditions in the vicinity of BZT by Kolodiazhnyi's research

Kolodiazhnyi [29] also found the highest *Qf* of 330 × 10^3–340 × 10^3 GHz positions deviated from the stoichiometric BMT composition which is located in the BMT-$Ba_5Ta_4O_{15}$-$Ba_3Ta_2O_8$ compositional triangle (CT) as shown in **Figure 24**. The positions located in the single-phase BMT, which was indicated by green line.
The position is close to the BMT-$Ba_5Ta_4O_{15}$ tie-line. A to H eight CTs are formed by BMT and five stable compounds, such as $Ba_5Ta_4O_{15}$, MgO, BaO, $Ba_9MgTa_{14}O_{45}$ and $Mg_4Ta_2O_9$, and three metastable compounds, $Ba_6Ta_2O_{11}$, $Ba_4Ta_2O_9$ and $Ba_3Ta_2O_8$. In A, B and C-CTs, although the samples demonstrated high density and a high degree

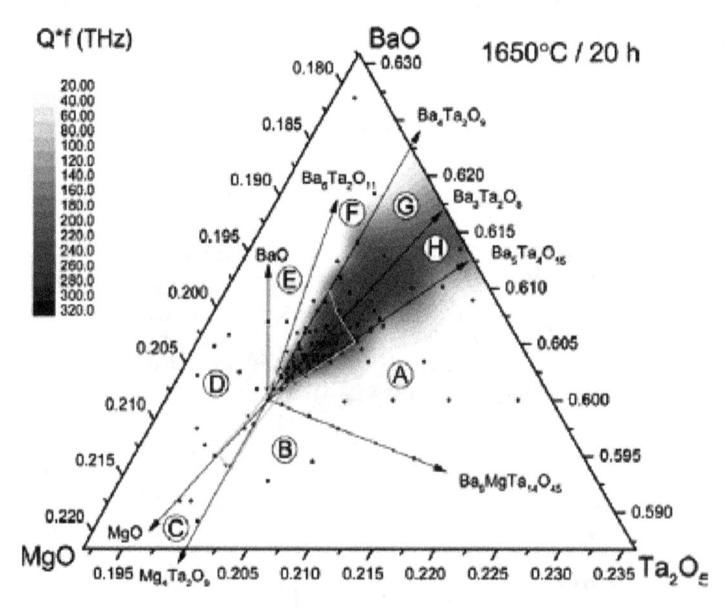

Figure 24.
Part of the BaO-MgO-Ta_2O_5 phase diagram in the vicinity of BMT divided into eight CTs. Small black dots indicate the target samples. Green line indicates an approximate boundary of the single-phase BMT.

of order, they showed low *Qf* values, attributed to the possible presence of the $Ba_9MgTa_{14}O_{45}$ second phase. Moreover, in D, E and F-CTs, as the samples were very low density, no electromagnetic resonance peaks were detected.

2.3.3.4 Koga's and Kolodiazhnyi's data comprehended in Kugimiya's data

Koga's data [24] and Kolodiazhnyi's [29] data are comparable with Kugimiya's BMT data [22]. The area (I) and the H-CT with the highest *Qf* as shown in **Figures 22** and **24**, respectively, are located on the opposite side of Kugimiya's data against the BMT-$Ba_5Ta_4O_{15}$ tie-line (**Figure 21**). These compositions will be comparable with that of the ideal crystal structure $Ba_{1+\alpha}(Mg_{1/3}Ta_{2/3+4\alpha/5}V_{\alpha/5})\ O_{3+3\alpha}$, as stated before in section (2.3.3.1) [22]. The formula is rewritten as $Ba(Mg_{1/3-\alpha/3}Ta_{2/3+2\alpha/15}V_{\alpha/5})O_3$ solid solutions on the tie-line BMT-$Ba_5Ta_4O_{15}$. The crystal structure in the composition region is ideal, without defects, and with an intrinsic high compositional density as described above. Surendran et al. [66] also reported a composition with high *Qf* deviated from stoichiometric BMT reviewed in detail in Intech Open Access Book [21].

2.3.4 Conclusions: important points concerning complex perovskites

- Ordering brings high *Qf* in the complex perovskite because of the long duration sintering. This situation has been bereaved for a long time. However, many examples contradicting this relation were presented.

- Koga et al. presented that *Qf* values of BZT did not depend on the ordering, preferably depending on the density and grain size.

- BZN with an order-disorder transition point at 1350°C (sample A) showed high *Qf* in the high-temperature disordered form. Moreover, annealing of the disordered sample B brings the ordered form, but the *Qf* does not improve. The both samples are analysed by the Rietveld method and HRTEM. The HRTEM presented the order form, disorder form and anti-phase domain by the FFT.
- Disordered samples with high density could not be synthesised by the solid-state reaction, but could be by SPS. The samples with disordered structure showed high *Q*. The ordering phenomenon is the only barometer of sintering in the solid-state reaction.

- Compositions deviated from stoichiometric complex perovskites such as BZT and BMT showed higher *Qf* and lower ordering than the stoichiometric composition. Based on these points, the ordering is not the reason for high *Qf*, and it is the only barometer of sintering.

- Intrinsic compositional density brings high *Qf*. On the BMT-$Ba_5Ta_4O_{15}$ tie-line, solid solutions are formed by the substitution Ta for Mg, which include high *Qf* compositions. The chemical composition with the highest *Qf* is $Ba_{1+\alpha}(Mg_{1/3}Ta_{2/3+4\alpha/5}V_{\alpha/5})O_{3+3\alpha}$, which is an ideal solid solutions without oxygen defects and neutralised vacancies (**Table 1**).

- Compositions deviated from stoichiometric BMT/BZT towards BaO and the Ta_2O_5 rich areas showing high *Qf*, as presented by Koga et al. [24], Kolodiazhny [29] and Surendran et al. [64], are comparable with intrinsic compositional density with high *Qf* as presented by Kugimiya [22].

3. Conclusions

The microwave dielectrics are the perfect, ideal and well-proportional crystal structures for low dielectric losses. Most of them belong to paraelectrics with inversion symmetry *i* and showing high symmetry and nondefects. In this chapter, the effects of ordering and symmetry were presented as follows: there are two types of ordering conditions. One is a case of nonphase transition such as pseudo tungsten-bronze solid solutions. These compounds show compositional ordering at a unique point of $x = 2/3$ on the $Ba_{6-3x}R_{8+2x}Ti_{18}O_{54}$ system, which shows the highest *Qf* without degradation of crystal symmetry. The other is a case of order-disorder phase transi-tion such as indialite/cordierite. Indialite with a disordered structure and a high symmetry of 6/*mmm* point group has a higher *Qf* than cordierite with an ordered structure and low symmetry of *mmm* point group. It is clarified that the effect of high symmetry is predominant for high *Qf*. In the case of complex perovskite, a long sintering time of more than 80 h brings a high *Qf* accompanying ordering. It was clarified that the ordering is not necessary for high *Qf* and only a barometer of sinter-ing in the solid-state reaction. Moreover, compositions deviated from stoichiometric complex perovskite showed higher *Qf* than the stoichiometric composition which has substituted Ta-ions for Mg/Zn-ions. The substitution brings a high density that is the compositional density. It was clarified that high compositional density brings high *Qf*.

Acknowledgements

The authors are grateful to Professors and graduate students of NIT, Meijo University and Hoseo University, and Doctors and researchers in the many companies, which collaborated with NIT. Visiting Professor Hitoshi Ohsato is grateful to the following projects: (1) support industries of Japan by Ministry of Economy, Trade and Industry (METI), Japan, (2) JSPS KAKENHI Grant Number 22560673, 25420721, JP16K06735 and (3) Nokia Foundation 2016 for Nokia Visiting Professors Project 201700003. Professor Heli Jantunen and Dr. Jobin Varghese are grateful to European Research Council Project No. 24001893 for financial assistance. The authors would also like to thank Honorary Research Professor Arthur E Hill of Salford University for valuable discussion and improving English during the preparation of this manuscript.

Author details

Hitoshi Ohsato[1,2]*, Jobin Varghese[1] and Heli Jantunen[1]

1 Microelectronics Research Unit, Faculty of Information Technology and Electrical Engineering, University of Oulu, Oulu, Finland

2 Department of Research, Nagoya Industrial Science Research Institute, Nagoya, Japan

*Address all correspondence to: ohsato.hitoshi@gmail.com

References

[1] Sebastian MT. Dielectric Materials for Wireless Communication. Amsterdam: Elsevier; 2008. ISBN 13:978-0-08-045330-9

[2] Sebastian MT, Ubic R, Jantunen H. Low-loss dielectric ceramic materials and their properties. International Materials Review. 2015;**60**:392-412

[3] Ohsato H. High frequency dielectric ceramics. In: Adachi G, editor. Materials Technology Hand Book for Rare-Earth Elements. Tokyo: NTS Inc; 2008. pp. 346-358 (Japanese)

[4] Ohsato H, Kagomiya I, Chae KW. Microwave dielectric ceramics with rare-earth elements (I). Journal of the Korean Physical Society. 2012;**61**:971-979

[5] Ohsato H, Kagomiya I, Kim JS. Microwave dielectric ceramics with rare-earth (II). Integrated Ferroelectrics. 2010;**115**:95-109

[6] Ohsato H. High frequency dielectrics. In: Shiosaki T, editor. Development and Applications of Ferroelectric Materials. Tokyo: CMC; 2001. pp. 135-147 (Japanese)

[7] Wakino K. Recent development of dielectric resonator materials and filters in Japan. Ferroelectrics. 1989;**91**:69-86

[8] Tamura H. Progress of ceramic dielectric materials for microwave and millimeterwave applications. In: MWE '99 Microwave Workshop Digest. 1999. pp. 175-180

[9] Kobayashi Y, Katoh M. Microwave measurement of dielectric properties of low-loss materials by the dielectric resonator rod method. IEEE Transaction on Microwave Theory and Technologies. 1985;**33**:586-592

[10] Ichinose N. High-frequency materials and their applications. New Ceramics & Electronic Ceramics. 1996;**9**(9):1-50

[11] Ohsato H. Microwave dielectrics. In: Fukunaga O, Haneda H, Makishima A, editors. Handbook of Multifunctional Ceramics. Tokyo: NTS; 2011. pp. 152-166 (Japanese)

[12] Ohsato H. Functional advances of microwave dielectrics for next generation. Ceramics International. 2012;**38**:S141-S146

[13] Ohsato H. Design of microwave dielectrics based on crystallography. In: Akedo J, Chen XM, Tseng T, editors. Advances in Multifunctional Materials and Systems II, Ceramic Transactions. Vol. 245. Hoboken, New Jersey: Wiley; 2014. pp. 87-100

[14] Kawashima S, Nishida M, Ueda I, Ouchi H. $Ba(Zn_{1/3}Ta_{2/3})O_3$ ceramics with low dielectric loss at microwave frequencies. Journal of the American Ceramic Society. 1983;**66**:421-423

[15] Ohsato H, Koga E, Kagomiya I, Kakimoto K. Origin of high *Q* for microwave complex perovskite. Key Engineering Materials. 2010;**421-422**:77-80

[16] Terada M, Kawamura K, Kagomiya I, Kakimoto K, Ohsato H. Effect of Ni substitution on the microwave dielectric properties of cordierite. Journal of the European Ceramic Society. 2007;**27**:3045-3148

[17] Ohsato H, Kim JS, Kim AY, Cheon CI, Chae KW. Millimeter-wave dielectric properties of cordierite/indialite glass ceramics. Japanese Journal of Applied Physics. 2011;**50**(9):09NF01-1-5

[18] Ohsato H. Millimeter-wave materials. In: Sebastian MT, Ubic R,

Jantunen H, editors. Microwave Materials and Applications. Wiley; 2017. pp. 203-265

[19] Ohsato H, Ohhashi T, Nishigaki S, Okuda T, Sumiya K, Suzuki S. Formation of solid solution of new tungsten bronze-type microwave dielectric compounds $Ba_{6-3x}R_{8+2x}Ti_{18}O_{54}$ (R = Nd and Sm, $0 \leq x \leq 1$). Japanese Journal of Applied Physics. 1993;**32**:4323-4326

[20] Ohsato H. Science of tungstenbronze-type like $Ba_{6-3x}R_{8+2x}Ti_{18}O_{54}$ (R = rare earth) microwave dielectric solid solutions. Journal of the European Ceramic Society. 2001;**21**:2703-2711

[21] Ohsato H. Microwave dielectrics with perovskite-type structure. In: Pan L, Zhu G, editors. Perovskite Materials —Synthesis, Characterization, Properties, and Applications. Rijeka, Croatia: INTECH; 2016. pp. 281-330. ISBN 978-953-51-4587-5. Available from: http://cdn.intechopen.com/pdfs-wm/49723.pdf [Accessed: 15-10-2018]

[22] Kugimiya K. Crystallographic study on the Q of $Ba(Mg_{1/3}Ta_{2/3})O_3$ dielectrics. In: Abstract for Kansai branch academic meeting. 5 September 2003; Senri-life science. B-20. In: Abstract for the 10th Meeting of Microwave/Millimeterwave Dielectrics and Related Materials on the Ceramic Soc. Japan: Nagoya Institute of Technology; 21st June 2004 (Japanese)

[23] Koga E, Moriwake H. Effects of superlattice ordering and ceramic microstructure on the microwave Q factor of complex Perovskite-type oxide $Ba(Zn_{1/3}Ta_{2/3})O_3$. Journal of the Ceramic Society of Japan. 2003;**111**:767-775 (Japanese)

[24] Koga E, Moriwake H, Kakimoto K, Ohsato H. Influence of composition deviation from stoichiometric $Ba(Zn_{1/3}Ta_{2/3})O_3$ on superlattice ordering and microwave quality factor Q. Journal of the Ceramic Society of Japan. 2005;**113**:172-178 (Japanese)

[25] Koga E, Yamagishi Y, Moriwake H, Kakimoto K, Ohsato H. Large Q factor variation within dense, highly ordered $Ba(Zn_{1/3}Ta_{2/3})O_3$ system. Journal of the European Ceramic Society. 2006;**26**:1961-1964

[26] Koga E, Yamagishi Y, Moriwake H, Kakimoto K, Ohsato H. Order-disorder transition and its effect on microwave quality factor Q in $Ba(Zn_{1/3}Nb_{2/3})O_3$ system. Journal of Electroceramics. 2006;**17**:375-379

[27] Fujimura T, Nishida M, Kugimiya K. Dielectric ceramics, United patent 5,246,898

[28] Ohsato H, Koga E, Kagomiya I, Kakimoto K. Dense composition with high-Q on the complex perovskite compounds. Ferroelectrics. 2009;**387**:28-35

[29] Kolodiazhnyi T. Origin of extrinsic dielectric loss in 1:2 ordered, single-phase $BaMg_{1/3}Ta_{2/3}O_3$. Journal of the European Ceramic Society. 2014;**34**:1741-1753

[30] Miyashiro A. Cordierite–indialite relations. American Journal of Science. 1957;**255**:43-62

[31] Gibbs GV. Polymorphism of cordierite. I. Crystal structure of low cordierite. American Mineralogist. 1966;**51**:1068-1087

[32] Wu JM, Huang HL. Effect of crystallization on microwave dielectric properties of stoichiometric cordierite glasses containing B_2O_3 and P_2O_5 glasses. Journal of Materials Research. 2000;**15**:222-227

[33] Izumi F, Ikeda T. A rietveld-analysis program RIETAN-98 and its applications to zeolites. Materials Science Forum. 2000;**321-324**:198-203

[34] Toraya H, Hibino H, Ohsumi K. A new powder diffractometer for synchrotron radiation with multiple-detector system. Journal of Synchrotron Radiation. 1996;**3**:75-83

[35] Ohsato H, Kagomiya I, Terada M, Kakimoto K. Origin of improvement of *Q* based on high symmetry accompanying Si–Al disordering in cordierite millimeter-wave ceramics. Journal of the European Ceramic Society. 2010;**30**:315-318

[36] Brown ID, Shannon RD. Empirical bond-strength–bond-length curves for oxides. Acta Cryst. 1973;**A29**:266-282

[37] Brown ID, Wu KK. Empirical parameters for calculating cation–oxygen bond valences. Acta Cryst. 1976;**B32**:1957-1959

[38] Ohsato H, Kim JS, Cheon CI, Kagomiya I. Crystallization of indialite/cordierite glass ceramics for millimeter-wave dielectrics. Ceramics International. 2015;**41**:S588-S595

[39] Ohsato H, Kim JS, Cheon CI, Kagomiya I. Millimeter-wave dielectrics of indialite/cordierite glass ceramics: Estimating Si/Al ordering by volume and covalency of Si/Al octahedron. Journal of the Ceramic Society of Japan. 2013;**121**:649-654

[40] Carvajal RJ. Fullprof Software. 2006. Available from: http://www-llb.cea.fr/fullweb/powder.htm [Accessed: 13-04-2018]

[41] Varfolomeev MB, Mironov AS, Kostomarov VS, Golubtsova LA, Zolotova TA. The synthesis and homogeneity ranges of the phases $Ba_{6-x}R_{8+2x/3}Ti_{18}O_{54}$. Russian Journal of Inorganic Chemistry. 1988;**33**:607-608

[42] Ohsato H, Mizuta M, Ikoma T, Onogi Z, Nishigaki S, Okuda T. Microwave dielectric properties of tungsten bronzetype $Ba_{6-3x}R_{8+2x}Ti_{18}O_{54}$ (*R* = La, Pr, Nd and Sm) solid solutions. Journal of the Ceramic Society of Japan International Edition. 1998;**106-185**:184-188

[43] Negas T, Davies PK. Influence of chemistry and processing on the electrical properties of $Ba_{6-3x}Ln_{8+2x}Ti_{18}O_{54}$ solid solutions. In: Material and Processes for Wireless Communications, Ceramic Transactions 53. Hoboken, New Jersey: Wiley; 1995. pp. 196-197

[44] Valant M, Suvorov D, Kolar D. X-ray investigations and dielectric property determination of the $Ba_{4.5}Gd_9Ti_{18}O_{54}$ compound. Japanese Journal of Applied Physics. 1996;**35**:144-150

[45] Ohsato H, Ohhashi T, Kato H, Nishigaki S, Okuda T. Microwave dielectric properties and structure of the $Ba_{6-3x}Sm_{8+2x}Ti_{18}O_{54}$ solid solutions. Japanese Journal of Applied Physics. 1995;**34**:187-191

[46] Fukuda K, Kitoh R, Awai I. Microwave characteristics of mixed phases of $BaPr_2Ti_4O_{12}$-$BaPr_2Ti_5O_{14}$ ceramics. Journal of Materials Research. 1995;**10**:312-319

[47] Valant M, Suvorov D, Rawn CJ. Intrinsic reasons for variations in dielectric properties of $Ba_{6-3x}R_{8+2x}Ti_{18}O_{54}$ (*R* = La-Gd) solid solutions. Japanese Journal of Applied Physics. 1999;**38**:2820-2826

[48] Matveeva RG, Varforomeev MB, Il'yuschenko LS. Refinement of the composition and crystal structure of $Ba_{3.75}Pr_{9.5}Ti_{18}O_{54}$. Zhurnal Neorganicheskoi Khimii. 1984;**29**:31- 34 (Trans. Russ: Journal of Inorganic Chemistry. 1984;**29**:17-19)

[49] Roth RS, Beach F, Antoro A, Davis K, Soubeyroux JL. Structural of the nonstoichiometric solid solutions $Ba_2RE_4[Ba_x+RE_{2/3-2/3x}]Ti_9O_{27}$ (*RE* = Nd, Sm). In: 14 Int Congress Crystallog: Mat

Sci. C-138: Collected Abstract 07. 9-9; Perth, Australia. 1987

[50] Ohsato H, Ohhashi T, Okuda T. Structure of $Ba_{6-3x}Sm_{8+2x}Ti_{18}O_{54}$ (0<x<1). In: Asian Crystallographic Association Conference (AsCA '92): Ext. Abstract 14U-50; November 1992; Singapore. 1992

[51] Kolar D, Gabrscek S, Suvorov D. Structural and dielectric properties of perovskite-like rare earth titanates. In: Duran P, Fernandes JF, editors. Third Euro-Ceramics V.2. Vol. 2. Spein: Faenza Editrice Lberica; 1993. pp. 229-234

[52] Lundberg M, Sundberg M, Magneli A. The "pentagonal column" as a building unit in crystal and defect structure of some groups of transition metal compounds. Journal of Solid State Chemistry. 1982;**44**:32-40

[53] Ohsato H, Nishigaki S, Okuda T. Superlattice and dielectric properties of dielectric compounds. Japanese Journal of Applied Physics. 1992;**31**(9B):3136-3138

[54] Ohsato H. Crystallography of dielectrics. In: Crystallography in Japan (II)–The glorious development. Tokyo, Japan: The Crystallographic Society of Japan; 2014. pp. 184-185 (Japanese)

[55] Ohsato H, Imaeda M, Komura A, Okuda T. Non-linear microwave quality factor change based on the site occupancy of cations on the tungstenbronze-type $Ba_{6-3x}R_{8+2x}Ti_{18}O_{54}$ (R = rare earth) solid solutions. In: Nair KM, Bhalla AS, editors. Dielectric Ceramic Materials, Ceramic Transactions 100. Hoboken, New Jersey: Wiley; 1998. pp. 41-50

[56] Ohsato H, Imaeda M, Takagi Y, Komura A, Okuda T. Microwave quality factor improved by ordering of Ba and rare-earth on the tungstenbronze-type $Ba_{6-3x}R_{8+2x}Ti_{18}O_{54}$ (R = La, Nd and Sm) solid solutions. In: Proceeding of the XIth IEEE International Symposium on Applications of Ferroelectrics; IEEE catalog number 98CH36245. 1998. pp. 509-512

[57] Ohsato H, Mizuta M, Okuda T. Crystal structure and microwave dielectric properties of Tungstenbronze-type $Ba_{6-3x}R_{8+2x}Ti_{18}O_{54}$ (R = La, Nd and Sm) solid solutions. In: Morawiec H, Stroz D, editors. Applied Crystallography. World Scientific Publishing; 1998. pp. 440-447

[58] Galasso F, Pyle J. Ordering in compounds of the $A(B'_{0.33}Ta_{0.67})O_3$ type. Inorganic Chemistry. 1963;**2**:482-484

[59] Yokotani Y, Tsuruta T, Okuyama K, Kugimiya K. Low-dielectric loss ceramics for microwave uses. National Technical Report. 1994;**40**:11-16 (Japanese)

[60] Tamura H, Konoike T, Sakabe Y, Wakino K. Microwave dielectric properties of Ba-Nd-Ti-O system doped with metal oxides. Journal of the American Ceramic Society. 1984;**67**:C-59

[61] Matsumoto H, Tamura H, Wakino K. $Ba(Mg,Ta)O_3$-$BaSnO_3$ high-Q dielectric resonator. Japanese Journal of Applied Physics. 1991;**30**:2347-2349

[62] Kageyama K. Crystal structure and microwave dielectric properties of $Ba(Zn_{1/3}Ta_{2/3})O_3$-$(Sr,Ba)(Ga_{1/3}Ta_{1/2})O_3$ ceramics. Journal of the American Ceramic Society. 1992;**75**:1767-1771

[63] Koga E, Moriwake H, Kakimoto K, Ohsato H. Synthesis of disordered $Ba(Zn_{1/3}Ta_{2/3})O_3$ by spark plasma sintering and its microwave Q factor. Japanese Journal of Applied Physics. 2006;**45**(9B):7484-7488

[64] Ohsato H, Azough F, Koga E, Kagomiya I, Kakimoto K, Freer R. High symmetry brings high Q instead of ordering in $Ba(Zn_{1/3}Nb_{2/3})O_3$: A HRTEM

study. In: Akedo J, Ohsato H, Shimada T, editors. Advances in Multifunctional Materials and Systems, Ceramic Transactions. Vol. 216. Hoboken, New Jersey: Wiley; 2010. pp. 129-136

[65] Wu H, Davies PK. Influence of non-stoichiometry on the structure and properties of $Ba(Zn_{1/3}Nb_{2/3})O_3$ microwave dielectrics: I. Substitution of $Ba_3W_2O_9$. Journal of the American Ceramic Society. 2006;**89**(7):2239-2249. DOI: 10.1111/j.1551-2916.2006.01007.x

[66] Surendran KP, Sebastian MT, Mohanan P, Moreira RL, Dias A. Effect of nonstoichiometry on the structure and microwave dielectric properties of $Ba(Mg_{0.33}Ta_{0.67})O_3$. Chemistry of Materials. 2005;**17**:142-151

5

Optical Propagation in Magneto-Optical Materials

Licinius Dimitri Sá de Alcantara

Abstract

Magneto-optical materials present anisotropy in the electrical permittivity controlled by a magnetic field, which affects the propagation characteristics of light and stands out in the design of nonreciprocal devices, such as optical isolators and circulators. Based on Maxwell's equations, this chapter focuses on the wave propagation in magneto-optical media. The following cases are covered: The propagation of a plane wave in an unbounded magneto-optical medium, where the phenomenon of Faraday rotation is discussed, and the guided propagation in planar magneto-optical waveguides with three and five layers, highlighting the phenomenon of nonreciprocal phase shift and its potential use on the design of nonreciprocal optical devices.

Keywords: magneto-optical media, light propagation, Faraday rotation, nonreciprocal phase shift, optical devices

1. Introduction

A material is classified as magneto-optical (MO) if it affects the propagation characteristics of light when an external magnetic field is applied on it. For ferromagnetic materials, which are composed by magnetically ordered domains, MO phenomena may also occur in the absence of an external magnetic field. A great number of magneto-optical phenomena are the direct or indirect outcome of the splitting of energy levels in an external or spontaneous magnetic field [1].

The MO effect depends on the polarization of the magnetic field. It also depends on the polarization of the light and on its propagation direction, so it is an anisotropic phenomenon, which has attracted great attention from researchers in optical devices. The MO materials can have their anisotropy controlled by a magnetostatic field (H_{DC}), and this behavior can be exploited on the design of nonreciprocal devices. By nonreciprocal devices or structures, it means that waves or guided modes supported by them have their propagation characteristics altered when the wave propagation sense is reversed. Optical isolators and circulators can be highlighted as examples of such devices. Isolators are designed to protect optical sources from reflected light and are present in optical amplification systems. The circulators are employed as signal routers and act in devices that extract wavelengths in WDM systems.

The design of optical devices with MO materials is addressed in several works such as [2–5]. The challenges for the design of such devices are the development of MO materials with high-induced anisotropy and high transparency at the optical spectrum. Therefore, research activities on the improvement of MO materials and structures have also great relevance and are covered in works such as [6–10]. Integration of MO materials and structures with other optical system components, with reduction of insertion losses, is also a target for researches in optical devices. Research of MO effects in optical structures such as photonic crystals has also been addressed [11–13].

This chapter presents analytical formalisms derived from Maxwell's and wave equations to analyze the propagation characteristics of transverse electromagnetic (TEM) waves in unbounded magneto-optical material. The guided propagation characteristics of transverse magnetic (TM) modes in three- and five-layered planar magneto-optical waveguides are also formalized and discussed. The analytical formalism is versatile so that each layer can be set as magneto-optical or isotropic in the mathematical model.

2. Wave propagation characteristics

This section focuses on the optical propagation analysis in magneto-optical media using Maxwell's equations as starting point. In a magnetized MO media, cyclotron resonances occur at optical frequencies, if the wave is properly polarized. This physical phenomenon induces a coupling between orthogonal electric field components in the plane perpendicular to the applied magnetostatic field H_{DC}, which affects the wave polarization. Depending on the orientation of the magneto-static field, the configuration of the electric permittivity tensor changes. If H_{DC} is oriented along one of the Cartesian axes, the relative elec tric permittivity assumes the form

$$\overline{\overline{\varepsilon}}_r = \begin{bmatrix} n^2 & 0 & 0 \\ 0 & n^2 & j\delta \\ 0 & -j\delta & n^2 \end{bmatrix}, \text{ for } H_{DC} \parallel x - \text{axis}; \tag{1}$$

$$\overline{\overline{\varepsilon}}_r = \begin{bmatrix} n^2 & 0 & j\delta \\ 0 & n^2 & 0 \\ -j\delta & 0 & n^2 \end{bmatrix}, \text{ for } H_{DC} \parallel y - \text{axis}; \tag{2}$$

$$\overline{\overline{\varepsilon}}_r = \begin{bmatrix} n^2 & j\delta & 0 \\ -j\delta & n^2 & 0 \\ 0 & 0 & n^2 \end{bmatrix}, \text{ for } H_{DC} \parallel z - \text{axis}. \tag{3}$$

where n is the refractive index of the material and δ is the magneto-optical constant. The MO constant is proportional to H_{DC}. If the sense of H_{DC} is reversed, $\delta(-H_{DC}) = -\delta(H_{DC})$, and for $H_{DC} = 0$, the off-diagonal components of the electric permittivity tensor are zero [14, 15].

2.1 TEM wave in an unbounded magneto-optical medium

Let us consider a TM wave propagating in an unbounded MO medium, as shown in **Figure 1**.

From Maxwell's equations, the vectorial Helmholtz equation for anisotropic media and for the electric field $\overline{E}(x,y,z)$ can be written as

$$\omega^2 \mu_0 \varepsilon_0 \overline{\overline{\varepsilon}}_r \overline{E} + \nabla^2 \overline{E} - \nabla(\nabla \cdot \overline{E}) = \overline{0}, \tag{4}$$

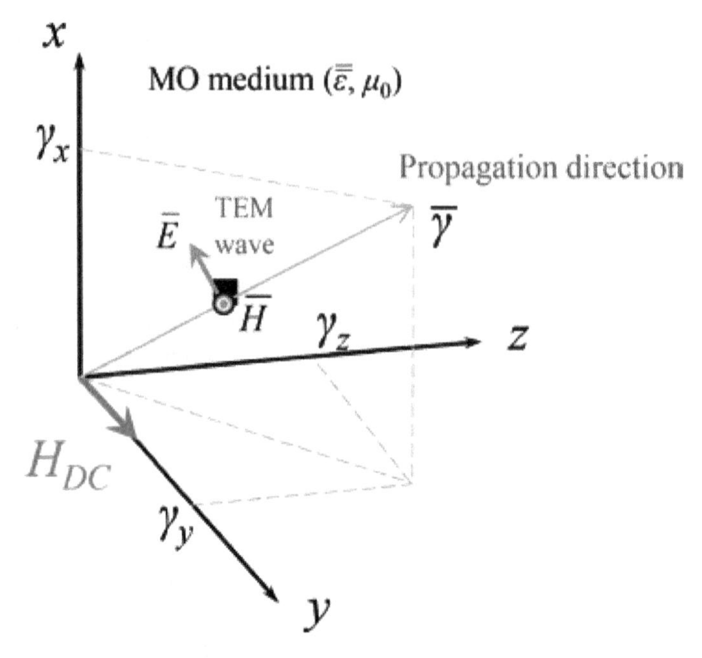

Figure 1.
TEM wave in an unbounded magneto-optical medium.

where ω is the angular frequency in rad/s, μ_0 is the magnetic permeability of the vacuum in H/m, and ε_0 is the electric permittivity of the vacuum in F/m.

To develop a plane wave solution for MO media, it is assumed that H_{DC} is parallel to the y-axis and $\bar{\bar{\varepsilon}}_r$ is given by Eq. (2) from now on. This assumption does not imply on lack of generality because it is assumed that the wave propagates at an arbitrary direction, with the electric field vector given by

$$\overline{E} = \overline{E}_0 \exp\left(j\omega t\right) \exp\left[-j\left(\gamma_x x + \gamma_y y + \gamma_z z\right)\right]. \tag{5}$$

where $\overline{\gamma} = \gamma_x \vec{\mathbf{i}} + \gamma_y \vec{\mathbf{j}} + \gamma_z \vec{\mathbf{k}}$ is the propagation constant vector.

From Gauss' law for a medium with equilibrium of charges, $\nabla.\left(\varepsilon_0 \bar{\bar{\varepsilon}}_r \overline{E}\right) = 0$, we obtain:

$$\nabla \cdot \overline{E} = j\frac{\delta}{n^2}\left(\frac{\partial E_x}{\partial z} - \frac{\partial E_z}{\partial x}\right). \tag{6}$$

Substituting Eq. (6) into Eq. (4) leads to

$$\omega^2 \mu_0 \varepsilon_0 \bar{\bar{\varepsilon}}_r \overline{E} + \nabla^2 \overline{E} - j\frac{\delta}{n^2}\nabla\left(\frac{\partial E_x}{\partial z} - \frac{\partial E_z}{\partial x}\right) = \overline{0}. \tag{7}$$

Expanding Eq. (7) in the Cartesian coordinates results in

$$\omega^2 \mu_0 \varepsilon_0 \left(n^2 E_x + j\delta E_z\right) + \frac{\partial^2 E_x}{\partial x^2} + \frac{\partial^2 E_x}{\partial y^2} + \frac{\partial^2 E_x}{\partial z^2} - j\frac{\delta}{n^2}\left(\frac{\partial^2 E_x}{\partial x \partial z} - \frac{\partial^2 E_z}{\partial x^2}\right) = 0, \tag{8}$$

$$\omega^2 \mu_0 \varepsilon_0 n^2 E_y + \frac{\partial^2 E_y}{\partial x^2} + \frac{\partial^2 E_y}{\partial y^2} + \frac{\partial^2 E_y}{\partial z^2} - j\frac{\delta}{n^2}\left(\frac{\partial^2 E_x}{\partial y \partial z} - \frac{\partial^2 E_z}{\partial x \partial y}\right) = 0, \tag{9}$$

$$\omega^2\mu_0\varepsilon_0(-j\delta E_x + n^2 E_z) + \frac{\partial^2 E_z}{\partial x^2} + \frac{\partial^2 E_z}{\partial y^2} + \frac{\partial^2 E_z}{\partial z^2} - j\frac{\delta}{n^2}\left(\frac{\partial^2 E_x}{\partial z^2} - \frac{\partial^2 E_z}{\partial x \partial z}\right) = 0. \quad (10)$$

The spatial derivatives in Eqs. (8)–(10) are now calculated by considering Eq. (5):

$$\left(\omega^2\mu_0\varepsilon_0 n^2 - |\gamma|^2 - j\frac{\delta}{n^2}\gamma_x\gamma_z\right)E_x + j\delta\left(\omega^2\mu_0\varepsilon_0 - \frac{1}{n^2}\gamma_x^2\right)E_z = 0 \quad (11)$$

$$\left(\omega^2\mu_0\varepsilon_0 n^2 - |\gamma|^2\right)E_y + j\frac{\delta}{n^2}\left(\gamma_y\gamma_z E_x - \gamma_x\gamma_y E_z\right) = 0, \quad (12)$$

$$-j\delta\left(\omega^2\mu_0\varepsilon_0 - \frac{1}{n^2}\gamma_z^2\right)E_x + \left(\omega^2\mu_0\varepsilon_0 n^2 - |\gamma|^2 - j\frac{\delta}{n^2}\gamma_x\gamma_z\right)E_z = 0, \quad (13)$$

where $|\gamma| = \sqrt{\gamma_x^2 + \gamma_y^2 + \gamma_z^2}$.

2.1.1 TEM wave with electric field vector parallel to H_{DC}

By observing Eqs. (11–13), we note that **when the electric field of the electromagnetic wave is polarized along the y-axis and is parallel to H_{DC}**, so that $E_x = E_z = 0$, the magneto-optical constant δ related to H_{DC} will have no effect on the propagation characteristics of the wave. In this case, from Eq. (12), the propagation constant modulus would be

$$|\gamma| = n\omega\sqrt{\mu_0\varepsilon_0}, \quad (14)$$

which is the same expression for a traveling wave in an isotropic material. Note that when the electric field is polarized along the y-axis, the wave is traveling in the plane xz, so that $\gamma_y = 0$.

2.1.2 The general expression for the propagation constant

In a general case, by solving the system formed by Eqs. (11) and (13), we obtain the following equation

$$\left(\omega^2\mu_0\varepsilon_0 n^2 - |\gamma|^2 - j\frac{\delta}{n^2}\gamma_x\gamma_z\right)^2 - \delta^2\left(\omega^2\mu_0\varepsilon_0 - \frac{1}{n^2}\gamma_x^2\right)\left(\omega^2\mu_0\varepsilon_0 - \frac{1}{n^2}\gamma_z^2\right) = 0. \quad (15)$$

Solving Eq. (15) for $|\gamma|$, we obtain:

$$|\gamma| = \sqrt{\omega^2\mu_0\varepsilon_0 n^2 - j\frac{\delta}{n^2}\gamma_x\gamma_z \pm \delta\sqrt{\left(\omega^2\mu_0\varepsilon_0 - \frac{1}{n^2}\gamma_x^2\right)\left(\omega^2\mu_0\varepsilon_0 - \frac{1}{n^2}\gamma_z^2\right)}}. \quad (16)$$

Note that when the MO constant $\delta = 0$, Eq. (16) reduces to Eq. (14).

The parameters γ_x and γ_z are projections of the propagation constant vector along the x and the y-axis, respectively.

2.1.3 TEM wave propagating parallel to H_{DC}

If the TEM wave is propagating along the H_{DC} direction (y-axis), so that $\gamma_x = \gamma_z = 0$, Eq. (16) assumes the simpler form:

$$|\gamma| = \omega\sqrt{\mu_0\varepsilon_0(n^2 \pm \delta)}, \tag{17}$$

and from Eq. (5), the electric field vector becomes

$$\overline{E} = \left(E_{0x}\vec{\mathbf{i}} + E_{0z}\vec{\mathbf{k}}\right)\exp\left[j\left(\omega t - y\omega\sqrt{\mu_0\varepsilon_0(n^2 \pm \delta)}\right)\right]. \tag{18}$$

From Eq. (11), we see that the electric field components are connected by

$$E_{0x} = -j\delta\frac{\omega^2\mu_0\varepsilon_0}{\left(\omega^2\mu_0\varepsilon_0 n^2 - |\gamma|^2\right)}E_{0z}. \tag{19}$$

Substituting Eq. (17) in Eq. (19), we obtain:

$$E_{0x} = \pm jE_{0z}. \tag{20}$$

Therefore, substituting Eq. (20) in Eq. (18), and given that $\pm j = \exp(\pm j\pi/2)$, the electric field components can be written as

$$E_x = E_{0z}\exp\left[j\left(\omega t - y\omega\sqrt{\mu_0\varepsilon_0(n^2 \pm \delta)} \pm \pi/2\right)\right], \tag{21}$$

$$E_z = E_{0z}\exp\left[j\left(\omega t - y\omega\sqrt{\mu_0\varepsilon_0(n^2 \pm \delta)}\right)\right]. \tag{22}$$

Eqs. (21) and (22) represent a circular polarized wave, which can be dismembered into two circular polarized eigenmodes propagating along the y-axis with different propagation constants. If the plus sign (in "±") is adopted for Eqs. (21) and (22), we obtain a counterclockwise (CCW) circular polarized eigenmode. Otherwise, if the minus sign is adopted, we obtain a clockwise (CW) circular polarized eigenmode, as shown in **Figure 2**. From Eq. (17), it is possible to associate an equivalent refractive index to each eigenmode:

$n^+ = \sqrt{n^2 + \delta}$, for the CCW circular polarized eigenmode;
$n^- = \sqrt{n^2 - \delta}$, for the CW circular polarized eigenmode.

A linear polarized wave propagating along the y-axis may be decomposed into two opposite circular polarized waves in the xz plane, as shown in **Figure 2**. Since these eigenmodes propagate with distinct propagation constants, the linear

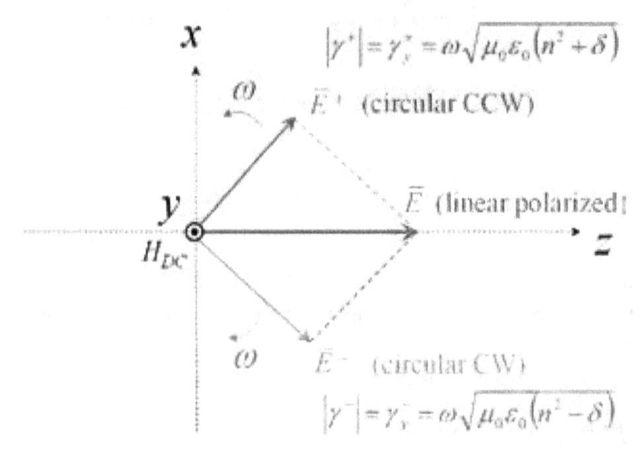

Figure 2.
Decomposition of a linear polarized TEM wave into two circular polarized components. The circular polarized components travel with distinct propagation constants in a MO medium.

polarization will rotate in the xz plane as the wave propagates along the y-axis, in a phenomenon known as **Faraday rotation**, which is depicted in **Figure 3**.

When the sense of the magnetostatic field H_{DC} is reversed, the magneto-optical constant δ changes its signal, and the values of n^+ and n^- are interchanged, and the sense of rotation of a linear polarized wave in the MO media will change.

The Faraday rotation angle (ϕ_F) may be calculated (in radians) as a function of the propagation distance y by

$$\phi_F = \frac{1}{2}(\phi^+ - \phi^-) = \frac{1}{2}\left(n^+ \frac{2\pi}{\lambda_0} y - n^- \frac{2\pi}{\lambda_0} y\right) = \frac{\pi}{\lambda_0}\left(\sqrt{n^2+\delta} - \sqrt{n^2-\delta}\right)y, \quad (23)$$

where λ_0 is the optical wavelength in vacuum. The Faraday rotation effect is responsible for a periodic power transfer between the transverse components, in this case, E_x and E_z. This phenomenon in MO materials may be exploited for the design of optical isolators based on Faraday rotation.

When a MO waveguide, with H_{DC} applied along its longitudinal direction, supports degenerate orthogonal quasi TEM modes, the power transfer between these modes will be maximized. **Figure 4** shows a MO rib waveguide [16], where layers 1 and 2 are composed of bismuth yttrium iron garnet (Bi-YIG) grown on top of a gadolinium gallium garnet (GGG) substrate with $n_{SR} = 1.94$. For the Bi-YIG layers, the relative permittivity tensor has the form of Eq. (2), with $\delta = 2.4 \times 10^{-4}$,

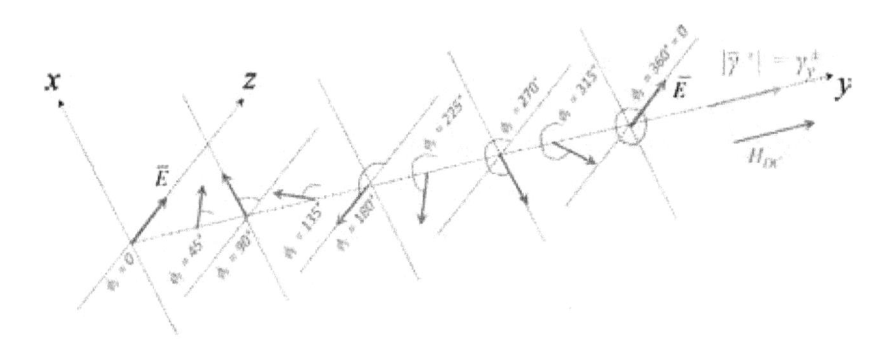

Figure 3.
Faraday rotation of a linear polarized TEM wave in a MO medium. The propagation direction is parallel to the magnetostatic field H_{DC}.

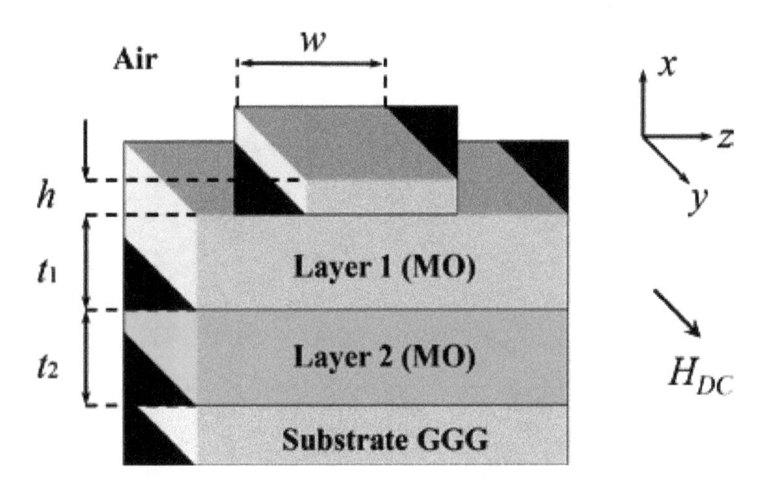

Figure 4.
Magneto-optical rib waveguide.

n_1 = 2.19, and n_2 = 2.18. The waveguide dimensions are w = 8 μm, h = 0.5 μm, t_1 = 3.1 μm, and t_2 = 3.4 μm. The optical wavelength is λ_0 = 1.485 μm.

Figure 5 shows numerical results for the power transfer between the transverse components along the propagation direction. These results were obtained using a finite difference vectorial beam propagation method (FD-VBPM) [17]. We observe that the length for maximum energy transfer is around 6800 μm In practice, as observed in [16], the device length must be set at half that length (~3400 μm) so that a 45° rotation is achieved at the output port. Therefore, if a reflection occurs at this point, the reflected field will complete a 90° rotation at the input port, which can then be blocked with a polarizer without affecting the input field, so that an optical isolator is obtained.

In Eq. (23), by adopting $\delta = 2.4 \times 10^{-4}$, $n = n_1 = 2.19$, $\lambda_0 = 1.485$ μm, and $\phi_F = \pi/4$ (45°), we obtain y = 3388 μm, which is a propagation length that converges with the FD-VBPM result.

2.1.4 TEM wave propagating along the diagonal of an imaginary cube

Before finishing this section, let us consider another particular case of propagation direction—suppose, in **Figure 1**, that $\gamma_x = \gamma_y = \gamma_z = \gamma_u$, with $\gamma_u \neq 0$. **This case corresponds to a TEM wave propagating along the diagonal of an imaginary cube, adjacent to the Cartesian axes**. From Eq. (16), we obtain:

$$|\gamma| = \sqrt{\omega^2\mu_0\varepsilon_0 n^2 - j\frac{\delta}{n^2}\gamma_u^2 \pm \delta\left(\omega^2\mu_0\varepsilon_0 - \frac{1}{n^2}\gamma_u^2\right)}. \tag{24}$$

From the relation $|\gamma| = \sqrt{\gamma_x^2 + \gamma_y^2 + \gamma_z^2}$ we can also obtain:

$$|\gamma| = \gamma_u\sqrt{3}. \tag{25}$$

Equaling Eqs. (24)–(25) and solving for γ_u result in

$$\gamma_u = \omega\sqrt{\frac{\mu_0\varepsilon_0(n^2 \pm \delta)}{3 \pm \frac{\delta}{n^2} + j\frac{\delta}{n^2}}}. \tag{26}$$

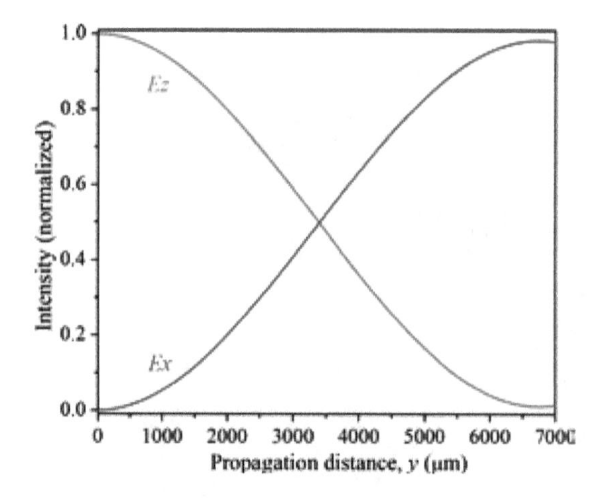

Figure 5.
Normalized intensity evolution of the transverse field components along the propagation direction (y-axis) of the MO waveguide.

Substituting Eq. (26) in Eq. (25), we obtain the propagation constant:

$$|\gamma| = \omega\sqrt{\frac{\mu_0\varepsilon_0(n^2 \pm \delta)}{1 \pm \frac{\delta}{3n^2} + j\frac{\delta}{3n^2}}}. \tag{27}$$

The corresponding electric field vector can be retrieved by substituting the results of Eqs. (26)–(27) in Eq. (11) to obtain

$$E_x = \pm jE_z. \tag{28}$$

However, for the considered propagation direction, the E_y component is not zero. From Eq. (12) we obtain:

$$E_y = -\frac{n^2 \pm \delta + j3(\delta \pm n^2)}{5n^2}E_z. \tag{29}$$

By using the results of Eqs. (26)–(29) in Eq. (5), we can express the electric field vector for this particular case by

$$\overline{E} = \left(\pm j\,\vec{\mathbf{i}} - \frac{n^2 \pm \delta + j3(\delta \pm n^2)}{5n^2}\,\vec{\mathbf{j}} + \vec{\mathbf{k}}\right)E_{0z}\exp\left[j\left(\omega t - \omega\sqrt{\frac{\mu_0\varepsilon_0(n^2 \pm \delta)}{3 \pm \frac{\delta}{n^2} + j\frac{\delta}{n^2}}}(x + y + z)\right)\right], \tag{30}$$

where **i**, **j**, and **k** are the unit vectors along the x-, y-, and z-axis, respectively.

As in the previous case of propagation, Eq. (30) provides two eigenmodes for TEM propagation. From Eq. (28) we can observe that, when projected in the xz plane, the electric field vector of each eigenmode is circular polarized. The combination of these eigenmodes will result in a wave with linear polarization progressively rotated as it propagates. The E_y component has the role of projecting the Faraday rotation to the plane perpendicular to the propagation direction (the diagonal of the cube), since the wave is TEM regarding this propagation direction. **Figure 6** shows a simulation of the TEM wave eigenmodes along the diagonal of an imaginary cube.

The simulations presented in **Figure 6** were performed for f = 193.4145 THz, n = 2, and δ = 0.2. Note that both eigenmodes present losses as they propagate. This is due the complex characteristic of the propagation constant expressed by Eq. (27), where the imaginary part depends on the magneto-optical constant δ. It was observed that increasing δ enhances the Faraday rotation but also increases the losses for diagonal propagation.

Equivalent refractive indexes for the circular polarized eigenmodes can be obtained from Eq. (27), which leads to the following equation to compute the Faraday rotation for diagonal propagation:

$$\phi_F = \frac{\pi}{\lambda_0}\mathrm{Re}\left(\sqrt{\frac{n^2 + \delta}{1 + \frac{\delta}{3n^2} + j\frac{\delta}{3n^2}}} - \sqrt{\frac{n^2 - \delta}{1 - \frac{\delta}{3n^2} + j\frac{\delta}{3n^2}}}\right)d, \tag{31}$$

where d is the propagation distance along the diagonal.

For n = 2, δ = 0.2, and λ_0 = 1.55 μm, we obtain ϕ_F/d = 0.27046 rads/μm. Comparing with the case for propagation along the y-axis (parallel to H_{DC}), by using Eq. (23), we obtain ϕ_F/y = 0.40549 rads/μm. These results show that we can obtain a better Faraday rotation when the propagation direction is aligned with the magnetostatic field, when considering TEM waves.

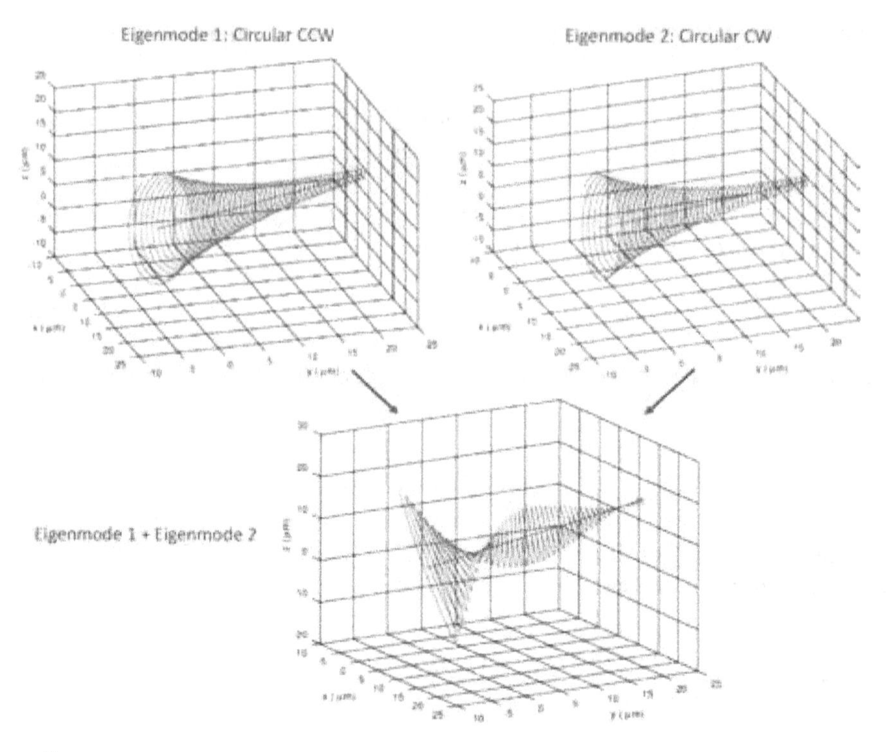

Figure 6.
TEM eigenmodes for diagonal propagation where $\gamma_x = \gamma_y = \gamma_z$. *The trajectory of the electric field vector is represented by red lines.*

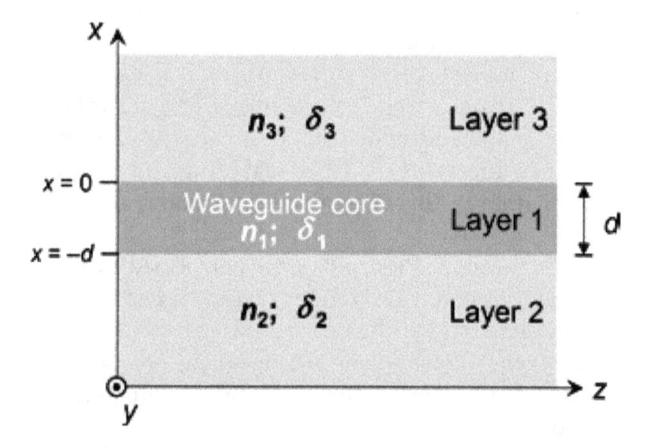

Figure 7.
Longitudinal section of a planar MO waveguide.

2.2 TM mode in a planar magneto-optical waveguide

Figure 7 presents a planar MO waveguide, which is composed by three MO layers. The magnetostatic field H_{DC} is applied along the *y*-axis. The propagation direction is now the *z*-axis. The planar waveguide supports transversal electric, TE, modes (H_x, E_y, H_z components) and transversal magnetic, TM, modes (E_x, H_y, E_z components). As discussed in Section 2.1.1, if H_{DC} is parallel to the electric field vector of the wave, then MO constant δ does not affect the propagation characteristics of the mode. Therefore, for the TE modes, no MO effect will be observed. For TM modes, however, the electric field components are perpendicular to H_{DC}, and nonreciprocal propagation characteristics will take place. In this section, mathematical expressions to calculate the propagation constants for TM modes in a MO planar waveguide will be derived. For the occurrence of guided modes in the structure shown in **Figure** 7, $n_1 > n_2$ and $n_1 > n_3$.

Defining $\bar{\bar{\xi}}$ as the inverse of the electric permittivity tensor of Eq. (2), we have:

$$\bar{\bar{\xi}} = \bar{\bar{\varepsilon}}_r^{-1} = \begin{bmatrix} \frac{n^2}{n^4-\delta^2} & 0 & -\frac{j\delta}{n^4-\delta^2} \\ 0 & \frac{n^2}{n^4-\delta^2} & 0 \\ \frac{j\delta}{n^4-\delta^2} & 0 & \frac{n^2}{n^4-\delta^2} \end{bmatrix} = \begin{bmatrix} \xi_{xx} & 0 & -\xi_{zx} \\ 0 & \xi_{yy} & 0 \\ \xi_{zx} & 0 & \xi_{zz} \end{bmatrix}. \tag{32}$$

From Maxwell's equations at the frequency domain, considering TM modes (E_x, H_y, E_z components) and no field spatial variations along the y-axis, we obtain:

$$j\omega\mu H_y = j\beta E_x + \frac{\partial E_z}{\partial x}, \tag{33}$$

$$E_x = \frac{1}{\varepsilon_0}\left(\xi_{xx}\frac{\beta}{\omega}H_y + j\frac{\xi_{zx}}{\omega}\frac{\partial H_y}{\partial x}\right), \tag{34}$$

$$E_z = \frac{1}{\varepsilon_0}\left(\xi_{zx}\frac{\beta}{\omega}H_y - j\frac{\xi_{zz}}{\omega}\frac{\partial H_y}{\partial x}\right), \tag{35}$$

where β is the propagation constant of the guided TM mode in radians per meter.

Substituting Eqs. (34)–(35) in Eq. (33), we obtain the following wave equation for nonreciprocal media in terms of the H_y component:

$$\frac{\partial^2 H_y}{\partial x^2} + \left(\frac{k_0^2 - \xi_{xx}\beta^2}{\xi_{zz}}\right)H_y = 0, \tag{36}$$

where $k_0 = \omega\sqrt{\mu\varepsilon_0}$.

The solution for H_y is expressed for each waveguide layer as.

$$H_y = C\exp(-\zeta x), \text{ for } x \geq 0. \tag{37}$$

$$H_y = C\cos(\kappa x) + D\,\text{sen}(\kappa x), \text{ for} -\text{d} \leq x \leq 0. \tag{38}$$

$$H_y = [C\cos(\kappa d) - D\,\text{sen}(\kappa d)]\exp[\gamma(x+d)], \text{ for } x \leq -\text{d}. \tag{39}$$

The solution for the component E_z at each layer is obtained by substituting the corresponding solution for H_y in Eq. (35), resulting in.

$$E_z = \frac{C}{\omega\varepsilon_0}\left(\xi_{zx}^{(3)}\beta + j\zeta\xi_{zz}^{(3)}\right)\exp(-\zeta x), \text{ for } x \geq 0. \tag{40}$$

$$E_z = \frac{1}{\omega\varepsilon_0}\left\{C\left[\xi_{zx}^{(1)}\beta\cos(\kappa x) + j\kappa\xi_{zz}^{(1)}\text{sen}(\kappa x)\right] + D\left[\xi_{zx}^{(1)}\beta\text{sen}(\kappa x) - j\kappa\xi_{zz}^{(1)}\cos(\kappa x)\right]\right\}, \text{ for} -\text{d} \leq x \leq 0. \tag{41}$$

$$E_z = \frac{C\cos(\kappa d) - D\text{sen}(\kappa d)}{\omega\varepsilon_0}\left(\xi_{zx}^{(2)}\beta - j\gamma\xi_{zz}^{(2)}\right)\exp[\gamma(x+d)], \text{ for } x \leq -\text{d}. \tag{42}$$

The superscripts between parentheses on the inverse permittivity tensor elements identify the corresponding waveguide layer, as specified in **Figure 7** . The continuity of E_z at $x = 0$ and at $x = -d$ leads to the following system:

$$C\left[\left(\xi_{zx}^{(3)} - \xi_{zx}^{(1)}\right)\beta + j\zeta\xi_{zz}^{(3)}\right] + D\left(j\kappa\xi_{zz}^{(1)}\right) = 0, \tag{43}$$

$$\begin{aligned} &C\left\{\left[\left(\xi_{zx}^{(1)} - \xi_{zx}^{(2)}\right)\beta + j\gamma\xi_{zz}^{(2)}\right]\cos(\kappa d) - j\kappa\xi_{zz}^{(1)}\,sen(\kappa d)\right\} \\ &+ D\left\{\left[\left(\xi_{zx}^{(2)} - \xi_{zx}^{(1)}\right)\beta - j\gamma\xi_{zz}^{(2)}\right]\mathrm{sen}(\kappa d) - j\kappa\xi_{zz}^{(1)}\cos(\kappa d)\right\} = 0. \end{aligned} \tag{44}$$

After solving this system formed by Eqs. (43)–(44), we obtain:

$$\tan(\kappa d) = \frac{\kappa\xi_{zz}^{(1)}\left[\zeta\xi_{zz}^{(3)} + \gamma\xi_{zz}^{(2)} - j\left(\xi_{zx}^{(3)} - \xi_{zx}^{(2)}\right)\beta\right]}{\left(\kappa\xi_{zz}^{(1)}\right)^2 - \left[\left(\xi_{zx}^{(3)} - \xi_{zx}^{(1)}\right)\beta + j\zeta\xi_{zz}^{(3)}\right]\left[\left(\xi_{zx}^{(2)} - \xi_{zx}^{(1)}\right)\beta - j\gamma\xi_{zz}^{(2)}\right]}. \tag{45}$$

The constants ζ, κ, and γ can be determined by substituting Eq. (37), Eq. (38), or Eq. (39), respectively, in Eq. (36), resulting in

$$\zeta = \sqrt{\frac{\xi_{xx}^{(3)}\beta^2 - k_0^2}{\xi_{zz}^{(3)}}}, \tag{46}$$

$$\kappa = \sqrt{\frac{k_0^2 - \xi_{xx}^{(1)}\beta^2}{\xi_{zz}^{(1)}}}, \tag{47}$$

$$\gamma = \sqrt{\frac{\xi_{xx}^{(2)}\beta^2 - k_0^2}{\xi_{zz}^{(2)}}}, \tag{48}$$

where $k_0 = 2\pi/\lambda_0$, and λ_0 is the optical wavelength.

From the roots of Eq. (45) for β, the dispersion curve for TM modes in MO waveguides can be retrieved. Assuming that $n_1 = 2.26$, $n_2 = 2.0$, $n_3 = 2.23$, $d = 1$ μm, and only the layer 3 is magneto-optical with $\delta = 0.019$, the dispersion curve for the fundamental and a superior TM mode is shown in **Figure 8**. We observe that the effective index profile changes when the propagation direction is reversed, which opens the possibility to the design of nonreciprocal devices. This phenomenon is known as *nonreciprocal phase shift*. If the magnetostatic field is not applied ($\delta = 0$), the effective index profile becomes reciprocal and converges to the dashed line

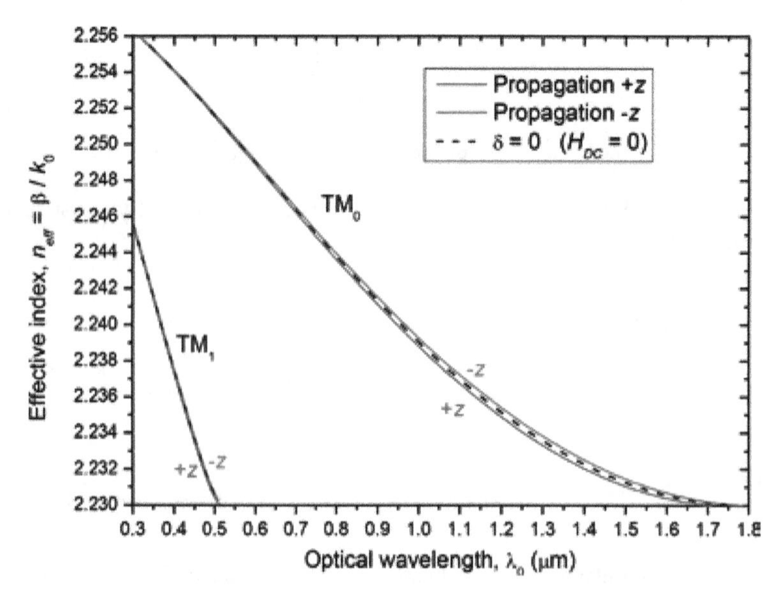

Figure 8.
Dispersion curves of the fundamental TM_0 mode and the superior TM_1 mode.

shown in **Figure 8**. The TM modes reach cutoff for optical wavelengths at which the effective index reaches the minimum value of 2.23. For greater optical wavelengths, the mode becomes irradiated and escapes through layer 3.

Figure 9 shows the transversal distributions of the H_y component at two distinct optical wavelengths. For this waveguide design, λ_0 = 1.55 μm is near cutoff, and the mode is highly distributed in the MO layer, which increases the nonreciprocal phase shift. Note from **Figure 8** that the difference between the effective indexes of the counter propagating TM modes are greater for optical wavelengths near cutoff, but as the wavelengths decreases, the mode becomes more confined at the waveguide core, and its interaction with the MO layer decreases, resulting in a decrease of the nonreciprocal phase shift effect, considering this waveguide configuration.

2.3 TM mode in a planar magneto-optical directional coupler

Now let us consider a five-layered MO planar structure as shown in **Figure 10**.

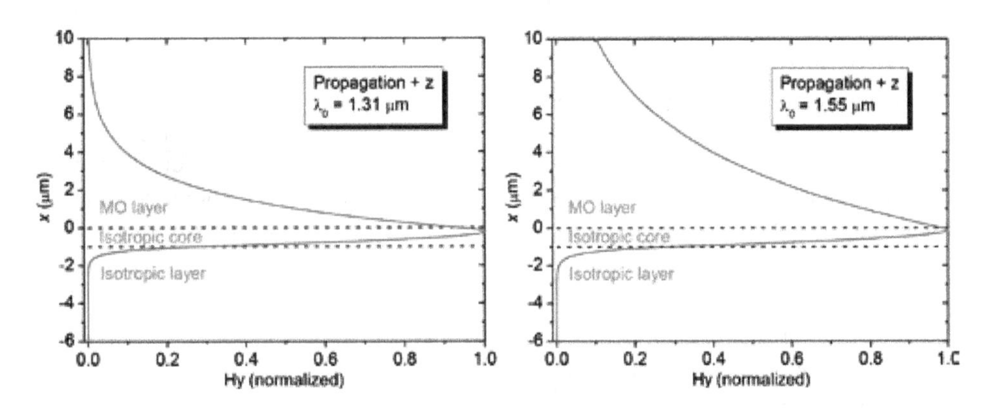

Figure 9.
Transversal distribution of the H_y *component of the fundamental* TM_0 *mode at* λ_0 = 1.31 μm *and at* λ_0 = 1.55 μm.

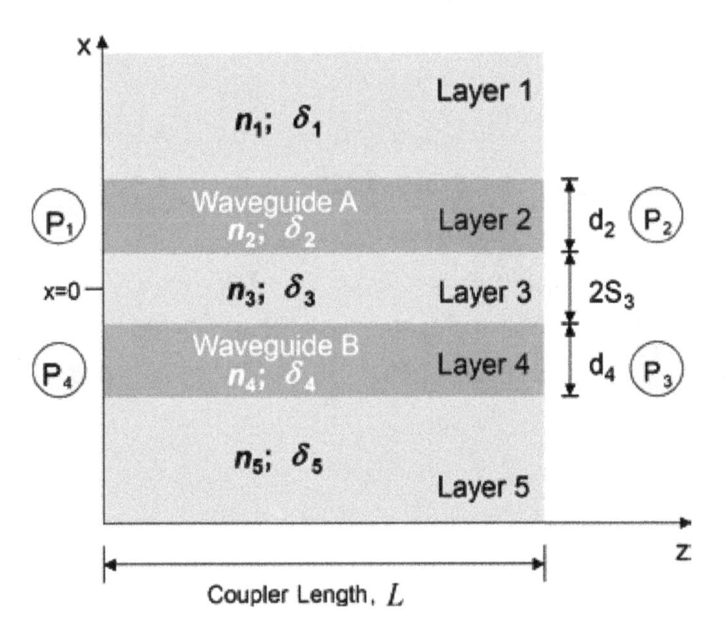

Figure 10.
Longitudinal section of the five-layered MO planar structure.

The solutions for Eq. (36) in each layer, making use of the proper radiation conditions, are [18]

$$H_y = A_1 \exp\left[-\gamma_1(x - S_3 - d_2)\right],\ \text{for } x \geq S_3 + d_2,$$
$$H_y = A_2 \cos\left[\kappa_2(x - S_3 - d_2/2)\right] + A_3 \sin\left[\kappa_2(x - S_3 - d_2/2)\right]\ \text{for } S_3 \leq x \leq S_3 + d_2,$$
$$H_y = A_4 \exp\left(-\gamma_3 x\right) + A_5 \exp\left(\gamma_3 x\right),\ \text{for } -S_3 \leq x \leq S_3,$$
$$H_y = A_6 \cos\left[\kappa_4(x + S_3 + d_4/2)\right] + A_7 \sin\left[\kappa_4(x + S_3 + d_4/2)\right],\ \text{for } -S_3-d_4 \leq x \leq -S_3,$$
$$H_y = A_8 \exp\left[\gamma_5(x + S_3 + d_4)\right],\ \text{for } x \leq -S_3-d_4,$$

where A_1 through A_8 are constants to be determined, κ_i and γ_j are given by.

$$\kappa_i = \sqrt{\frac{k_0^2 - \xi_{xx}^{(i)}\beta^2}{\xi_{zz}^{(i)}}},\ i = 2, 4, \tag{49}$$

$$\gamma_j = \sqrt{\frac{\xi_{xx}^{(j)}\beta^2 - k_0^2}{\xi_{zz}^{(j)}}},\ j = 1, 3, 5, \tag{50}$$

where $k_0 = 2\pi/\lambda_0$, and λ_0 is the optical wavelength.

The electric field components E_x and E_z can be directly obtained with Eq. (34) and Eq. (35), respectively. Applying the boundary conditions for the tangential components H_y and E_z, one obtains a system of eight equations and eight unknowns, which can be conveniently written in matrix form as follows:

$$[M(\beta)]\mathbf{A} = \mathbf{0}. \tag{51}$$

Here, $[M(\beta)]$ is an 8×8 matrix that depends on the unknown longitudinal propagation constant β and $\mathbf{A} = [A_1\ A_2\ \ldots\ A_8]^T$. The propagation constant can be easily found by solving the equation $Det([M(\beta)]) = 0$. The nonzero elements of the matrix $[M(\beta)]$ are listed below:

$$M_{11} = 1;\ M_{12} = -\cos\left(\kappa_2 d_2/2\right);\ M_{13} = -\sin\left(\kappa_2 d_2/2\right);$$
$$M_{21} = -j\xi_{zx}^{(1)}\beta + \gamma_1\xi_{zz}^{(1)};\ M_{22} = j\xi_{zx}^{(2)}\beta\cos\left(\kappa_2 d_2/2\right) - \kappa_2\xi_{zz}^{(2)}\sin\left(\kappa_2 d_2/2\right);$$
$$M_{23} = -j\xi_{zx}^{(2)}\beta\sin\left(k_2 d_2/2\right) + k_2\xi_{zz}^{(2)}\cos\left(k_2 d_2/2\right);$$
$$M_{32} = \cos\left(\kappa_2 d_2/2\right);\ M_{33} = -\sin\left(\kappa_2 d_2/2\right);\ M_{34} = -\exp\left(-\gamma_3 S_3\right);\ M_{35} = -\exp\left(\gamma_3 S_3\right);$$
$$M_{42} = -j\xi_{zx}^{(2)}\beta\cos\left(\kappa_2 d_2/2\right) - \kappa_2\xi_{zz}^{(2)}\sin\left(\kappa_2 d_2/2\right);$$
$$M_{43} = j\xi_{zx}^{(2)}\beta\sin\left(\kappa_2 d_2/2\right) - \kappa_2\xi_{zz}^{(2)}\cos\left(\kappa_2 d_2/2\right);$$
$$M_{44} = \left(j\xi_{zx}^{(3)}\beta - \xi_{zz}^{(3)}\gamma_3\right)\exp\left(-\gamma_3 S_3\right);\ M_{45} = \left(j\xi_{zx}^{(3)}\beta + \xi_{zz}^{(3)}\gamma_3\right)\exp\left(\gamma_3 S_3\right);$$
$$M_{54} = \exp\left[\gamma_3 S_3\right];\ M_{55} = \exp\left[-\gamma_3 S_3\right];\ M_{56} = -\cos\left(\kappa_4 d_4/2\right);\ M_{57} = -\sin\left(\kappa_4 d_4/2\right);$$
$$M_{64} = -\left(j\xi_{zx}^{(3)}\beta + \xi_{zz}^{(3)}\gamma_3\right)\exp\left[\gamma_3 S_3\right];\ M_{65} = -\left(j\xi_{zx}^{(3)}\beta + \xi_{zz}^{(3)}\gamma_3\right)\exp\left[-\gamma_3 S_3\right];$$
$$M_{66} = j\xi_{zx}^{(4)}\beta\cos\left(\kappa_4 d_4/2\right) - \xi_{zz}^{(4)}\kappa_4\sin\left(\kappa_4 d_4/2\right);$$
$$M_{67} = j\xi_{zx}^{(4)}\beta\sin\left(\kappa_4 d_4/2\right) + \xi_{zz}^{(4)}\kappa_4\cos\left(\kappa_4 d_4/2\right);$$
$$M_{76} = \cos\left(\kappa_4 d_4/2\right);\ M_{77} = -\sin\left(\kappa_4 d_4/2\right);\ M_{78} = -1;$$
$$M_{86} = -j\xi_{zx}^{(4)}\beta\cos\left(\kappa_4 d_4/2\right) - \xi_{zz}^{(4)}\kappa_4\sin\left(\kappa_4 d_4/2\right);$$
$$M_{87} = j\xi_{zx}^{(4)}\beta\sin\left(\kappa_4 d_4/2\right) - \xi_{zz}^{(4)}\kappa_4\cos\left(\kappa_4 d_4/2\right);$$
$$M_{88} = j\xi_{zx}^{(5)}\beta + \xi_{zz}^{(5)}\gamma_5;$$

As an example, **Table 1** shows the material parameters and layer thicknesses for each layer. Layers 1 and 5 are unbounded, and their thicknesses are theoretically infinite for the analytical model. The optical wavelength is $\lambda_0 = 1.55$ μm.

Figure 11 shows a plot of guided supermodes that occurs in the planar structure for forward propagation (along +*z*). The guided propagation along the five-layered structure, as well the periodical energy exchange of light between the two waveguides, can be expressed as a linear combination of these supermodes. The coupling length for the structure is given by $L_\pi = \pi/|\beta_1 - \beta_2|$, where β_1 and β_2 are the propagation constants of the supermodes obtained from the roots of $Det([M(\beta)]) = 0$. The computed coupling length, which refers to the propagation along the $+z$ axis, is $L_\pi^+ = 1389.84\,\mu m$.

Figure 12 shows the plot of the supermodes, now considering backward propagation of the TM mode (along -*z*). The computed coupling length, which refers to the backward propagation along the *z*-axis, is $L_\pi^- = 689\,\mu m$.

Layer	Parameters		
	n	δ	Thickness (μm)
1	2.23	−0.019	∞
2	2.26	0	1.20
3	2.00	0	0.75
4	2.26	0	1.23
5	2.23	−0.019	∞

Table 1.
Material and geometric parameters of the MO directional coupler.

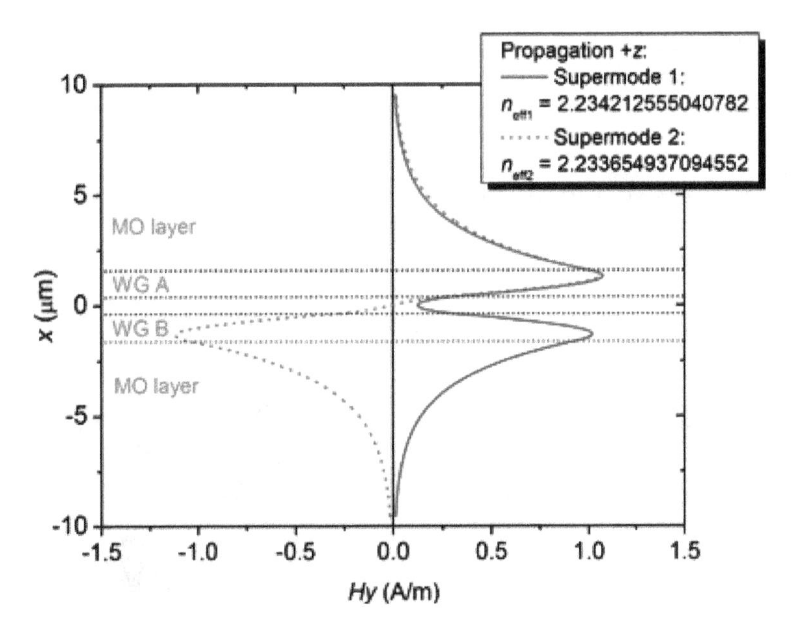

Figure 11.
Transversal distribution of the supermodes (H_y component) for forward propagation (+z).

Considering both propagation senses, when the condition $L = L_\pi^+ = 2L_\pi$ for the length of the directional coupler is achieved, we obtain an optical isolator calibrated for the given optical wavelength. The operation of the optical isolator is depicted in **Figure 13**. If an optical source is placed at the port 1 of the waveguide A, all optical power will be coupled into port 3 of the waveguide B, if the length of the directional coupler is $L = L_\pi^+$. If some light is reflected at port 3, since $L = 2L_\pi^-$, all optical power is directed to the port 4. Therefore, the optical source at port 1 becomes isolated from the reflected light. **Figures 14**, **15** show simulations of the forward and backward optical propagation in the MO directional coupler via a propagation projection of a linear combination of the corresponding supermodes.

The MO directional coupler of **Figure 10** also acts as an optical circulator, considering the following sequence of input and output ports: 1 to 3; 3 to 4; 4 to 2; and 2 to 1.

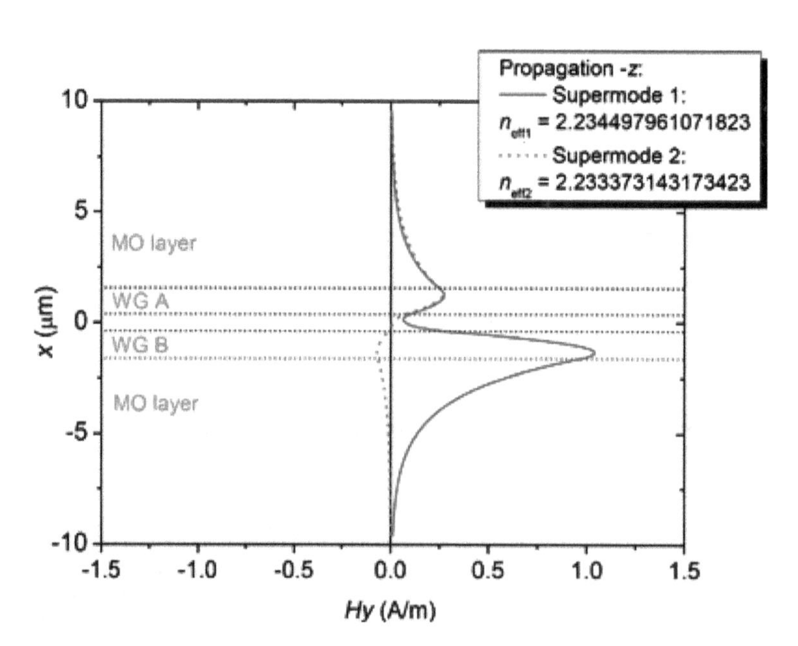

Figure 12.
Transversal distribution of the supermodes (H_y component) for backward propagation (−z).

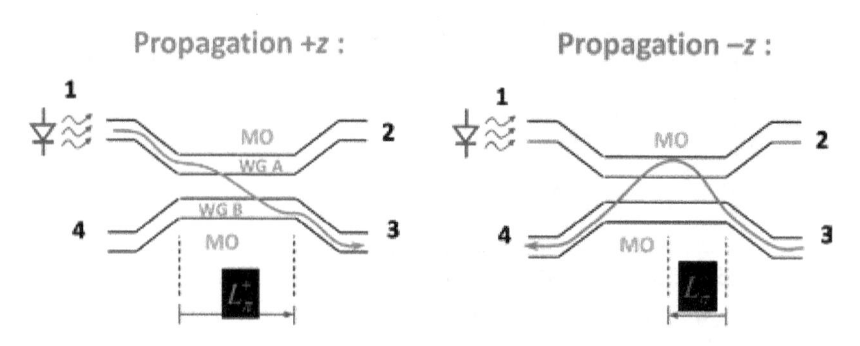

Figure 13.
Operation of an optical isolator based on nonreciprocal phase shift.

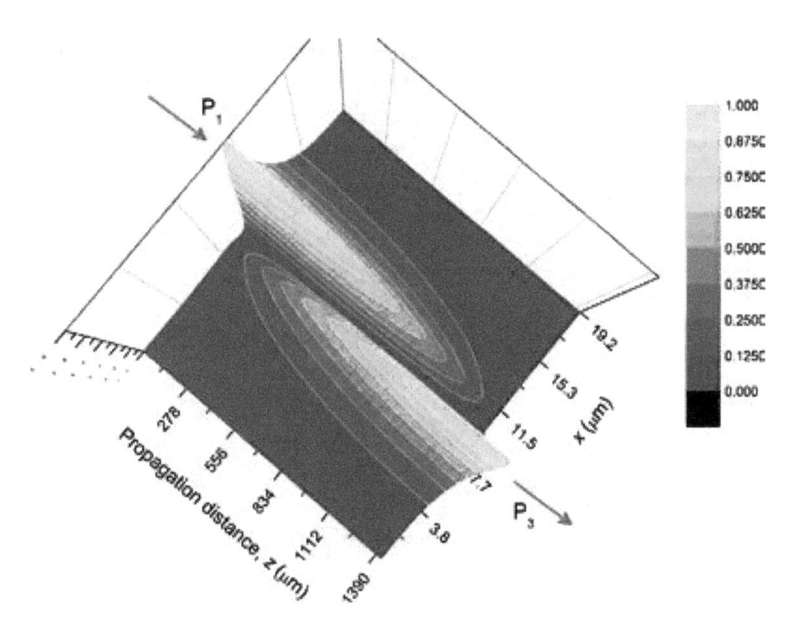

Figure 14.
Forward propagation simulation of the TM mode component H_y *excited at port 1 (*P_1*) of the five-layered structure. The light exits through port 3 (*P_3*). The starting transversal* H_y *field was supermode 1 plus supermode 2 of* ***Figure 11***.

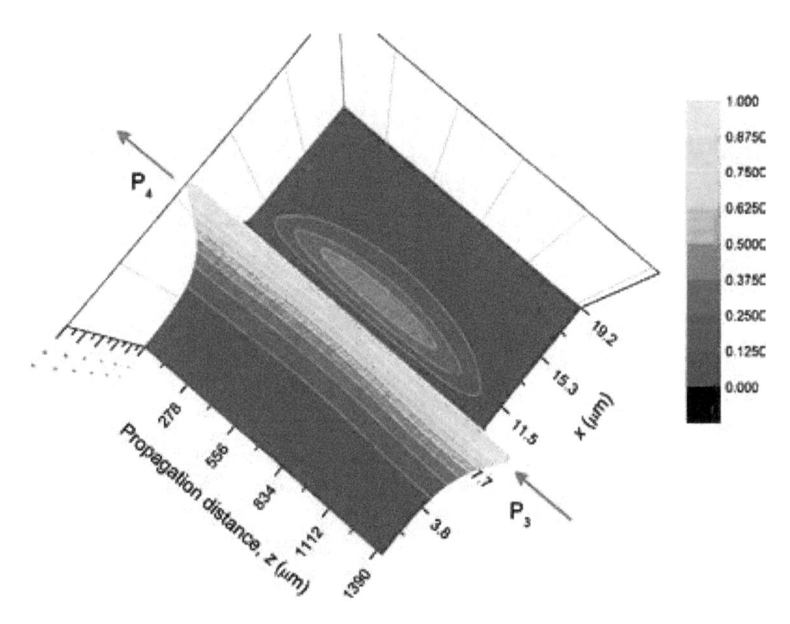

Figure 15.
Backward propagation simulation of the TM mode component H_y *excited at port 3 (*P_3*) of the five-layered structure. The light exits through port 4 (*P_4*). The starting transversal* H_y *field was supermode 1 minus supermode 2 of* ***Figure 12***.

3. Conclusions

The propagation characteristics of optical waves in magneto-optical media and in planar waveguides with three and five MO layers were exposed. The effects of Faraday rotation and nonreciprocal phase shift were discussed with mathematical background to support the analyses. The propagation of TEM waves in unbounded MO media was discussed, where it was shown that the Faraday rotation is maximized when the propagation occurs in the same direction of the applied magnetostatic field. It was also mathematically shown that if there is no such alignment, losses may be added to the wave propagation. A planar MO waveguide and a directional coupler were also analyzed in the context of their nonreciprocity. For these structures, nonreciprocity is observed for TM-guided modes. The theoretical analyses confirm that magneto-optical materials have great potential to be employed on the design of nonreciprocal optical devices, such as isolators and circulators.

Author details

Licinius Dimitri Sá de Alcantara
Cyberspatial Institute, Federal Rural University of Amazon, Belém, Brazil

*Address all correspondence to: licinius@ufra.edu.br

References

[1] Antonov V, Harmon B, Yaresko A. Electronic Structure and Magneto-Optical Properties of Solids. Dordrecht: Springer; 2004. 528 p. DOI: 10.1007/1-4020-1906-8

[2] Levy M. The on-chip integration of magnetooptic waveguide isolators. IEEE Journal of Selected Topics in Quantum Electronics. 2002;**8**(6):1300-1306. DOI: 10.1109/JSTQE.2002.806691

[3] Ando K. Waveguide optical isolator: A new design. Applied Optics. 1991; **30**(9):1080-1095. DOI: 10.1364/AO.30.001080

[4] Bahlmann N, Chandrasekhara V, Erdmann A, Gerhardt R, Hertel P, Lehmann R, et al. Improved design of magnetooptic rib waveguides for optical isolators. Journal of Lightwave Technology. 1998;**16**(5):818-823. DOI: 10.1109/50.669010

[5] Li TF, Guo TJ, Cui HX, Yang M, Kang M, Guo QH, et al. Guided modes in magneto-optical waveguides and the role in resonant transmission. Optics Express. 2013;**21**(8):9563-9572. DOI: 10.1364/OE.21.009563

[6] Bolduc M, Taussig AR, Rajamani A, Dionne GF, Ross CA. Magnetism and magnetooptical effects in Ce-Fe oxides. IEEE Transactions on Magnetics. 2006; **42**(10):3093-3095. DOI: 10.1109/TMAG.2006.880514

[7] Fratello VJ, Licht SJ, Brandle CD. Innovative improvements in bismuth doped rare-earth iron garnet Faraday rotators. IEEE Transactions on Magnetics. 1996;**32**(5):4102-4107. DOI: 10.1109/20.539312

[8] Pedroso CB, Munin E, Villaverde AG, Medeiros Neto JA, Aranha N, Barbosa LC. High Verdet constant Ga:S:La:O chalcogenide glasses for magneto-optical devices. Optical Engineering. 1999;**38**(2). DOI: 10.1117/1.602080

[9] Kalandadze L. Influence of implantation on the magneto-optical properties of garnet surface. IEEE Transactions on Magnetics. 2008; **44**(11):3293-3295. DOI: 10.1109/TMAG.2008.2001624

[10] Nomura T, Kishida M, Hayashi N, Ishibashi T. Evaluation of garnet film as magneto-optic transfer readout film. IEEE Transactions on Magnetics. 2011; **47**(8):2081-2086. DOI: 10.1109/TMAG.2011.2123103

[11] Inoue M, Arai K, Fujii T, Abe M. One-dimensional magnetophotonic crystals. Journal of Applied Physics. 1999;**85**(8):5768-5770. DOI: 10.1063/1.370120

[12] Koerdt C, Rikken GL, Petrov EP. Faraday effect of photonic crystals. Applied Physics Letters. 2003;**82**(10): 1538-1540. DOI: 10.1063/1.1558954

[13] Zvezdin AK, Belotelov VI. Magnetooptical properties of two-dimensional photonic crystals. The European Physical Journal B–Condensed Matter and Complex Systems. 2004;**37**: 479-487. DOI: 10.1140/epjb/e2004-00084-2

[14] Haider T. A review of magneto-optic effects and its application. International Journal of Electromagnetics and Applications. 2017;**7**(1):17-24. DOI: 10.5923/j.ijea.20170701.03

[15] Zak J, Moog ER, Liu C, Bader SD. Magneto-optics of multilayers with arbitrary magnetization directions. Physical Review B. 1991;**43**:6423-6429. DOI: 10.1103/PhysRevB.43.6423

[16] Wolfe R, Lieberman R, Fratello V, Scotti R, Kopylov N. Etch-tuned ridged waveguide magneto-optic isolator. Applied Physics Letter. 1990;**56**: 426-428. DOI: 10.1063/1.102778

[17] Alcantara LDS, Teixeira FL, Cesar AC, Borges BH. A new full-vectorial FD-BPM scheme: Application to the analysis of magnetooptic and nonlinear saturable media. Journal of Lightwave Technology. 2005;**23**(8):2579-2585. DOI: 10.1109/JLT.2005.850811

[18] Alcantara LDS, De Francisco CA, Borges BH. Analytical model for magnetooptic five-layered planar waveguides. In: SBMO/IEEE MTT-S International Microwave and Optoelectronics Conference (IMOC'17); 27–30 August 2017; Águas de Lindóia, Brazil. pp. 1-5. DOI: 10.1109/IMOC.2017.8121106

6

Ferromagnetism in Multiferroic $BaTiO_3$, Spinel MFe_2O_4 (M = Mn, Co, Ni, Zn) Ferrite and DMS ZnO Nanostructures

Kuldeep Chand Verma, Ashish Sharma, Navdeep Goyal and Ravinder Kumar Kotnala

Abstract

Multiferroic magnetoelectric material has significance for new design nano-scale spintronic devices. In single-phase multiferroic $BaTiO_3$, the magnetism occurs with doping of transition metals, TM ions, which has partially filled d-orbitals. Interestingly, the magnetic ordering is strongly related with oxygen vacancies, and thus, it is thought to be a source of ferromagnetism of TM:$BaTiO_3$. The nanostructural MFe_2O_4 (M = Mn, Co, Ni, Cu, Zn, etc.) ferrite has an inverse spinel structure, for which M^{2+} ions in octahedral site and Fe^{3+} ions are equally distributed between tetrahedral and octahedral sites. These antiparallel sub-lattices (cations M^{2+} and Fe^{3+} occupy either tetrahedral or octahedral sites) are coupled with O_{2-} ion due to superexchange interaction to form ferrimagnetic structure. Moreover, the future spintronic technologies using diluted magnetic semiconductors, DMS materials might have realized ferromagnetic origin. A simultaneous doping from TM and rare earth ions in ZnO nanoparticles could increase the antiferromagnetic ordering to achieve high-Tc ferromagnetism. The role of the oxygen vacancies as the dominant defects in doped ZnO that must involve bound magnetic polarons as the origin of ferromagnetism.

Keywords: magnetically ordered material, oxygen vacancies, spin glass

1. Introduction

Multiferroics are combining multiple order parameters, offer an exciting way of coupling phenomena such as electronic and magnetic order. However, because simultaneous electric and magnetic order is difficult to achieve, multiferroics-especially those that function at or approaching room temperature-are extremely rare. For a crystal, when electrons are surrounded atomic nuclei to orient themselves in a same fashion then the crystal induces macroscopic ferromagnetism and electrical polarization. Since induction of ferromagnetism is essential in the technologies that involved sensors, computer hard drives, power generation, etc. Recently, the diluted magnetic semiconductors (DMSs) such as ZnO, SnO_2, TiO_2, etc. have generated potential in spintronics because DMS has the collective ordering that mediated via semiconductor charge carriers, as well as electron scattering at localized magnetic impurities and electron–electron interactions. To use DMSs for practical spintronic devices, a relatively high concentration of magnetic elements needed in the semiconductor host, and a large ferromagnetism is required with a Curie temperature (T_c) above room temperature. The transition metal (TM) ferrites with a spinel structure (MFe_2O_4; M = Co^{2+}, Ni^{2+}, Cu^{2+}, Zn^{2+}, etc.) are used in a wide variety of technological applications such as magnetic memory devices and biomedicine. However, these spinel ferrites are the candidate materials for multifer-roic heterostructure due to their excellent magnetic response. For such multiferroic heterostructures, the perovskite ($BaTiO_3$, $PbTiO_3$, $BiFeO_3$, etc.) has an opportunity of higher piezoelectric coefficient that may pool with magnetostrictive materials ($CoFe_2O_4$, $NiFe_2O_4$, $ZnFe_2O_4$, etc.) via lattice strain effect.

1.1 Multiferroic $BaTiO_3$

The magnetoelectric (ME) effect—the induction of magnetization by an electric field and the induction of electric polarization by a magnetic field [1]. Multiferroic nanostructures have given recent advances in new type of memory devices including multistate data storage and spintronics [2]. $BaTiO_3$ (BTO) is a rare single-phase multiferroic. In multiferroics, the magnetic order is due to exchange interactions between magnetic dipoles, which themselves originate from unfilled shells of electron orbitals. Similarly, the electric order is due to the ordering of local electric dipoles, elastic order is due to the ordering of atomic displacements due to strain. The three crystal-lographic phases of BTO are ferroelectric: rhombohedral <190 K, orthorhombic for 190 K< T < 278 K and tetragonal for 278 K < T < 395 K. At higher temperatures, BTO is a paraelectric. For tetragonal BTO (**Figure 1(b)**) having lattice constants: $a(Å) = b(Å) = 3.99$; $c(Å) = 4.03$; space group = P4*mm*, the displacement of Ti^{4+} ion along *c*-axis might be induced electrical polarization (ferroelectricity). It involved hybridization of charge among Ti cation (3d states) with O anions (2p states). In cubic phase, the Ba^{2+} is located at the centre of the cube with coordination number 12.

Since to the formation of ME random access memories (MERAMs), the main thing that required ME coupling via interfacial exchange coupling among a multiferroic and a ferromagnet, which can change the magnetization of the ferromagnetic coating with respect to a voltage (**Figure 1(a)**) [3]. For such MERAMs, an electric field is enabled by ME coupling could control the exchange coupling between multiferroic and ferromagnetic at the interface. This exchange coupling at the interface reins the magnetization of the ferromagnetic layer, and therefore the magnetization might be change with multiferroic electrical polarization. Therefore, for perovskite (ABO_3) BTO, Ba^{2+} (A-site cation) induce the required distortion for ferroelectricity, while magnetism can be achieved by the doping such as TM = Cr, Mn, Fe, Co, Ni, Cu along B-site cation [4, 5].

1.2 Spinel ferrites

The spinel structure typically represented as AB_2O_4, where 'A' indicates four-fold coordinated tetrahedral sub-lattice sites and 'B' indicates six-fold coordinated octahedral sub-lattice (**Figure 1(c)**). There are 8 A-sites in which the metal cations are tetrahedrally coordinated with oxygen, and 16 B-sites, which possess octahedral coordination. Normal ferrites have divalent cations residing solely as the central ion on the tetrahedral sub-lattice with only trivalent Fe cations occupying octahedral sub-lattice sites. Harrisa and Sepelak [6] suggested superexchange interaction, the J_{BB} is strong and negative indicating antiferromagnetic coupling between Fe^{3+}–O^{2-}–Fe^{3+} with octahedral sub-lattice cation spins largely canceling out. When

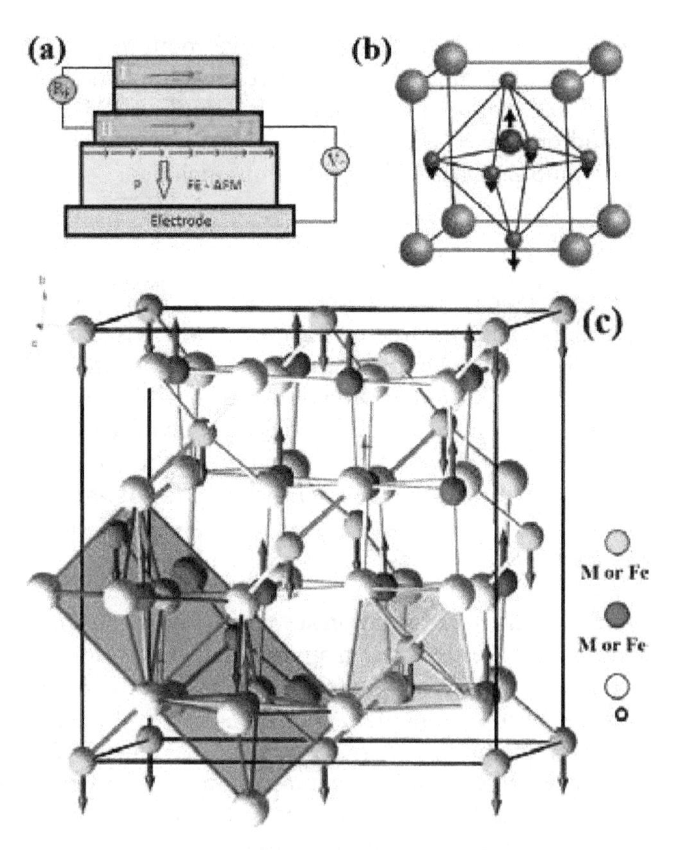

Figure 1.
(a) Schematic MERAM device in which the binary information is stored by the magnetization direction of the ferromagnetic layer (I & II), read by the resistance, R_p of the magnetic trilayer, and written by applying a voltage across the multiferroic ferroelectric-antiferromagnetic layer (FE-AFM). (b) Tetragonal crystalline structure of BTO, with a lattice ferroelectric distortion. (c) Spinel unit cell. Arrows represent magnetic moments and their antiparallel alignment. Adopted from Harrisa and Sepelak [6].

the divalent cation resides on the [B] site and A and B ions share the balance of [B] and the totality of (A), the spinel is inverse: $(A_{1-\delta}B_{\delta})[A_{\delta}B_{2-\delta}]O_4$, where δ is the inversion degree, *i.e.*, $NiFe_2O_4$. Ferrite of the type $NiFe_2O_4$ (NFO), $CoFe_2O_4$ (CFO) and $MnFe_2O_4$ (MFO) with the spinel structure are magnetic ceramics which have potential in electronic and magnetic components [7]. The NFO has an inverse spinel structure for which Fe^{3+} ions occupied tetrahedral A-sites; whereas Fe^{3+} and Ni^{2+} ions are sit on octahedral B-sites. This NFO is ferrimagnetic material has magneti-zation originated by antiparallel spins on A- and B-sites. However, for normal spinel structure of CFO, Co is a divalent atom, occupying tetrahedral A-sites, while Fe is a trivalent atom, sitting on the octahedral B-sites. For MFO, the inverted spinel struc-ture is partial for which the Mn^{2+} and Fe^{3+} ions with half-filled 3d shell (ground state is singlet S^6 with spin = 5/2 with zero orbital momentum) and the crystal field is not sufficient to split it [8]. The magnetic moment of MFO agrees well with Neel's coupling scheme and has lower resistivity than CFO and NFO ferrites [9].

1.3 Diluted magnetic semiconductors

Recently, the realization of spin in DMS, semiconductor with substituted magnetic impurities (Fe, Co, Ni, Cu) has attracted great interest in design of spin-tronics devices like spin field-effect transistors, non-volatile memory devices, and programmable logic gates [10–13]. Among various DMS, ZnO is a promising spin source, since it epitomizes DMS with T_c well above room temperature. The DMS ZnO is a wide band gap that gained recent research in spintronics, due to an ability to change its optical and magnetic behavior with doping of TM = Fe, Co, Mn, Ni, Cu, Cr, or V ions and/or by intrinsic defects, such as oxygen vacancy (V_O) and zinc vacancy (V_{Zn}). The DMS ZnO has wide applications included:

1.3.1 Magnetic recording

For read heads in magnetic disk recorders (computer components), read head senses the magnetic bits that are stored on the media, which is stored as magnetized regions of the media, called magnetic domains, along tracks (**Figure 2(a)**) [13]. Magnetization is stored as a "0" in one direction and as a "1" in the other. Although, there is no magnetic field emanating from the interior of a magnetized domain itself, uncompensated magnetic poles in the vicinity of the domain walls generate magnetic fields (sensed by the GMR element) that extend out of the media.

1.3.2 Nonvolatile memories

The "Nonvolatile" refers to information storage that does not "evaporate" when power is removed from a system, *i.e.*, magnetic disks and tapes. Prinz [13] has recently demonstrated that GMR elements can be fabricate in arrays with standard lithographic processes to obtain memory that has speed and density approaching that of semiconductor memory, but is nonvolatile. A schematic representation of RAM that is constructed of GMR elements is shown in **Figure 2(b)**. The spin-dependent scattering of the carriers (electrons & holes) is minimized for parallel magnetic moment of the ferromagnetic layer, to induce lowest value of resistance. However, the highest resistance is the result of maximized spin-dependent scattering carriers via anti-aligned ferromagnetic layers. An external magnetic field could give the direction of magnetic moments, applied to the materials. The spin-valve structures of GMR set into series using lithographic (wires) termed as a sense line. This sense line has information storage due to resistance (resistance of elements). The sense line runs the current which is detected at the end by amplifiers because resistance changes in the elements.

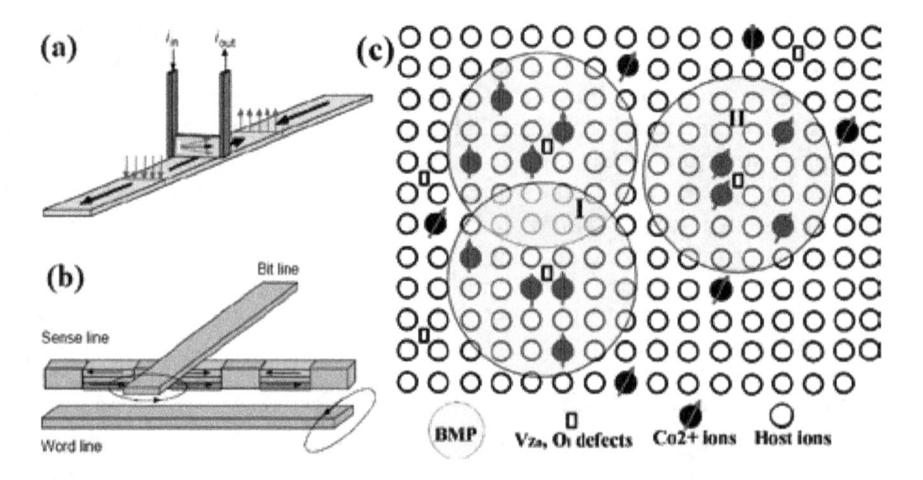

Figure 2.
A schematic representation of (a) GMR read head (I) that passes over recording media containing magnetized regions, (b) RAM that is constructed of GMR elements. Adopted from Prinz [13], (c) bound magnetic polaron (BMP), V_{Zn}, O_i trapped carriers couple with the 3d shell spins of TM ions within its hydrogenic orbit.

1.3.3 Ferromagnetism in DMS ZnO due to bound magnetic polarons

Generally, the room temperature ferromagnetism (RTFM) in DMS ZnO is given because of bound magnetic polaron (BMP) model [14]. The localized spins of the dopant ion interact with the charge carriers that are bound to a small number of defects such as oxygen vacancies, resulting into a magnetic polarization of the surrounding local moments. Since the magnetism in TM ions doped ZnO nanoparticles relates with exchange interactions between unpaired electron spins, that arising from the lattice imperfections such as oxygen vacancies, V_O at the surface of the nanoparticles. Pal et al. [15] described BMP formation in Co doped ZnO and shown in **Figure 2(c)**. The electrons trapped in the defect vacancies undergo orbital coupling with the *d* shells of the adjacent divalent dopant ion and form BMP. In BMP model, the bound electrons (holes) hold in defect states that coupled through TM ions to overlap ferromagnetic regions, which responsible into high T_C (due to forma-tion of long-range ferromagnetic ordering) [16]. When the dopant ions are donors or acceptors, the exchange interactions are sp-d that would lead BMPs formation.

2. Experimental methods

The multiferroic, DMS and ferrites materials might be synthesized by methods such as a sol-gel [17], chemical combustion [18], hydrothermal [19], metallo-organic decomposition (MOD) [20], conventional solid-state reaction [21], sol-gel precipitation [22], thermal evaporation [23], etc.

3. Result and discussion

3.1 Multiferroic systems of $BaTiO_3$

3.1.1 X-ray diffraction for $BaTiO_3$ and $BaTM_{0.01}Ti_{0.99}O_3$ nanoparticles

The sol-gel method is used to prepare pure $BaTiO_3$ and $BaTM_{0.01}Ti_{0.99}O$ [TM = Cr (BTO:Cr), Mn (BTO:Mn), Fe (BTO:Fe), Co (BTO:Co), Ni (BTO:Ni), Cu (BTO:Cu)] nanoparticles [4]. The TM ions in perovskite BTO structure highly influenced lattice constants to induce lattice strain and unit cell expansion, which responsible into defects vacancies formation. **Figure 3** shows the Rietveld refinement (Full-Prof program) of X-ray diffraction (XRD) patterns for pure and $BaTM_{0.01}Ti_{0.99}O$ nanoparticles measured at room temperature. A polycrystalline with tetragonal BTO phase (space group:P4*mm*) is detected. The fitting parameters, $R_p(\%)$ = 6.7, 7.1, 6.7, 4.9, 9.1, 8.2 and 9.8, $R_{wp}(\%)$ = 3.2, 9.9, 9.2, 10.0, 12.5, 1.2 and 14.6, χ^2 = 1.1, 1.4, 0.81, 1.4, 3.3, 3.3 and 0.9 and distortion ratio, (c/a) = 1.00959, 1.00932, 1.00688, 1.00909, 1.00776, 1.00625 and 1.00508, respectively, refined for pure BTO and BTO with Cr, Mn, Fe, Co, Ni, Cu doping. Also, **Figure 3(a)** shows the tetragonal splitting of (200) diffraction peak at 2θ = 44.3–45.7°. The diffraction peak of pure BTO has shifted towards a lower diffraction angle with TM doping which supports the lattice strain in BTO. Such splitting of (200) peak might be con-firmed the tetragonal phase formation. It is due to electrostatic repulsions between 3d electrons of Ti^{4+} ions and 2p electrons of O^{2-} ions, the structure becomes dis-torted. It is also reported in Ref. [4] that the average particles size, from TEM, D_{TEM} (nm) = 20 ± 3, 13 ± 1, 33 ± 5, 35 ± 3, 17 ± 1 and 47 ± 7, the value of saturation magne-tization, M_s (emu g^{-1}) = 0.056, 0.042, 0.066, 0.035, 0.013 and 0.021, and the ME coupling constant, α_{ME} (mV cm^{-1} Oe^{-1}) = 25.91, 11.27, 31.15, 16.58, 11.61 and 16.48, respectively, measured with Cr, Mn, Fe, Co, Ni, Cu doping into BTO.

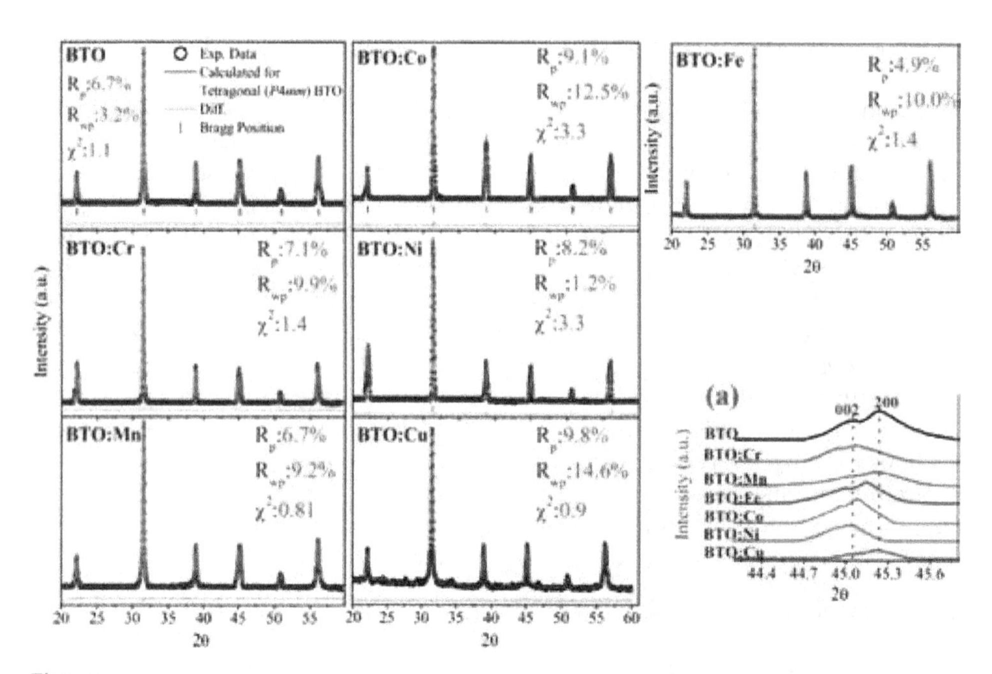

Figure 3.
XRD pattern for pure $BaTiO_3$ and $BaTM_{0.01}Ti_{0.99}O_3$ nanoparticles. (a) Showing the splitting of the (200) peak. Adopted from Verma and Kotnala [4].

3.1.2 Nanostructure of $BaTiO_3$ (BTO) and $BaFe_{0.01}Ti_{0.99}O_3$ (BFTO) multiferroic

The BTO and BFTO were prepared by a hydrothermal method of processing temperature 180°C/48 h [5]. The XRD pattern shows the coexistence of cubic/tetragonal/hexagonal phases of BTO and cubic/tetragonal of BFTO. **Figure 4(a** and **b)** reveals the TEM images of BTO and BFTO nanostructure. It shown that BTO (**Figure 4(a)**) is the product that consist of nanorods structure having hexagonal like face of average diameter 50 nm and length 75 nm. However, **Figure 4(b)** shows the nanowires forma-tion of BFTO with average diameter ~45 nm and the length >1.5 μm. It is also reported in Ref. [5] that the room temperature *M-H* hysteresis shows diamagnetism in BTO and ferromagnetism in BFTO with M_s ~ 82.23 memu g^{-1}, M_r ~ 31.91 memu g^{-1} with H_c ~ 122.68 Oe and ME coupling coefficient, α_{ME} = 16 mV Oe^{-1} cm^{-1}.

3.1.3 Ferromagnetism/ferroelectricity in Ce,La:$BaFe_{0.01}Ti_{0.99}O_3$ nanostructures

The nano-aggregation type $Ba(Fe_{0.67}Ce_{0.33})_{0.01}Ti_{0.99}O_3$ (BFTO:Ce) and $Ba(Fe_{0.67}La_{0.33})_{0.01}Ti_{0.99}O_3$ (BFTO:La) product is synthesized by a hydrothermal process [2]. Rietveld refinement of XRD pattern indicates polycrystalline phase with tetragonal BFTO. It is reported that the Ce and La ions in BFTO improved lattice distortion, c/a ratio. These dopant Ce and La in BFTO forms nano-aggregation type product with aver-age value of aggregation diameter, D = 40 and 22 nm, respectively, for BFTO:Ce and BFTO:La. The formation of tetragonal BTO phase and lattice defects due to vacancies is attributed by Raman active modes. These defects and vacancies are also confirmed with photoluminescence measurement that might be altered due to higher surface-to-volume ratio in nano-aggregation. **Figure 4(d** and **e)** shows the ferromagnetic behavior of Ce, La doped BFTO, respectively, by magnetization versus field (M-H_{dc} hysteresis) measured at room temperature [2]. The values of M_s (emu g^{-1}) = 0.15 and 0.08, and M_r (emu g $^{-1}$) = 0.039 and 0.015 with H_c (Oe) = 242 and 201, respectively, for BFTO:Ce and BFTO:La. It is well predicted that the formation of Fe^{4+}–O^{2-}–Fe^{4+} interaction is ferromagnetic, which dominate in BFTO over the antiferromagnetic Fe^{3+}–O^{2-}–Fe^{4+}

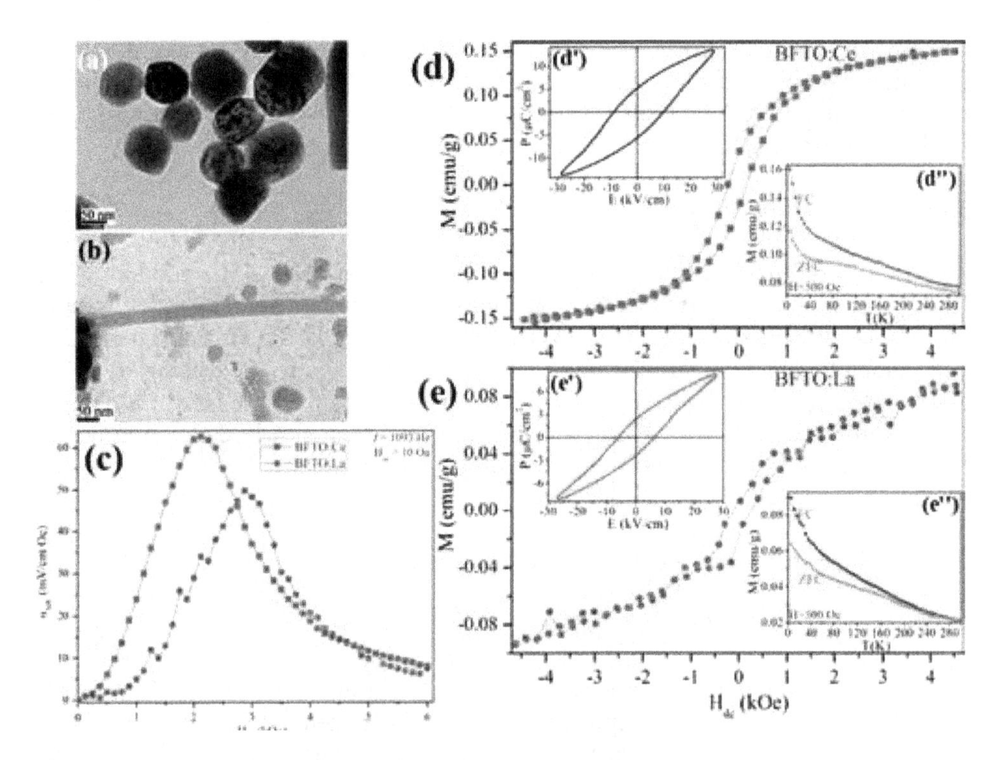

Figure 4.
Transmission electron microscopy (TEM) of (a) pure $BaTiO_3$ (b) $Ba(Fe_{0.01}Ti_{0.99})O_3$. (c) ME voltage coefficient (α_{ME}), (d and e) M-H_{dc} hysteresis at room temperature, (d' and e') P-E hysteresis, and (d" and e") M(T) following ZFC/FC at H = 500 Oe for Ce, La doped BFTO. Adopted from Refs. [2, 5].

and Fe^{3+}–O^{2-}–Fe^{3+} interactions, producing weak ferromagnetism. Generally, the ferromagnetism in Fe-doped $BaTiO_3$ is explained into two ways: the partially filled inner shells (d- or f-levels) and formation of nanostructures. An F-centre exchange (FCE) mechanism describes the required ferromagnetism [24, 25]. Such a mechanism expects Fe^{3+}–V_O^{2-}–Fe^{3+} transition that generally exists in the structure where an electron trapped in oxygen vacancies, V_O to form F-centre. For this, the electron occupies an orbital, p_z that overlaps d_z^2 of d shells in iron neighbors. The Fe^{3+} ions have $3d^5$ electronic configurations for which spin down trapped electron and spin up in two iron Neighbors. Therefore, the F-centre has the exchange interaction among two iron ions that would leads to ferromagnetism. The insets of **Figure 4(d′** and **e′)** show the polarization-electric field (P-E) hysteresis at room temperature. With Ce doping into BFTO, the value of spontaneous polarization, P_s ($\mu C\ cm^{-2}$) = 13.83 and remanent polarization, P_r ($\mu C\ cm^{-2}$) = 5.45 with electric coercivity, E_c ($kV\ cm^{-1}$) = 9.53. However, with La doping, P_s ($\mu C\ cm^{-2}$) = 8.28, and P_r ($\mu C\ cm^{-2}$) = 2.46, with E_c ($kV\ cm^{-1}$) = 6.14. These values of polarization have an improvement over reported work [26–28]. This is due to nano-aggregation formation and lattice distortion enhancement in BTO lattice. The smaller polarization in BFTO:La is the nano-size effect that involved compensation of polarization-induced surface charges [29].

3.1.4 Magnetization at low temperature measurement

The origin of observed ferromagnetism at room temperature in Ce- and La-doped BFTO is described by measuring their magnetization from zero-field cooling (ZFC) and field cooling (FC) at 500 Oe (**Figure 4(d″** and **e″)**). A clear sep-aration between FC and ZFC retains up to low temperature without blocking tem-perature is observed. This is an indication of weak antiferromagnetic interactions.

An upward curvature observed in M-T curve suggests a Curie-Weiss like behavior. It is attributed with short-range ferromagnetism, or a spin-cluster within a matrix of spin disorder [30]. Li et al. [31] suggested that the oxygen vacancy might be medi-ate antiferromagnetic-ferromagnetic interactions in multiferroics.

3.1.5 ME coupling in Ce, La-substituted BFTO

The longitudinal ME coupling coefficient, α_{ME} is measured for Ce and La-doped BFTO and shown in **Figure 4(c)**. The samples are biased with *ac* magnetizing field, H_{ac} = 10 Oe at 1093 Hz, and a *dc* magnetic field, H_{dc} is applied collinear to it. The value of α_{ME} was determined as a function of *dc* magnetic field using: $\alpha_{ME} = \frac{\partial E}{\partial H} = \frac{1}{t}\frac{\partial V}{\partial H} = \frac{Vout}{t \times Hac}$. In **Figure 4(c)**, the value of α_{ME} increases rapidly to a maximal value (due to an enhancement of elastic interactions) and then slowly decreasing in the higher region of H_{dc}. The maximum value of α_{ME} (mV cm^{-1} Oe^{-1}) is 62.65 and 49.79, respectively, for BFTO:Ce and BFTO:La.

3.2 Spinel ferrites MFe_2O_4 (M = Mn, Co, Ni, Zn, Cu, etc.)

3.2.1 Structural studies of $Co_{0.65}Zn_{0.35}Fe_2O_4$

Figure 5 (a) shows the Rietveld refinement (space group: $Fd\bar{3}m$) of $Co_{0.65}Zn_{0.35}Fe_2O_4$ (CZFO) ferrite, which indicate spinel phase [32]. The value of χ^2 is 1.648. The lattice parameter of CZFO is found to be 8.4183 Å. A three-dimensional sketch of CZFO unit cell projected along the *c*-axis (**Figure 5(b)**). The inset of **Figure 5(a)** shows the SEM image of a CZFO pellet sample to displays densely packed grains with few scattered pores and voids. It is also reported in Ref. [32] that the high dielectric permittivity value is obtained. The ferrimagnetic-paramagnetic phase transition is ~640 K.

3.2.2 Lattice constant, grains size and magnetism in pure NFO, CFO and MFO thin films

Table 1 shows the experimental results of lattice constant (*a*), grain's size [*x* (SEM) & *x′* (AFM)], M_s, M_r and H_c measured at room temperature (300 K) and 10 K, and magnetic phase transition temperature (T_{pm}) of $NiFe_2O_4$ (NFO),

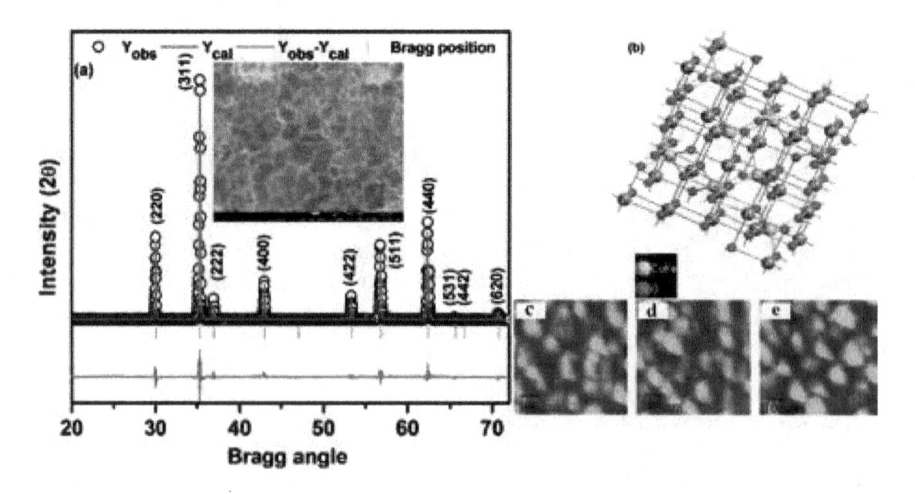

Figure 5.
(a) XRD patterns and SEM image (inset) of $Co_{0.65}Zn_{0.35}Fe_2O_4$ (CZFO). (b) CZFO unit cell with a spinel structure. AFM of (c) NFO, (d) CFO (e) MFO thin films. Adopted from Refs. [9, 32].

Sample	a (Å)	x (nm)	x' (nm)	M_s (emu cc^{-1})		M_r (emu cc^{-1})		H_c (Oe)		T_{pm} (K)
				300 K	10 K	300 K	10 K	300 K	10 K	
NFO	8.161	44	46	50.60	76.42	14.33	63.31	265.33	73.90	687
CFO	8.312	60	61	33.50	54.57	15.50	8.90	1292.00	69.33	693
MFO	8.425	74	75	5.40	13.35	1.10	7.20	113.30	39.28	581

Adopted from Verma et al. [9].

Table 1.
*Values of lattice constant (**a**), grain's size (**x**) (SEM), grain's size (**x'**) (AFM), M_s, M_r and H_c at 300 K and 10 K, and magnetic phase transition (T_{pm}) of pure NFO, CFO and MFO thin films.*

$CoFe_2O_4$ (CFO) and $MnFe_2O_4$ (MFO) thin films [9]. These ferrites were prepared by a MOD method using spin coating. The thickness of all the film is ~700 nm. It is also reported that the miller indices of cubic spinel ferrites structure are (2 2 0), (3 1 1), (2 2 2), (4 0 0), (3 3 1), (4 2 2) and (5 1 1), respectively, detected with diffraction angle 2θ = 30.48, 34.99, 37.48, 42.58, 48.02, 51.23 and 55.85° for NFO, for CFO, 2θ = 30.25, 34.87, 37.36, 42.58, 47.20, 51.23 and 55.84° and 2θ = 29.77, 34.64, 37.36, 41.75, 47.08, 51.23 and 57.04° for MFO. The higher coercivity value of CFO than NFO and MFO is the effect of higher magneto-crystalline anisotropy of Co cations than Ni and Mn. The observed values of M_s, M_r and H_c of the NFO, CFO and MFO films are quite smaller than bulk form [9]. For this, the decrease in M_s in ferrite nanoparticles is the canted spins in the surface layers [33]. Also, it observed very smaller values of M_r, M_s and H_c of MFO because lowering number of magnetic domains and the rate of alignment of the spins with the applied field [34, 35].

3.2.3 AFM images of NFO, CFO and MFO thin films

The value of average grain's size from AFM (**Figure 5(c–e)**) is 46, 61 and 75 nm, and the surface roughness 2.5, 4 and 2 nm, respectively, for NFO, CFO and MFO ferrite thin films, were synthesized by MOD method [9].

3.2.4 Magnetic ordering of ferrites in $MFe_2O_4/BaTiO_3$ nanocomposite

The multiferroic $MFe_2O_4/BaTiO_3$ (MFO/BTO) thin films were fabricated by a MOD method [20]. The addition of ferrite MFe_2O_4 in BTO results into lattice strain due to tetragonal distortion, unit cell expansion/contraction and lattice mismatch. It enhances ME coupling. The lattice distortion, *c*/*a*, tetragonal phase of BTO, spinel ferrite MFO and the lattice strain in composite phases is studied by XRD pattern [20]. These reported results also shown the AFM images to evaluate the value of average grain size for MnFO/BTO, CFO/BTO, NFO/BTO and ZFO/BTO, which is 25, 102, 24 and 133 nm, respectively. XPS analysis indicate that Fe exist into mixed +2/+3 valence states and O in both O^{2-} and deficient states. **Figure 6(a)** shows the room temperature ferromagnetic behavior of MFO/BTO thin films by measuring ferromagnetic hyster-esis (M-H_{dc}). The measured value of saturation magnetization, M_s (kJ T^{-1} m^{-3}) = 1.29, 20.25, 27.64 and 6.77, remanent magnetization M_r (kJ T^{-1} m $^{-3}$) = 0.03, 3.75, 6.76 and 1.49 and with magnetic coercivity, H_c (× 10^5 Am^{-1}) = 0.013, 0.079, 0.167 and 0.135, respectively, for MnFO/BTO, CFO/BTO, NFO/BTO and ZFO/BTO nano-composite. An enlarged view of M-H_{dc} as the inset of **Figure 6(a)** is also shown for more clarification. This magnetic measurement of MFO/BTO is compared with magnetization of a single phase MnFO (5.40 kJ T^{-1} m^{-3}), CFO (33.50 kJ T^{-1} m^{-3}), NFO (50.60 kJ T^{-1} m^{-3}) and ZFO (230 kJ T^{-1} m^{-3}), which indicate abrupt reduc-tion in nanocomposite sample. This is because non-magnetic BTO phase exist in nanocomposite, which reduced the magnetization of ferrite. Since the non-magnetic elements weaken the A-B superexchange interaction which result into an increase the distance between the magnetic moments in A and B sites in spinel structure. Since the magnetization decreases/increases in ferrites is a relaxation process that might be concerned with the redistribution of oxygen vacancies reported by Wang et al. [36]. The presence of a magnetic dead or antiferromagnetic coating on the nanostructural surface have reducible value of saturation magnetization of ferrite nanoparticles [37]. Vamvakidis et al. [38] suggested that the reduction in inversion degree (~0.22) of $MnFe_2O_4$ due to the partial oxidation of Mn^{2+} to Mn^{3+} ions results into weaker superexchange interactions between the tetrahedral and octahedral sites within the spinel structure. Bullita et al. [39] reported the cation distribution of $ZnFe_2O_4$ at the nanoscale level, which is contributed by partial inverted spinel structure results into

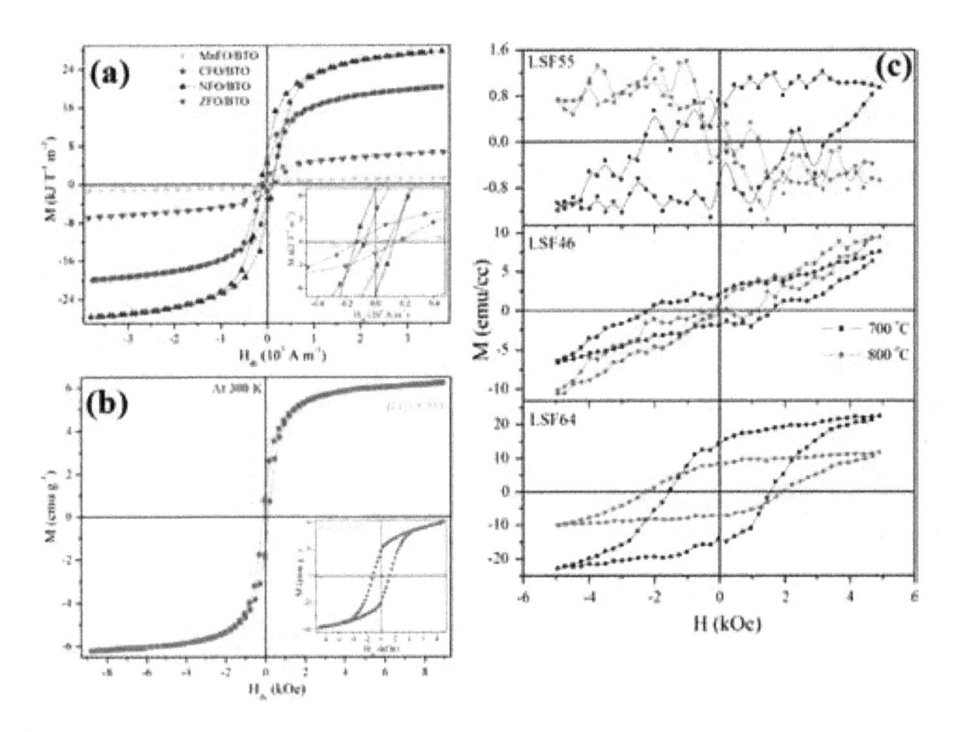

Figure 6.
(a) M-H_{dc} hysteresis of $MFe_2O_4/BaTiO_3$ nanocomposite thin films at 300 K. (b) Room temperature M-H_{dc} measurement for pure CFO and BTO-CFO nanoparticles. (c) the M-H measurement of LSF thin films. Adopted from Refs. [19, 20, 35].

an increase magnetization. Peddis et al. [40] indicates inversion degree of $CoFe_2O_4$ nanoparticles, which has a better correlation between spin canting and cationic distribution to get competitively higher M_s.

3.2.5 Magnetization in CFO and 0.25BTO–0.75CFO nanoparticles

The multiferroic 0.25BTO–0.75CFO nanocomposite is prepared by hydrothermal process at 180°C/48 h [19]. The spinel CFO and tetragonal BTO structure is studied by XRD pattern. The 0.25BTO–0.75CFO nanocomposite has nanoparticles formation with average particles size from FESEM is 80 ± 10 nm and from TEM is 76 ± 13 nm. **Figure 6(b)** shows the M-H_{dc} hysteresis for pure CFO and 0.25BTO–0.75CFO nanoparticles, measured at room temperature. The value of M_s (emu g^{-1}) = 39.027 and 6.199, M_r (emu g^{-1}) = 19 724 and 1.545 with H_c (Oe) = 1226 and 149, respectively, for CFO and 0.25BTO–0.75CFO nanoparticles. These mea-sured values of M_s and H_c for pure CFO are smaller than reported for bulk $CoFe_2O_4$ (M_s (emu g^{-1}) = 73 and H_c(kOe) = 5, at 300 K). However, the nanocomposite of 0.25BTO–0.75CFO has abrupt decrement in M_s value, which is explained on the basis of BTO coating over CFO. For nanostructural CFO, the inversion degree must be swung with surface spin-canting and finite size-effect, to execute magnetic disorder. The change in bond angle and bond length may also influence the magne-tization of CFO in composite. Since the spin coupling due to 3d unpaired electrons in Co^{2+} and Fe^{3+} along A and B-sites of spinel structure would lead to the ferrimag-netism of $CoFe_2O_4$. Due to such spin arrangement by unpaired electrons, there exist Fe_B–O–Fe_B, Fe_B–O–Co_B, Fe_B–O–Fe_A, Co_B–O–Fe_A and Co_A–O–Fe_B superexchange interactions to induce magnetic ordering. In this, the antiferromagnetic interactions are strongest via A–O–B superexchange (inter-sublattice) and the weak ferromag-netism is attributed with A–O–A and B–O–B superexchange (intra-sublattice). It is reported for an ideal spinel CFO, the bond angle is 90° of B–O–B, 120° of A–O–B and 80° of A–O–A magnetic exchange interactions. In the present case, the Rietveld analysis [19], calculated the values of bond-angle/length between A and B sites of CFO in 0.25BTO–0.75CFO nanocomposite, *i.e.*, bond angle = 91.07° for Fe_B–O–Fe_B, 133.57° for Fe_A–O–Co_B and 76.13° of Fe_A–O–Fe_A. The values of bond length (l), $l_{A\text{-}A}$ = 0.3626(1) nm, $l_{B\text{-}B}$ = 0.2961(2) nm and $l_{A\text{-}B}$ = 0.34713 (2) nm. It is reported that the superexchange interactions might be related with bond-angle/length among bonds in Fe, Co and O atoms [41]. For the bond angle of 90° (Fe_B—O–Fe_B bonds), the ferromagnetic super-exchange interaction are resulted. When this bond angle is increased, the antiferromagnetic super-exchange interactions are there and generally, bond 120° of super-exchange Fe_A–O–Fe_B is the dominant interaction of magnetite. Above 120° of bond angle, the antiferromagnetic strength is increased which is maximum at 180° [42]. Hence the calculated values of bond angles for Co and Fe ions along A and B-sites are deviated with theoretical one that might be indicated the distorted spinel lattice to influence resulting magnetic behavior.

3.2.6 Ferromagnetism in $xLi_{0.5}Fe_{2.5}O_4$-$(1 - x)SrFe_2O_4$ thin films

The role of hard/soft ferrites composite of $xLi_{0.5}Fe_{2.5}O_4$-$(1 - x)SrFe_2O_4$ (LSF) (x = 0.4, 0.5, and 0.6) on the magnetic exchange-spring systems have been studied [35]. LSF thin films were prepared by MOD method. XRD result shows the poly-crystalline behavior of ferrite with cubic phase of $Li_{0.5}Fe_{2.5}O_4$ and orthorhombic $SrFe_2O_4$. The magnetic behavior of the three series of LSF films is investigated by measuring M-H hysteresis (**Figure 6(c)**). The Li doping into LSF enhanced the value of M_s and M_r and decreased coercivity value. At equal content of Li:Sr., the magnetic behavior is unpredicted. At annealing temperature of 800°C, the LSF55 thin film have anti S-type (diamagnetism), LSF64 has reducible magnetization, while LSF46 has a similar magnetism that seen at 700°C of annealing temperature. The values of saturation magnetization, M_s (emu cc^{-1}) = 1.21, 8.01, and 22.78, remanent magnetization, M_r (emu cc^{-1}) = 0.64, 2.13, and 14.20, with magnetic coercivity, H_c(kOe) = 2.32, 2.18, and 1.54, respectively, measured for LSF55, LSF46, and LSF64 sample, annealed at 700°C. Also from **Figure 6(c)**, the magnetic hys-teresis is involved two-step processes, which might have typically exchange-spring

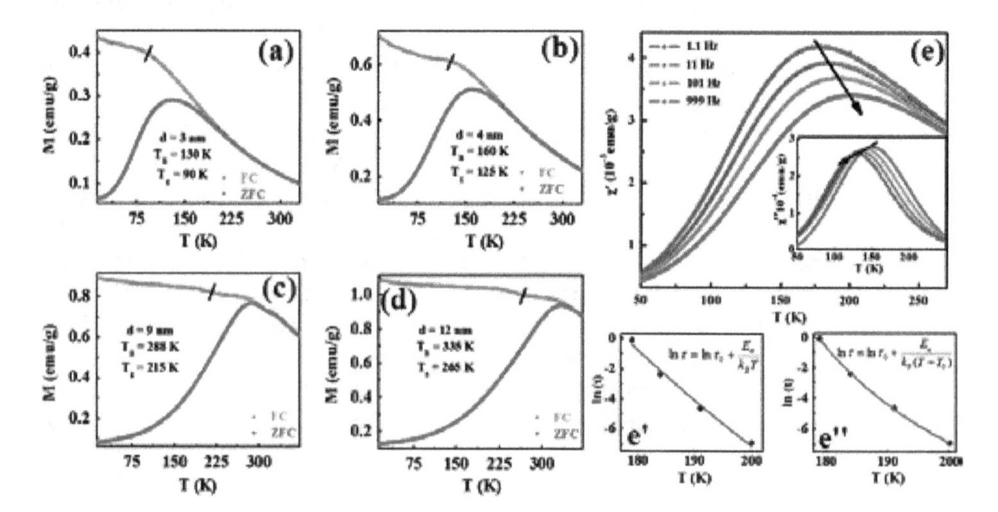

Figure 7.
(a–d) ZFC-FC magnetization of $CoFe_2O_4$ nanoparticles. (e) Temperature dependent $\chi'(T)/\chi''(T)$ of ac susceptibility for 4 nm $CoFe_2O_4$. (e') Arrhenius law (e'') Vogel-Fulcher law of $\chi'(T)$. Adopted from Mohapatra et al. [45].

regime. The positive nucleation field is increased with Li doping. This positive nucleation field occurrence is predicted with a micro-magnetic model, where the perpendicular bilayer of shape anisotropy contribution is there [43].

3.2.7 Super-spin glass formation in $CoFe_2O_4$ nanoparticles

The magnetic $CoFe_2O_4$ nanoparticles of mean size 2–16 nm have been synthesized through a solventless thermolysis technique [44]. **Figure 7 (a–d)** shows the ZFC/FC magnetization of 3, 4, 9 and 12 nm $CoFe_2O_4$ nanoparticles in a field of 5 Oe [45]. The ZFC magnetization of these $CoFe_2O_4$ nanoparticles has observed maxima, which represents the blocking temperature, T_B. However, the FC magnetization is continuously increased below T_B, but at a very low temperature, the FC magnetization slightly become constant. Expectedly, the FC magnetization below T_B is abruptly increased with a decrease in temperature, which is because the superparamagnetic nanoparticles have no inter-particle interactions. This is typically a signature of super-spin glass transition [44]. The temperature dependent real $\chi'(T)$ and the imaginary $\chi''(T)$ components of the *ac* magnetic susceptibility for 4 nm $CoFe_2O_4$ nanoparticles are shown in **Figure**.7**(e)** The $\chi'(T)$ curve for 1.1 Hz displays a sharp peak at the blocking temperature, T_B = 179 K and the position of peak is frequency dependent. By increasing frequency from 1.1 to 999 Hz, there is shifting of peak position from 179 to 200 K that calculated the value of $\Delta\ _mT$ = 21 K. However, $\chi''(T)$ in the inset of **Figure 7(e)** has also peak shifting behavior by changing peak position from 130 to 151 K with frequency 1.1–999 Hz. This is analyzed with relation: $\varphi = \Delta T_m / T_m \Delta \log(f)$, where ΔT_m is the difference between T_m measured in $\Delta\log(f)$ frequency interval. For non-interacting nanoparticles, this parameter 'φ' is usually more than 0.13, for nanoparticle based superspin-glasses, the range is $0.005 < \varphi < 0.05$ and for intermediate interactions, $0.05 < \varphi < 0.13$) [46]. From **Figure 7**, the φ values calculated to be ~0.015, lies for super-spin-glass systems. The super-spin glass behavior is further studied with Neel-Arrhenius law and Vogel-Fulcher model as shown in **Figure 7(e′** and **e″**).

3.3 ZnO: diluted magnetic semiconductor

Recently, DMS ZnO has a technological potential due to its large direct band gap (3.37 eV) and exciton binding energy about 60 meV, which is comparably high [47]. The room temperature ferromagnetism, RTFM in 3d ions doped ZnO is reported by Dietl et al. [48]. But the pure ZnO is also give RTFM when the crystalline product have small sized nanoparticles [49, 50]. The oxygen vacancies are located on the nanoparticles surface and responsible in major participation of RTFM [51]. Garcia et al. [49] found that the ZnO nanoparticles had absorbed certain organic molecules to modify the electronic structure to give RTFM without any magnetic impurity. Xu et al. [50] suggested singly charged oxygen vacancies that depend upon nano-size and heating condition and located mainly near on ZnO nanoparticles surface to induce ferromagnetism. During last decade, the magnetism of ZnO with TM doping is extensively studied. Sato et al. [52] have used local density approximation (LDA) to discuss ferromagnetic ordering which is more stable in Fe-doped ZnO. Karmakar et al. [53] have found antiferromagnetism, which prefers to stabilize Fe-doped ZnO without any native defects. Spaldin [54] had found ferromagnetic ordering which is not possible when Zn sites are substituted with Co or Mn, unless additional hole carriers are incorporated. However, rare earth (RE) atoms have partially filled *f*-orbitals which carry high magnetic moments and form magnetic coupling as for TM ions with partially filled *d*-orbitals [55]. Deng et al. [56] investigated the effect of La doping on the electronic structure and optical properties of ZnO using the first-principle calculation. It is also recently reported that the RE Ce ions are responsible in support antiferromagnetic interactions in TM = Co, Fe doped ZnO [57].

3.3.1 XRD for La, Gd substituted $Zn_{0.95}Co_{0.05}O$ nanostructure

The $Zn_{0.95}Co_{0.05}O$ (ZCO5), $Zn_{0.92}Co_{0.05}La_{0.03}O$ (ZCLO53) and $Zn_{0.92}Co_{0.05}Gd_{0.03}O$ (ZCGO53) nanostructures were synthesized by a sol-gel process [11]. The structural information of La and Gd doped ZCO5 nanostructure using the Rietveld method for XRD pattern of wurtzite structure (space group $P6_3mc$) refined and shown in **Figure 8(a)**. The miller indices of wurtzite structure (100), (002), (101), (102) and (110), respectively, observed with diffraction angle $2\theta = 31.88$, 34.49, 36.34, 47.59 and 49.65° of ZnO. The XRD peak intensity of $Zn_{0.95}Co_{0.05}O$ is reduced with La and Gd doping into it. This is because the ionic radii of La^{3+} and Gd^{3+} ions is much larger than TM Zn^{2+} and Co^{2+} ions, which results into lattice deformation of ZnO. Due to this, the lattice parameters such as lattice distortion, Zn–O bond length/angle and per unit cell volume are affected to induce lattice defects. The value of lattice constant, $a = 3.252(1)$ Å, 3.253(2) Å and 3.255(1) Å, $c = 5.204(1)$ Å, 5.218(3) Å and 5.212(3) Å, $V = 47.660(1)$ Å^3, 47.818(3) Å^3 and 47.823(2) Å^3, and $\chi^2 = 3.37$, 0.893 and 2.9, respectively, extracted for ZCO5, ZCLO53 and ZCGO53. It is also reported that the average value of particles size, D of nano-aggregation is 142 and 86 nm, respectively, for ZCLO53 and ZCGO53 [12]. The wurtzite ZnO structure and lattice defects (vacancies/interstitials) are also found with Raman study. Photoluminescence spectra have near band edge emission (shown energy band gap, $E_g = 3.26$ eV for ZCLO53 and for ZCGO53, $E_g = 3.27$ eV) and the defects/

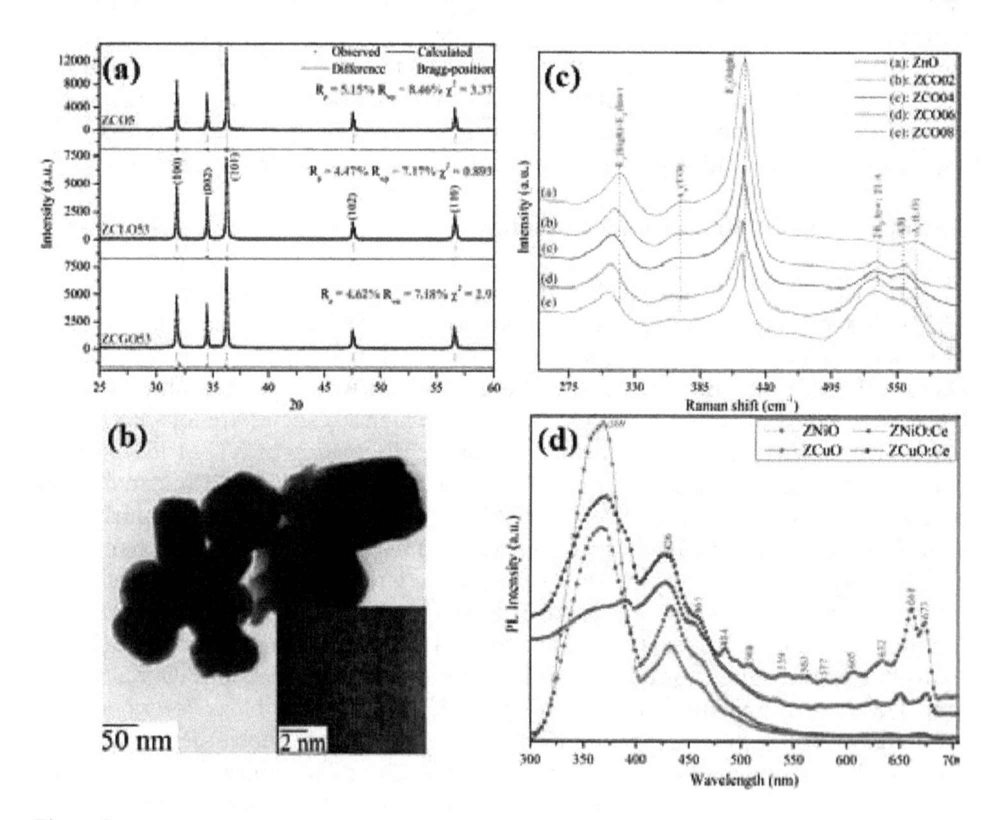

Figure 8.
(a) XRD pattern of Co, La and Gd doped ZnO nanostructure. (b) TEM image of $Zn_{0.94}Fe_{0.03}Ce_{0.03}O$ nanoparticles. (c) Raman spectra of pure and low Co concentrated ZnO nanoparticles. (d) Photoluminescence (PL) of Ni, Cu, Ce substituted ZnO nanoparticles. Adopted from Refs. [10, 11, 57, 58].

V_O evolution with variation in visible PL spectra. The $Zn_{0.95}Co_{0.05}O$ nanoparticles are paramagnetic at room temperature and involved superparamagnetic transition at low temperature. The antiferromagnetic interactions are enhanced with La and Gd doping into $Zn_{0.95}Co_{0.05}O$, is confirmed with ZFC/FC magnetization.

3.3.2 TEM of $Zn_{0.94}Fe_{0.03}Ce_{0.03}O$ (ZFCeO) nanoparticles

The DMS ZFCeO nanoparticles were synthesized by a sol-gel process [57]. **Figure 8(b)** shows the TEM image and the average particles size, D, of nanoparticles is 97 ± 4 nm. The inset of TEM is the HRTEM to show the crystalline formation and lattice spacing after doping into ZnO. It can be noted that the distorted lattice has an enhanced interplanar spacing d [corresponding to (101) planes] ~ 0.247 nm. For pure ZnO, the value of d ~ 0.237 nm. In HRTEM image, some little spots that covered the lattice fringes of spacing are also observed. This is an indication towards ferromagnetic clusters or structural inclusions formation.

3.3.3 Raman study evaluated lattice structure inducing defects

The $Zn_{1-x}Co_xO$ [x = 0.002 (ZCO02), 0.004 (ZCO04), 0.006 (ZCO06) and 0.008 (ZCO08)] nanoparticles were synthesized by a sol-gel process [58]. The XRD pattern results into wurtzite ZnO structure. The ZnO with Co doping has nanorods type morphology with diameter, D (nm) 18 ± 2, 23 ± 3, 41 ± 5 and 53 ± 3, and length (nm) = 39 ± 3, 57 ± 5, 95 ± 3 and 127 ± 5, respectively, for ZCO02, ZCO04, ZCO06 and ZCO08. **Figure 8(c)** shows Raman vibrational modes, which are located at around 314, 368, 422 and 533 cm^{-1} attributed to E_2(high)-E_2(low), A_1(TO), E_2(high) and ($2B_1$ low; 2LA) phonon modes, respectively [59]. The sharpest and strongest peak at about 422 cm^{-1} can be attributed to nonpolar high frequency mode, E_2(high), involved motion of oxygen, which is the characteristic of wurtzite lattice. With increasing Co concentration, a pronounced weakening in peak height, E_2(high) mode, than pure ZnO, has been observed. The intensity of E_2 mode of pure ZnO is shifted towards lower frequencies with increasing Co doping. This happens because decreasing binding energy of Zn-O bonds and a tensile strain in nanograins. An additional strong peak known as additional mode (AM) is observed at 554 cm^{-1} whose intensity is increased with Co concentration. This AM mode is the quasi-longitudinal optical mode formed with abundant shallow donor defects (Zn interstitial, oxygen vacancies, etc.). It is also reported in Ref. [59] that the RTFM is enhanced in low Co concentrated ZnO nanoparticles due to lattice defects. The low temperature ZFC/FC magnetic measurement indicates long-range antiferromagnetic-ferromagnetic ordering to form BMPs.

3.3.4 Photoluminescence spectra extract ZnO lattice defects

The $Zn_{0.95}Ni_{0.05}O$ (ZNiO), $Zn_{0.91}Ni_{0.05}Ce_{0.04}O$ (ZNiO:Ce), $Zn_{0.95}Cu_{0.05}O$ (ZCuO) and $Zn_{0.91}Cu_{0.05}Ce_{0.04}O$ (ZCuO:Ce) nanoparticles were synthesized by a sol-gel process [10]. **Figure 8(d)** show the PL emission for Ni, Cu, Ce substituted ZnO nanoparticles at room temperature. The peak at 369 (3.36 eV) is correlated with surface exciton recombination, which is the near band edge emission of ZnO [59]. The visible emission is formed due to radiative recombination of a photogenerated hole for which an electron occupied oxygen vacancies. The violet emission at 426 nm is the effect of radiative defects related oxygen and Zn vacancies. The peak at 461 and 484 nm is the blue emission, which have two defect level formed due to transition from Zn_i to valance band or bottom of the conduction band to O interstitial. The peak at 632, 661 and 673 nm is the red emission formed with intrinsic defect of O [60]. The Ce-doping into ZNiO and ZCuO is the production of oxygen vacancies defects in PL spectra [61].

3.3.5 Ferromagnetism at 300 K, 10 K of $Zn_{0.95}Ni_{0.05}O$ and $Zn_{0.91}Ni_{0.05}Ce_{0.04}O$ nanoparticles

Figure 9 shows M-H hysteresis at 300 and 10 K, respectively, for ZNiO, ZNiO:Ce nanoparticles [10]. At 300 K, the values of M_s (emu g^{-1}) = 0.073 and 0.085, and M_r (emu g^{-1}) = 0.0066 and 0.0187 with H_c (Oe) = 150 and 558, respectively, for ZNiO and ZNiO:Ce. At 10 K, the value of M_s (emu g^{-1}) = 0.096 and 0.198 and M_r (emu g^{-1}) = 0.0129 and 0.0642 with H_c (Oe) = 76 and 165, respectively, for ZNiO and ZNiO:Ce. The RTFM in TM substituted ZnO has a great influence of the intrinsic defects such as oxygen vacancies. This is since the BMP model ascribed that the bound electrons (holes) in the defect states can couple with TM ions and cause the ferromagnetic regions to overlap, giving long range ferromagnetic order-ing. A theoretical prediction based on first principle calculation in Ni-doped ZnO shows that the ferromagnetic ordering is energetically favorable through double or superexchange mechanisms [10]. The defects such as oxygen vacancies in DMS are responsible in generating carriers for ferromagnetic ordering. The observed values of M_s and M_r at 10 K, are enhanced than at room temperature. However, at low temperature (~10 K), the value of H_c is varies abruptly. This is due to the existence of some ferromagnetic cluster assemblies in the samples, which enlarged H_c, may an indication towards superparamagnetic/spin-glass formation [62].

3.3.6 FC/ZFC magnetization of $Zn_{0.95}Ni_{0.05}O$ and $Zn_{0.91}Ni_{0.05}Ce_{0.04}O$ nanoparticles

The origin of observed RTFM in Ni, Ce substituted ZnO is evaluated by the temperature dependent magnetization [M(T)] with FC at 500 Oe and ZFC measurements (**Figure 9**) [47]. The FC/ZFC curves are separated with decrease in temperature, which usually appears with a coexistent system of antiferromagnetic and ferromagnetic phases. The steep increase of magnetization in FC curve with decreasing temperatures below 50 K is the characteristic of DMS [63]. Decreasing

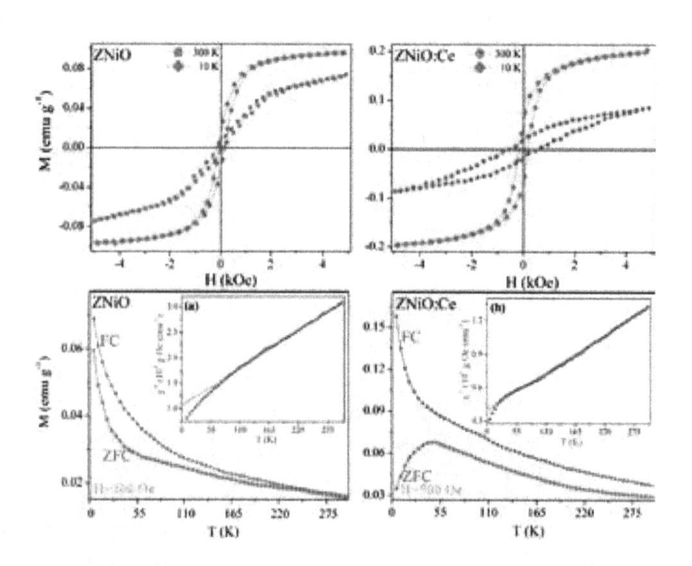

Figure 9.
M-H hysteresis at 300 and 10 K, and M(T) measurement following FC/ZFC at H = 500 Oe for ZNiO, ZNiO:Ce nanoparticles. Inset shows their χ^{-1}-T of Curie-Weiss law. Adopted from Verma and Kotnala [10].

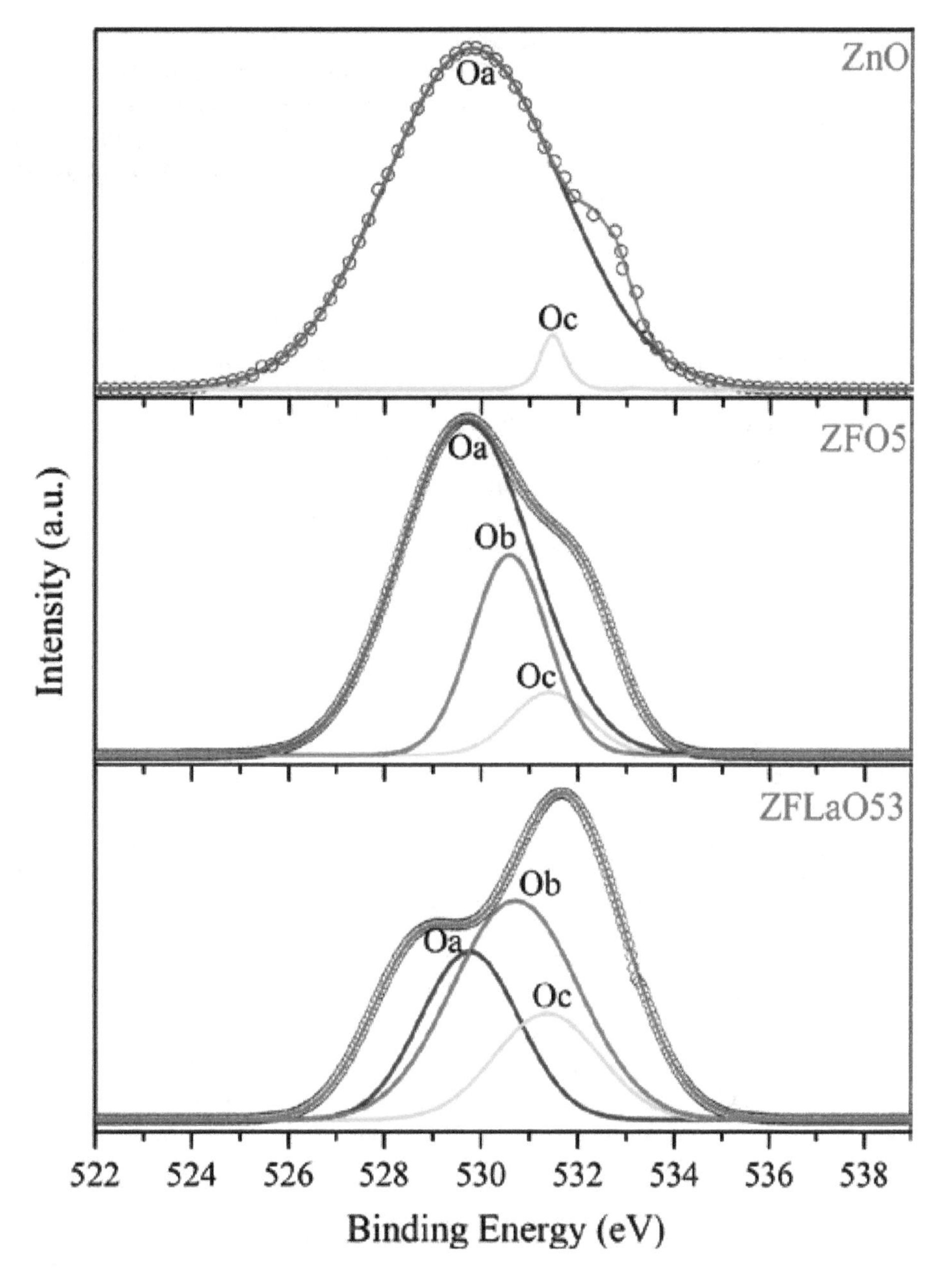

Figure 10.
O 1s XPS spectra of pure ZnO, $Zn_{0.95}Fe_{0.05}O$ and $Zn_{0.92}Fe_{0.05}La_{0.03}O$ nanoparticles. Adopted from Verma and Kotnala [47].

temperatures result into an increase in polarons interaction distance, leading to overlap between neighboring polarons and so allowing them to interact through the magnetic impurities, forming correlated polarons clusters. Though the direct interaction among localized carriers induced antiferromagnetism, and the ferromagnetic interaction due to BMP is possible with large magnetic impurities. When the unoccupied 3d states overlap the impurity band might be attributed high T_c value of ZnO. This is explained on the basis of donor impurity band for which oxy-gen vacancies involved F-centers into host ZnO [64]. The dip in ZFC curve indicates the existence of the blocking temperature, T_B. The upward curvature observed in the FC M(T) measurements (inset of **Figure 9**) of Ni and Cesubstituted ZnO nanoparticles, involve a Curie–Weiss behavior related with susceptibility (χ) as: $\chi = \frac{C}{T-\theta}$; where C is the material specific Curie constant, T is the absolute temperature, and θ is the Weiss constant (in K) [10]. The estimated value of θ is found to be negative, which indicate antiferromagnetic interactions. But χ^{-1}(T) displays a notable deviation from the Curie-Weiss law that might be attributed by short-range ferromagnetism, antiferromagnetism, or a spin-glass system [65].

3.3.7 O 1s XPS spectra

In order to examined the presence of oxygen vacancies, V_O, the O 1s XPS spectra of pure ZnO, $Zn_{0.95}Fe_{0.05}O$ (ZFO5) and $Zn_{0.92}Fe_{0.05}La_{0.03}O$ (ZFLaO53) nanoparticles are given in**Figure 10** [47]. It deconvoluted into three peaks (O_a, O_b, O_c). The peak located on the low binding energy side (O_a) around 529.73 eV is attributed to O^{2-} ions on the wurtzite structure. The O_b peak at about 530.65 eV is associated with O^{2-} ions in oxygen-deficient regions within ZnO matrix, which indicate the formation of defects. The O_c peak at about 531.37 eV is usually attributed to chemisorbed oxygen on the nanostructural surface of the ZnO, such as adsorbed $-CO_3$, H_2O or adsorbed O_2 [66, 67].

4. Conclusion

The ferromagnetism in multiferroic, DMS and ferrites is a topic of intensive research due to their superficial potential in spintronic and magnetic memory devices. The lattice defects related to these multiferroic, DMS and ferrites have an important role in mediating observed ferromagnetism. The lattice defects mainly are oxygen vacancies and their formation might be related with crystalline structure, heating condition during syntheses and types of nanostructural product.

Acknowledgements

This work is financially supported by Dr. D.S. Kothari Postdoctoral Fellowship (No.F.4-2/2006 (BSR)/PH/16-17/0066) under University Grants Commission, India.

Author details

Kuldeep Chand Verma[1*], Ashish Sharma[1], Navdeep Goyal[1]
and Ravinder Kumar Kotnala[2]

1 Department of Physics, Panjab University, Chandigarh, India

2 CSIR-National Physical Laboratory, New Delhi, India

*Address all correspondence to: dkuldeep.physics@gmail.com

References

[1] Buurma AJC, Blake GR, Palstra TTM. Multiferroic Materials: Physics and Properties, from Encyclopedia of Materials: Science and Technology (2nd edn.), 2006. Reference Module in Materials Science and Materials Engineering. Amsterdam: Elsevier; 2016. pp. 1-7

[2] Verma KC, Kotnala RK. Lattice defects induce multiferroic responses in Ce, La-substituted $BaFe_{0.01}Ti_{0.99}O_3$ nanostructures. Journal of the American Ceramic Society. 2016;**99**:1601-1608

[3] Bibes M, Barthelemy A. Towards a magnetoelectric memory. Nature Materials. 2008;**7**:425-426

[4] Verma KC, Kotnala RK. Multiferroic approach for Cr, Mn, Fe, Co, Ni, Cu substituted $BaTiO_3$ nanoparticles. Materials Research Express. 2016;**3**:055006

[5] Verma KC, Vinay G, Kaur J, Kotnala RK. Raman spectra, photoluminescence, magnetism and magnetoelectric coupling in pure and Fe doped $BaTiO_3$ nanostructures. Journal of Alloys and Compounds. 2013;**578**:5-11

[6] Harrisa VG, Sepelak V. Mechanochemically processed zinc ferrite nanoparticles: Evolution of structure and impact of induced cation inversion. Journal of Magnetism and Magnetic Materials. 2018;**465**:603-610

[7] Sugimoto M. The past, present, and future of ferrites. Journal of the American Ceramic Society. 1999;**82**:269-279

[8] Price GD, Price SL, Burdett JK. The factors influencing cation site-preferences in spinels a new mendelyevian approach. Physics and Chemistry of Minerals. 1982;**8**:69-76

[9] Verma KC, Singh VP, Ram M, Shah J, Kotnala RK. Structural, microstructural and magnetic properties of $NiFe_2O_4$, $CoFe_2O_4$ and $MnFe_2O_4$ nanoferrite thin films. Journal of Magnetism and Magnetic Materials. 2011;**323**:3271-3275

[10] Verma KC, Kotnala RK. Understanding lattice defects to influence ferromagnetic order of ZnO nanoparticles by Ni, Cu, Ce ions. Journal of Solid State Chemistry. 2017;**246**:150-159

[11] Verma KC, Kotnala RK. Defects due to lattice distortion and nano-size intermediate ferromagnetism in La, Gd substituted $Zn_{0.95}Co_{0.05}O$. Current Applied Physics. 2016;**16**:175-182

[12] Kaur J, Kotnala RK, Gupta V, Verma KC. Anionic polymerization in Co and Fe doped ZnO: Nanorods, magnetism and photoactivity. Current Applied Physics. 2014;**14**:749-756

[13] Prinz GA. Magnetoelectronics. Science. 1998;**282**:1660-1663

[14] Verma KC, Kotnala RK. Realizing ferromagnetic ordering in SnO_2 and ZnO nanostructures with Fe, Co, Ce ions. Physical Chemistry Chemical Physics. 2016;**18**:17565-17574

[15] Pal B, Dhara S, Giri PK, Sarkar D. Room temperature ferromagnetism with high magnetic moment and optical properties of Co doped ZnO nanorods synthesized by a solvothermal route. Journal of Alloys and Compounds. 2014;**615**:378-385

[16] McCabe GH, Fries T, Liu MT, Shapira Y, Ram-Mohan LR, Kershaw R, et al. Bound magnetic polarons in p-type $Cu_2Mn_{0.9}Zn_{0.1}SnS_4$. Physical Review B. 1997;**56**(**11**):6673-6680

[17] Verma KC, Kotnala RK, Negi NS. Improved dielectric and ferromagnetic properties in Fe-doped $PbTiO_3$ nanoparticles at room temperature.

Applied Physics Letters. 2008;**92**:152902-152903

[18] Verma KC, Kotnala RK. Tailoring the multiferroic behavior in $BiFeO_3$ nanostructures by Pb doping. RSC Advances. 2016;**6**:57727-57738

[19] Verma KC, Singh M, Kotnala RK, Goyal N. Magnetic field control of polarization/capacitance/voltage/resistance through lattice strain in $BaTiO_3$-$CoFe_2O_4$ multiferroic nanocomposite. Journal of Magnetism and Magnetic Materials. 2019;**469**:483-493

[20] Verma KC, Singh D, Kumar S, Kotnala RK. Multiferroic effects in $MFe_2O_4/BaTiO_3$ (M = Mn, Co, Ni, Zn) nanocomposites. Journal of Alloys and Compounds. 2017;**709**:344-355

[21] Shen Y, Sun J, Li L, Yao Y, Zhou C, Su R, et al. The enhanced magnetodielectric interaction of $(1 - x)BaTiO_3$–$xCoFe_2O_4$ multiferroic composites. Journal of Materials Chemistry C. 2014;**2**:2545-2551

[22] Jiang K, Zhu JJ, Wu JD, Sun J, Hu ZG, Chu JH. Influences of oxygen pressure on optical properties and interband electronic transitions in multiferroic bismuth ferrite nanocrystalline films grown by pulsed laser deposition. ACS Applied Materials & Interfaces. 2011;**3**:4844-4852

[23] Li H, Huang Y, Zhang Q, Qiao Y, Gu Y, Liu J, et al. Facile synthesis of highly uniform Mn/Co-codoped ZnO nanowires: Optical, electrical, and magnetic properties. Nanoscale. 2011;**3**:654-660

[24] Kaur J, Kotnala RK, Verma KC. Multiferroic properties of $Ba(Fe_xTi_{1-x}) O_3$ nanorods. Materials Letters. 2011;**65**:3160-3163

[25] Verma KC, Kaur J, Negi NS, Kotnala RK. Multiferroic and magnetoelectric properties of nanostructured $BaFe_{0.01}Ti_{0.99}O_3$ thin films obtained under polyethylene glycol conditions. Solid State Communications. 2014;**178**:11-15

[26] Liu S, Akbashev AR, Yang X, Liu X, Li W, Zhao L, et al. Hollandites as a new class of multiferroics. Scientific Reports. 2014;**4**:6203

[27] Zhao H, Kimura H, Cheng Z, Osada M, Wang J, Wang X, et al. Large magnetoelectric coupling in magnetically short-range ordered $Bi_5Ti_3FeO_{15}$ film. SC Reports. 2014;**4**:5255

[28] Bretos I, Jimenez R, Lazaro CG, Montero I, Calzada ML. Defect-mediated ferroelectric domain depinning of polycrystalline $BiFeO_3$ multiferroic thin films. Applied Physics Letters. 2014;**104**:092905

[29] Spanier JE, Kolpak AM, Urban JJ, Grinberg I, Ouyang L, Yun WS, et al. Ferroelectric phase transition in individual single-crystalline $BaTiO_3$ nanowires. Nano Letters. 2006;**6**:735-739

[30] Beltran JJ, Barrero CA, Punnoose A. Evidence of ferromagnetic signal enhancement in Fe and Co codoped ZnO nanoparticles by increasing superficial Co^{3+} content. Journal of Physical Chemistry C. 2014;**118**:13203-13217

[31] Li W et al. Oxygen-vacancy-induced antiferromagnetism to ferromagnetism transformation in $Eu_{0.5}Ba_{0.5}TiO_3$-delta multiferroic thin films. Scientific Reports. 2013;**3**:618-616

[32] Pradhan DK, Kumari S, Puli VS, Das PT, Pradhan DK, Kumar A, et al. Correlation of dielectric, electrical and magnetic properties near the magnetic phase transition temperature of cobalt zinc ferrite. Physical Chemistry Chemical Physics. 2017;**19**:210-218

[33] Clavel G, Marichy C, Willinger MG, Ravaine S, Zitoun D, Pinna N. $CoFe_2O_4$-TiO_2 and $CoFe_2O_4$-ZnO thin film nanostructures elaborated from colloidal chemistry and atomic layer deposition. Langmuir. 2010;**26**(23):18400

[34] Ahmed MA, Okasha N, Mansour SF, El-dek SI. Bi-modal improvement of the physico-chemical characteristics of PEG and MFe_2O_4 subnanoferrite. Journal of Alloys and Compounds. 2010;**496**:345-350

[35] Verma KC, Kotnala RK. Spring like ferromagnetic behavior of $x Li_{0.5}Fe_{2.5}O_4$-$(1 - x)SrFe_2O_4$ nanoferrite thin films. Journal of Nanoparticle Research. 2011;**13**:4437-4444

[36] Wang P, Jin C, Zheng D, Li D, Gong J, Li P, et al. Strain and ferroelectric-field effects co-mediated magnetism in (011)-$CoFe_2O_4$/Pb $(Mg_{1/3}Nb_{2/3})_{0.7}Ti_{0.3}O_3$ multiferroic heterostructures. ACS Applied Materials & Interfaces. 2016;**8**:24198-24204

[37] Bateer B, Tian C, Qu Y, Du S, Yang Y, Ren Z, et al. Synthesis, size and magnetic properties of controllable $MnFe_2O_4$ nanoparticles with versatile surface functionalities. Dalton Transactions. 2014;**43**:9885-9891

[38] Vamvakidis K, Katsikini M, Sakellari D, Paloura EC, Kalogirou O, Samara CD. Reducing the inversion degree of $MnFe_2O_4$ nanoparticles through synthesis to enhance magnetization: Evaluation of their ^1H NMR relaxation and heating efficiency. Dalton Transactions. 2014;**43**:12754-12765

[39] Bullita S, Casu A, Casula MF, Concas G, Congiu F, Corrias A, et al. $ZnFe_2O_4$ nanoparticles dispersed in a highly porous silica aerogel matrix: A magnetic study. Physical Chemistry Chemical Physics. 2014;**16**: 4843-4852

[40] Peddis D, Yaacoub N, Ferretti M, Martinelli A, Piccaluga G, Musinu A, et al. Cationic distribution and spin canting in $CoFe_2O_4$ nanoparticles. Journal of Physics. Condensed Matter. 2011;**23**:426004-426008

[41] Nedelkoski Z, Kepaptsoglou D, Lari L, Wen T, Booth RA, Oberdick SD, et al. Origin of reduced magnetization and domain formation in small magnetite nanoparticles. Scientific Reports. 2017;**7**:45997

[42] Robinson DW. Magnetism and the chemical bond. In: John B, editor. Goodenough Interscience. New York: Wiley; 1963

[43] Casoli F, Nasi L, Albertini F, Fabbrici S, Bocchi C, Germini F, et al. Morphology evolution and magnetic properties improvement in FePt epitaxial films by in situ annealing after growth. Journal of Applied Physics. 2008;**103**:043912

[44] Mohapatra J, Mitra A, Bahadur D, Aslam M. Surface controlled synthesis of MFe_2O_4 (M = Mn, Fe, Co, Ni and Zn) nanoparticles and their magnetic characteristics. CrystEngComm. 2013;**15**:524-532

[45] Mohapatra J, Mitra A, Bahadur D, Aslam M. Superspin glass behavior of self-interacting $CoFe_2O_4$ nanoparticles. Journal of Alloys and Compounds. 2015;**628**:416-423

[46] Jonsson PE. Superparamagnetism and spin glass dynamics of interacting magnetic nanoparticle systems. Advances in Chemical Physics. 2004;**128**:191-248

[47] Verma KC, Kotnala RK. Oxygen vacancy induced by La and Fe into ZnO nanoparticles to modify ferromagnetic ordering. Journal of Solid State Chemistry. 2016;**237**:211-218

[48] Dietl T, Ohno H, Matsukura F, Cibert J, Ferrand D. Zener model description of ferromagnetism in zinc-blende magnetic semiconductors. Science. 2000;**287**:1019-1022

[49] Garcia MA, Merino JM, Pinel EF, Quesada A, dela Venta J, Ruız Gonzalez ML, et al. Magnetic properties of ZnO nanoparticles. Nano Letters. 2007;**7**:1489-1492

[50] Xu X, Xu C, Dai J, Hu J, Li F, Zhang S. Size dependence of defect-induced room temperature ferromagnetism in undoped ZnO nanoparticles. Journal of Physical Chemistry C. 2012;**116**:8813-8818

[51] Gao D, Zhang Z, Fu J, Xu Y, Qi J, Xue D. Room temperature ferromagnetism of pure ZnO nanoparticles. Journal of Applied Physics. 2009;**105**:113928-113924

[52] Sato K, Katayama-Yoshida H. Magnetic interactions in transition-metal-doped ZnO: An ab initio study. Ferromagnetism in a transition metal atom doped ZnO. Physica E. 2001;**10**:251-255

[53] Karmakar D, Mandal SK, Kadam RM, Paulose PL, Rajarajan AK, Nath TK, et al. Ferromagnetism in Fe-doped ZnO nanocrystals: Experiment and theory. Physical Review B. 2007;**75**:144404-144414

[54] Spaldin NA. Search for ferromagnetism in transition-metal-doped piezoelectric ZnO. Physical Review B. 2004;**69**:125201-125207

[55] El Hachimi AG, Zaari H, Benyoussef A, El Yadari M, El Kenz A. First-principles prediction of the magnetism of 4f rare-earth-metal-doped wurtzite zinc oxide. Journal of Rare Earths. 2014;**32**:715-721

[56] Deng SH, Duan MY, Xu M, He L. Effect of La doping on the electronic structure and optical properties of ZnO. Physica B. 2011;**406**:2314-2318

[57] Verma KC, Kotnala RK. Defects-assisted ferromagnetism due to bound magnetic polarons in Ce into Fe, Co:ZnO nanoparticles and first-principle calculations. Physical Chemistry Chemical Physics. 2016;**18**:5647-5657

[58] Verma KC, Bhatia R, Kumar S, Kotnala RK. Vacancies driven magnetic ordering in ZnO nanoparticles due to low concentrated Co ions. Materials Research Express. 2016;**3**:076103

[59] Zeng H, Duan G, Li Y, Yang S, Xu X, Cai W. Blue luminescence of ZnO nanoparticles based on non-equilibrium processes: Defect origins and emission controls. Advanced Functional Materials. 2010;**20**:561-572

[60] Kumar S, Basu S, Rana B, Barman A, Chatterjee S, Jha SN, et al. Structural, optical and magnetic properties of sol-gel derived ZnO:Co diluted magnetic semiconductor nanocrystals: An EXAFS study. Journal of Materials Chemistry C. 2014;**2**:481-495

[61] Dar MI, Arora N, Singh NP, Sampath S, Shivashankar SA. Role of spectator ions in influencing the properties of dopant-free ZnO nanocrystals. New Journal of Chemistry. 2014;**38**:4783-4790

[62] Kovaleva NN, Kugel KI, Bazhenov AV, Fursova TN, Loser W, Xu Y, et al. Formation of metallic magnetic clusters in a Kondo-lattice metal: Evidence from an optical study. Scientific Reports. 2012;**2**:890-897

[63] Singhal A, Achary SN, Manjanna J, Chatterjee S, Ayyub P, Tyagi AK. Chemical synthesis and structural and magnetic properties of dispersible cobalt- and nickel-doped ZnO

Nanocrystals. Journal of Physical Chemistry C. 2010;**114**:3422-3430

[64] Lu ZL, Hsu HS, Tzeng YH, Zhang FM, Du YW, Huang JCA. The origins of ferromagnetism in Co-doped ZnO single crystalline films: From bound magnetic polaron to free carrier-mediated exchange interaction. Applied Physics Letters. 2009;**95**:102501-102503

[65] Yang CY, Lu YH, Lin WH, Lee MH, Hsu YJ, Tseng YC. Structural imperfections and attendant localized/itinerant ferromagnetism in ZnO nanoparticles. Journal of Physics D: Applied Physics. 2014;**47**:345003

[66] Janotti A, Van de Walle CG. Oxygen vacancies in ZnO. Applied Physics Letters. 2005;**87**:122102-122103

[67] Verma KC, Kotnala RK, Goyal N. Multi-Functionality of Spintronic Materials, from Nanoelectronics Devices, Circuits and Systems. Amsterdam, Netherlands: Elsevier; 2019. pp. 153-215

7

Dielectric Responses in Multilayer C_f/Si_3N_4 as High-Temperature Microwave-Absorbing Materials

Heng Luo, Lianwen Deng and Peng Xiao

Abstract

High-temperature microwave-absorbing materials are in great demand in military and aerospace vehicles. The high-temperature dielectric behavior of multilayer C_f/Si_3N_4 composites fabricated by gelcasting has been intensively investigated at temperature coverage up to 800°C in the X-band (8.2–12.4 GHz). Experimental results show that the permittivity of Si_3N_4 matrix exhibits excellent thermo-stability with temperature coefficient lower than 10^{-3}°C^{-1}. Taking temperature-dependent polarized bound charge and damping coefficient into consideration, a revised dielectric relaxation model with Lorentz correction for Si_3N_4 ceramics has been established and validated by experimental results. Besides, a general model with respect to permittivity as a function of temperature and frequency has been established with the help of nonlinear numerical analysis to reveal mechanisms of temperature-dependent dielectric responses in C_f/Si_3N_4 composites. Temperature-dependent permittivity has been demonstrated to be well distributed on circular arcs with centers actually kept around the real (ε') axis in the Cole-Cole plane. Furthermore, space charge polarization and relaxation are discussed. These find-ings point to important guidelines to reveal the mechanism of dielectric behavior for carbon fiber functionalized composites including but not limited to C_f/Si_3N_4 composites at high temperatures, and pave the way for the development of high-temperature radar absorbing materials.

Keywords: high temperature, microwave-absorbing material, dielectric, relaxation

1. Introduction

Wireless electronic devices and communication instruments have found wide application in our daily life. Their efficient operation depends strongly on transmission behavior of alternating electromagnetic wave with frequency ranging from kilohertz (KHz) to gigahertz (GHz), and vice versa, are very sensitive to interference from external electromagnetic wave. Driven by the demand for both adequate interference rejection and controlled radiation, more and more efforts have been devoted to high-performance electromagnetic compatibility/interference (EMC/EMI) materials. As we all know, the propagation behavior of electromagnetic wave when encountering a material could be divided into three types in principle: reflection, absorption, and transmission. As typical EMI materials, metals or materials with high electrical conductivity could prevent external electromagnetic wave from penetration due to the large amount of free electrons. Last decades have witnessed intensive efforts toward exploring lightweight and cost-effective electromagnetic interference (EMI) materials with adequate shielding effectiveness [1–5], involving carbonaceous fillers-enabled polymers, novel lightweight metal composites, etc.

However, the primary function of EMI shielding is to reflect radiation using charge carriers that interact directly with incident electromagnetic field, and the back-radiation would in turn affect the surrounding environment and devices. What is more, the reflected radiation may also be caught by radar observation systems and lead to exposure of moving trace, which is extremely undesirable from the defense-oriented point of view. As a result, electromagnetic wave-absorbing materials with reduced reflection on the surface as well as enhanced internal attenuation are more favorable candidates for EMI shielding, especially in the GHz range. Polymers modified by carbon nanomaterials (e.g., carbon nanotubes [6–8], carbon nanofibers [9, 10], graphene [11, 12], etc.), metal powders [13, 14], and ferrite [15] have been demonstrated to be excellent microwave-absorbing/shielding materials especially in the X-band (8.2–12.4 GHz) [16, 17], and have achieved successful application [18–20]. However, due to their inferior temperature stability and mechanical properties, their application is limited toward applica-tion under high temperature. For example, the temperature on the windward side of high-speed aircraft (>3 Ma) could reach up to 1000°C due to the aerodynamic heating effect. As a result, ceramics and their derivative architecture (r-GOs/SiO_2, CNT/SiO_2, ZnO/$ZrSiO_4$, SiC_f/SiC, etc.) [21–30] with the integration of desirable dielectric responses, high strength, oxidation resistance, thermo-stability, and low density have attracted growing attention for high-temperature-absorbing materials. Besides, Si_3N_4 ceramics are one of the most intensively studied ceramics in high-temperature applications due to their superior antioxidation (>1200°C) and mechanical and chemical stabilization properties [31–38]. More importantly, owing to the excellent electrical insulation property and low dielectric constant, Si_3N_4 ceramics are expected to be a promising candidate matrix as high-temperature microwave-absorbing materials [25, 29, 30, 39–42]. However, previous work mainly focused on experimental evolution of complex dielectric responses with temperature and qualitative analyses according to the Debye theory. Still, model-ing for high-temperature dielectric behaviors is relatively limited and remains a great challenge due to the complexity of the components and microstructures for high-temperature microwave-absorbing materials, as well as high-temperature measurement system. It also should be noted that microwave dissipation capacity of a composite is strongly dependent on the structural design. Many investigations have shown that incorporation of reasonable structural design, involving multilayer structure [43–45] and periodic structure in metamaterials [46, 47], is an effective way to regulate dielectric response and guarantee desirable attenuation perfor-mance. Moreover, taking full advantage of tunable electromagnetic parameters in each layer, optimal microwave impedance matching as well as absorbing capability could be achieved. This fact means that it is essential to explore the mechanism of dielectric behavior of laminate-structure materials from new viewpoints.

In this chapter, we mainly focus on the microwave dielectric responses in laminate-structure or multilayer-structure C_f/Si_3N_4 composites from both experimental and theoretical points of view. Furthermore, a general model with respect to permittivity as a function of temperature and frequency would be established to reveal mechanisms of temperature-dependent dielectric responses for C_f/Si_3N_4 composites. These findings point to important guidelines to reveal the mechanism of dielectric behavior for carbon fiber functionalized composites including but not limited to C_f/Si_3N_4 composites at high temperatures, and pave the way for the development of high-temperature radar absorbing materials.

2. Experiments

2.1 Preparation of multilayer C_f/Si_3N_4 composites

Commercially available carbon fibers (T700, 12 K, TohoTenax Inc., Japan) were used as starting materials in this work. In order to avoid damage at high-temperature sintering, pyrolytic carbon(PyC)/SiC dual-coating on carbon fibers was prepared by chemical vapor deposition based on Methyltrichlorosilane (MTS)-H2-Ar system at 1150°C. The powder mixture of 85 wt% α-Si_3N_4 (purity > 93%, d_{50} = 0.5 μm, Beijing Unisplendor Founder High Technology Ceramics Co. Ltd., China), 5 wt% Al_2O_3, and 10 wt% Y_2O_3 was mixed with solvent-based acrylamide-N,N′-methylenebisacrylamide (AM-MBAM) system, and consolidated via gelcasting and pressureless-sintering route (as illustrated in **Figure 1**). Details of the multilayer C_f/Si_3N_4 samples' preparation are given in our previous work [35]. Note that each layer of carbon fiber plays dominant role to attenuate microwave energy, which could be adjudged the surface density of short carbon fiber.

2.2 High-temperature electromagnetic measurements

In order to evaluate the high-temperature permittivity, specimens with the size of 22.86 × 10.16 × 1.5 mm^3 were polished and determined in X-band through the wave-guide method with a vector network analyzer (Agilent N5230A, USA). As shown in **Figure 2**, the as-prepared Si_3N_4 ceramic sample was heated by an inner heater with a ramp rate of 10°C/min up to 800°C in air. For accuracy of measurement, the device was carefully calibrated with the through-reflect-line (TRL) approach, and a period of 10 min was applied to guarantee system stability at each evaluated temperature.

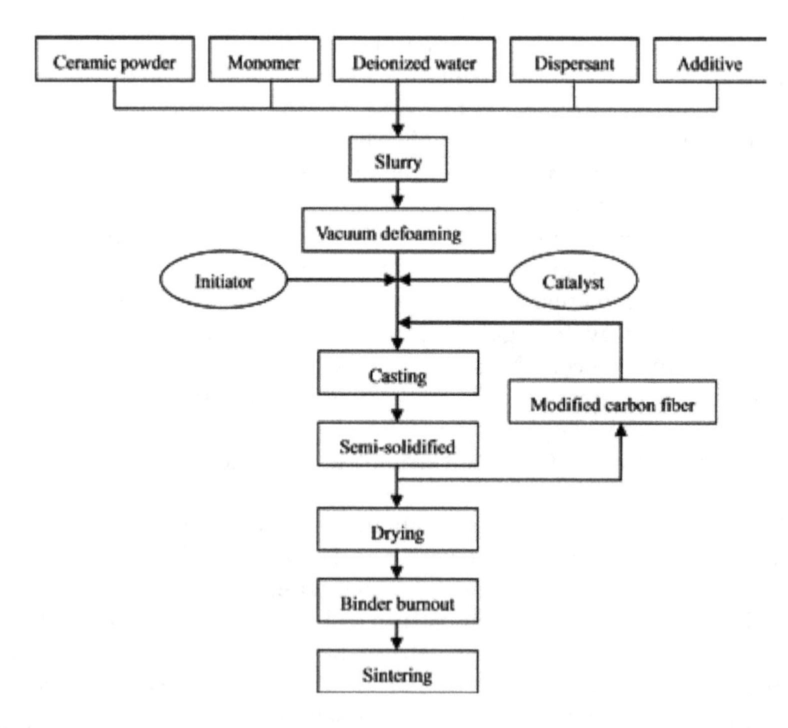

Figure 1.
The gelcasting process for preparation of multilayer C_f/Si_3N_4 composites (reprinted with permission from Ref. [35]).

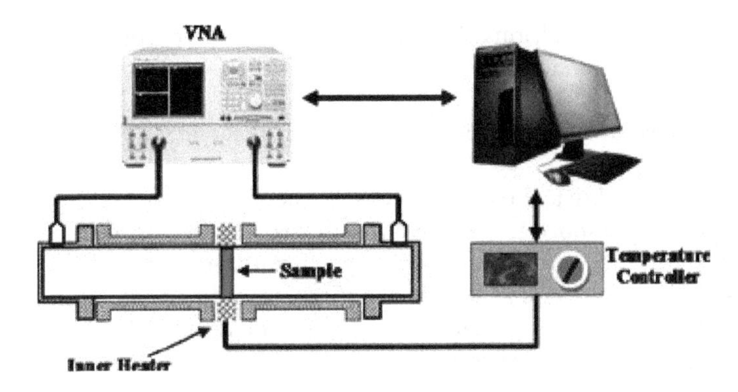

Figure 2.
Schematic diagram of complex permittivity test apparatus (reprinted with permission from Ref. [48]).

3. Microwave dielectric properties

3.1 Structure of multilayer C_f/Si_3N_4 composites

The optical image of cross-section of multilayer C_f/Si_3N_4 composites is shown in **Figure 3(a)**. As expected, three layers filled with short carbon fibers are uniformly embedded in the Si_3N_4 matrix. The microstructure of Si_3N_4 ceramic was formed by rod-like particles, which are evenly distributed and intercross with each other to form the main pores. Energy dispersive spectroscopy (EDS) analysis at spots A and B in **Figure 3(c)** demonstrates that the PyC/SiC interphase could effectively promote the chemical compatibility between carbon fibers and Si_3N_4 ceramic at high-temperature circumstance, which could be further proved by the XRD investigations (see **Figure 4**). As seen in **Figure 4**, in addition to the main β-Si_3N_4 peaks

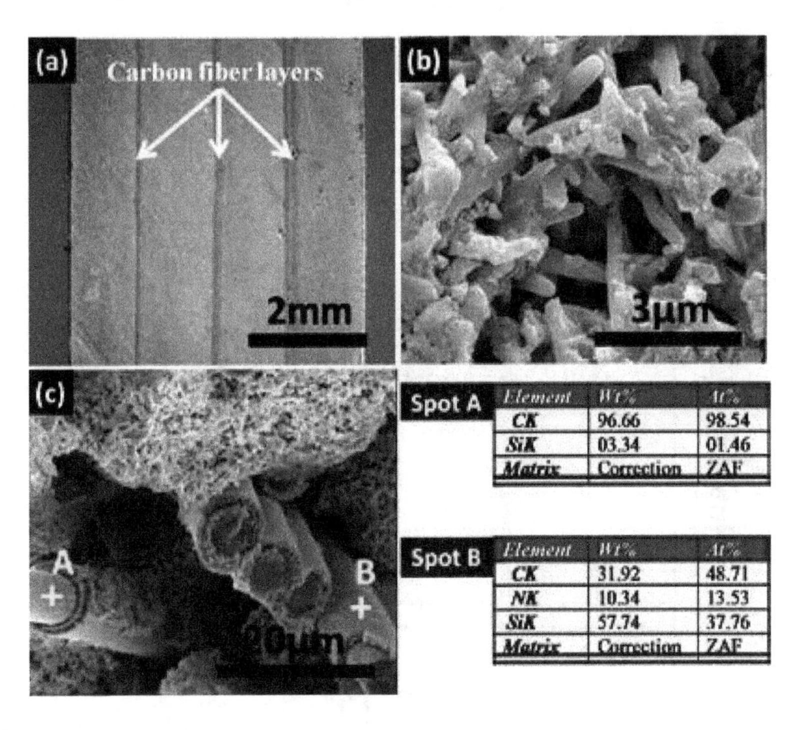

Spot A

Element	Wt%	At%
CK	96.66	98.54
SiK	03.34	01.46
Matrix	Correction	ZAF

Spot B

Element	Wt%	At%
CK	31.92	48.71
NK	10.34	13.53
SiK	57.74	37.76
Matrix	Correction	ZAF

Figure 3.
(a) Cross-section of multilayer C_f/Si_3N_4 composites, facture surface located at (b) Si_3N_4 matrix and (c) carbon fibers (reprinted with permission from Ref. [39]).

and Y-Al oxide peaks, additional C and β-SiC peaks, corresponding to the carbon fiber and modification coating, were detected for C_f/Si_3N_4 composites.

3.2 Room-temperature dielectric properties of multilayer C_f/Si_3N_4 composites

According to the classical transmission line theory, microwave complex permittivity ($\varepsilon = \varepsilon' - j\varepsilon''$) is an important parameter to determine the absorbing performance. **Figure 5(a)** shows the real and imaginary permittivity of multilayer C_f/Si_3N_4 composites at X-band, as well as as-prepared Si_3N_4 ceramics. Clearly, the dielectric constant of Si_3N_4 ceramics presents frequency-independent behavior. The mean real and imaginary parts of permittivity and dielectric loss($\tan\delta = \varepsilon''/\varepsilon'$) of pure Si_3N_4 ceramic were 7.7, 0.04, a n d 5.3×10^{-3}, respectively. The relatively low dielectric constant is considered to be helpful for microwave impedance matching with free space, which tends to reduce reflection of electromagnetic wave from the surface of material and enhance energy propagating in the material.

However, both the real permittivity and imaginary permittivity of C_f/Si_3N_4 sandwich composites decrease markedly as frequency increases at X-band, varying from 12.3 and 5.1 to 7.9 and 1.2, respectively. This phenomenon is usually called frequency dispersion characteristic, whichisacknowledged to be beneficial to broaden the microwave absorption bandwidth. The reflection loss (R) of Si_3N_4 and C_f/Si_3N_4 sandwich composites was calculated according to the formula as follows:

$$R(dB) = 20 \log\left|\frac{Z_{in} - 1}{Z_{in} + 1}\right| \quad (1)$$

and

$$Z_{in} = \sqrt{\frac{\mu_r}{\varepsilon_r}} \tanh\left[j\left(\frac{2\pi}{c}\right)\sqrt{\mu_r \varepsilon_r} fd\right] \quad (2)$$

where Z_{in} refers to input impedance, j is the imaginary unit (i.e., equals to $\sqrt{-1}$), c is the velocity of electromagnetic waves in free space, f is the microwave frequency,

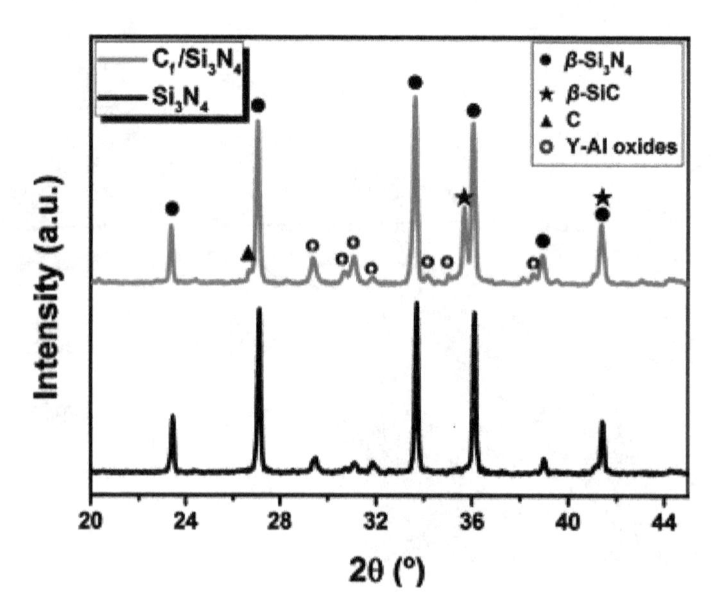

Figure 4.
XRD patterns of as-prepared Si_3N_4 ceramics and C_f/Si_3N_4 composites (reprinted with permission from Ref. [39]).

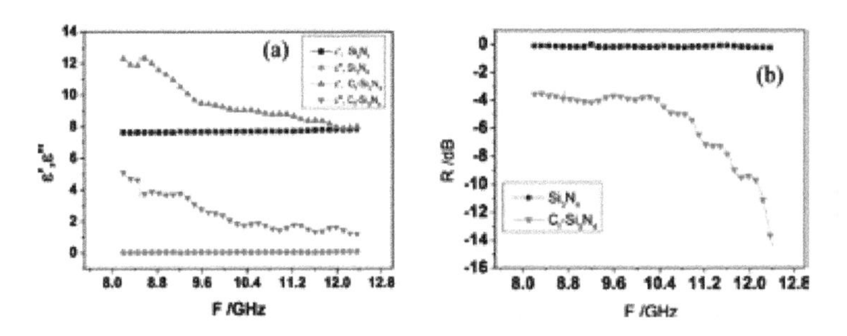

Figure 5.
(a) The permittivity and (b) reflection loss curves of Si_3N_4 and C_f/Si_3N_4 sandwich composites over X-band (reprinted with permission from Ref. [35]).

and *d* is the thickness of the samples. As depicted in **Figure 5(b)**, the microwave absorption ability of the C_f/Si_3N_4 sandwich composites was significantly enhanced compared with pure Si_3N_4 ceramic. The reflection loss of the C_f/Si_3N_4 sandwich composites gradually decreases from −3.5 dB to −14.4 dB with the increase of frequency, while that of the pure Si_3N_4 ceramic remains at −0.1 dB.

The enhanced microwave-absorbing performance could be mainly attributed to polarization relaxation. As we know, there exists migration of free electrons inside the electro-conductive carbon fibers, as well as charge accumulation at interfaces between short carbon fibers and insulated matrix when subjected to external electric field. As a result, the chopped carbon fibers are more inclined to be equiva-lent to micro-dipoles. With increase of frequency, the orientation of these dipoles could not keep up with change of electric field gradually, resulting in the real part of permittivity (ε') of C_f/Si_3N_4 sandwich composites decrease gradually. Furthermore, the scattering effect from defects and the crystal lattice on the back-and-forth movement of electrons under alternating electromagnetic waves predominately contributes to the dissipation of EM energy, which results in thermal energy.

For a deep-seated investigation of frequency-dependent dielectric responses of multilayer C_f/Si_3N_4 composites (**Figure 6(a)**), here we proposed an equivalent RC circuit model, where each layer of carbon fiber plays a role of one electrode in a plane-parallel capacitor, while each layer of Si_3N_4 ceramic plays the role of the dielectric (**Figure 6(b)**). Considering the existence of leakage current, leakage resistances were applied in equivalent circuit (**Figure 6(c)**).

According to the circuit theory knowledge, the relationship between permittivity and frequency ω follows:

$$\begin{aligned} \varepsilon' &= \frac{Q(C_1 + C_2)^2}{R_1^2 C_1 C_2 (C_1 + C_2)} \cdot \frac{1}{\omega^2} + \frac{Q R_1^2 C_1^2 C_2^2}{R_1^2 C_1 C_2 (C_1 + C_2)} \\ \omega \cdot \varepsilon'' &= P \frac{(C_1 + C_2)^2 + R_1^2 C_1^2 C_2^2 \cdot \omega^2}{(R_1 + R_2)(C_1 + C_2)^2 + R_1^2 R_2 C_1^2 C_2^2 \cdot \omega^2} \end{aligned} \quad (3)$$

where *P*, *Q*, and *C* are constants and are determined by the surface density of carbon fiber layers and thickness of Si_3N_4 layers. We have analyzed our experimental permittivity based on Eq. (3) using Trust-Region algorithm, which is illustrated in **Figure 7**. The points in **Figure 7** indicate the experimental data, while the results predicted by equivalent circuit model are given as solid line. Clearly, for multilayer C_f/Si_3N_4 composites, both ε' and ($\omega \cdot \varepsilon''$) are inversely proportional to the frequency square ω^2 , and the predicted results agree quite well with the measured data. Additionally, the experimental data show oscillation phenomena at high frequency, which may result from charge and discharge processes between C_1 and C_2 (**Figure 6**) with the increase of frequency. Note that, even though the imaginary part of

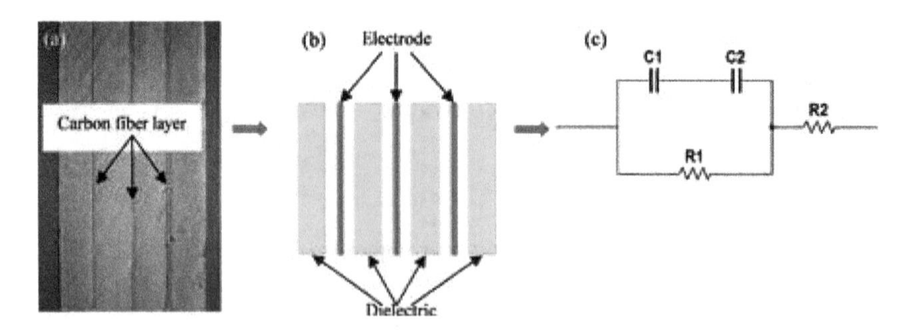

Figure 6.
(a) Cross-section morphology, (b) structural schematic diagram, and (c) equivalent circuit diagram of C_f/Si_3N_4 sandwich composites (reprinted with permission from Ref. [35]).

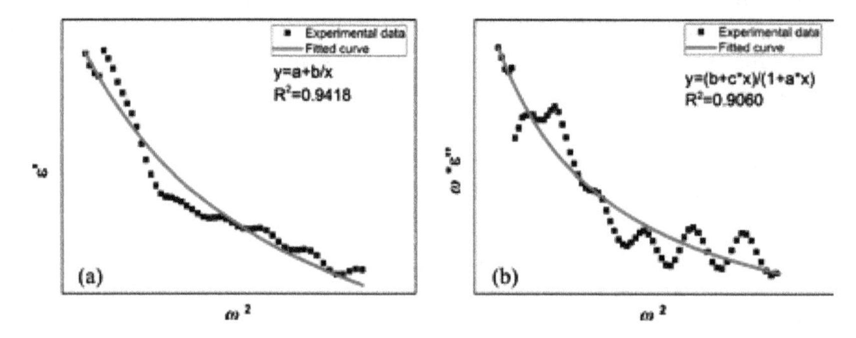

Figure 7.
Experimental data and curves of (a) ε ′ and (b) ω · ε ″ versus ω 2 (reprinted with permission from Ref. [35]).

permittivity declines faster than the real part with frequency increase, the reflection loss presents enhanced trend with frequency increase. This phenomenon is mainly attributed to the fact that microwave-absorbing efficiency is the combination of reflection from the material surface and attenuation inside the material. The lower the permittivity, the better its impedance matching between air and absorber. As a result, in order to achieve optimal reflection loss, one must lower reflection as much as possible and keep a modest loss tangent simultaneously.

3.3 High-temperature dielectric behaviors of Si_3N_4 ceramics

Due to the fact that there will inevitably be some variation of electromagnetic performance or even the mechanical property of materials served in high-temperature condition, the dielectric property would be supposed to dynamically change with temperature. How and to what extent does the permittivity dynami-cally change with temperature (increase or decrease)? All these are quite critical in parameter modification strategy for improving the accuracy of radar detection and guidance. Consequently, it is of utmost importance to explore the evolution of dielectric properties of Si_3N_4 ceramics used in high-temperature circumstances. Three-dimensional (3D) plots of the effect of temperature on permittivity of Si_3N_4 ceramics over X-band are shown in **Figure 8**. Clearly, the real permittivity (ε') shows no obvious change even though temperature rises up to 800° C. Likewise, thanks to the excellent electrical insulation of Si_3N_4 ceramics, the loss tangent (as shown in **Figure 8(b)**) is almost independent of frequency and temperature and remains lower than 0.06.

Herein, temperature coefficient κ is used to explicate the impact of temperature on dielectric response of as-prepared Si_3N_4 ceramics:

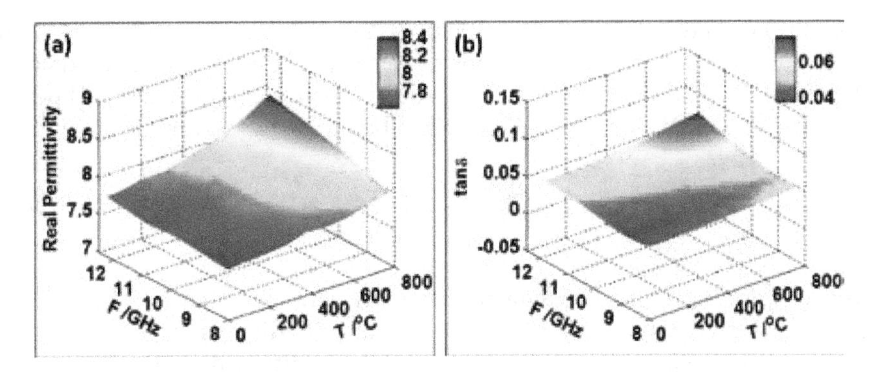

Figure 8.
Three-dimensional plots of complex permittivity of Si_3N_4 ceramics versus frequency and temperature (reprinted with permission from Ref. [39]).

$$\kappa = \frac{1}{\varphi} \cdot \frac{\Delta\varphi}{\Delta T}, \tag{4}$$

where T is the temperature and φ refers to either the dielectric constant or loss tangent. As summarized in **Table 1**, the temperature coefficients of both dielectric constant and loss tangent remain around 10^{-4}°C^{-1}. From this perspective, the as-prepared Si_3N_4 ceramics exhibit excellent thermo-stability of dielectric response within the range of evaluated temperatures. This weak temperature dependence further corroborates as-prepared Si_3N_4 ceramics to be a competitive candidate as the matrix of high-temperature microwave-absorbing materials.

It should be noted that the real permittivity increases slightly with frequency increase, which is contrary to the ordinary frequency dispersion effect described by the Debye model [49–55]. In order to further expound this peculiar frequency dispersion characteristic, it is essential to explore the details of electronic polarizing processing of Si_3N_4 ceramics. Considering the covalent bonding, the electronic polarization in Si_3N_4 ceramics mainly results from the bound charge's displacement deviated from the equilibrium position. The motion equation of bound charge driven by an external electric field $E_0 e^{j\omega t}$ can be expressed as:

$$m\frac{\partial^2 x}{\partial t^2} = qE_0 e^{j\omega t} - fx - 2\eta\frac{\partial x}{\partial t} \tag{5}$$

Temperature (°C)	$\kappa_{\varepsilon'}$ (×10^{-4}°C^{-1})		$\kappa_{\tan\delta}$ (×10^{-4}°C^{-1})	
	8.2 GHz	12.4 GHz	8.2 GHz	12.4 GHz
25	0.46	0.45	2.08	2.84
100	0.22	0.93	1.79	3.11
200	0.45	0.031	2.83	4.83
300	1.04	1.02	3.02	3.91
400	0.16	1.31	2.87	4.89
500	0.40	1.11	4.98	6.92
600	0.97	1.70	4.68	5.97
700	1.16	1.94	3.49	4.34

Table 1.
Temperature coefficient of permittivity and loss tangent at selected frequency (reprinted with permission from Ref. [39]).

where m, q, and x are the mass, charge, and displacement of single bound charge, respectively; f is the coefficient of restoring force; and η is the damping coefficient. Taking Lorentz correction [56–58] into consideration, the real permittivity of Si_3N_4 ceramic containing N polarized bound charges could be obtained as:

$$\varepsilon = \varepsilon_s + \frac{Nq^2}{\varepsilon_0 m} \cdot \frac{\omega_0{}^2 - \omega^2}{\left(\omega_0{}^2 - \omega^2\right)^2 + 4\eta^2\omega^2} \tag{6}$$

where ω_0 is the resonant frequency of Si_3N_4 ceramics, and ε_0 and ε_s are the vacuum permittivity and static dielectric constant of Si_3N_4 ceramics, respectively. Theoretical results have shown that the order of magnitude of resonant frequency ω_0 is around 10 eV (~10^{15} Hz) [59, 60] which is considerably larger than the tested frequency (~10^{10}Hz). Combining with Eq. (4), the real permittivity increases slightly with frequency increase, which is closely coincident with the experimental results. Furthermore, taking the effect of temperature into consideration, N and η should follow:

$$N \propto \exp\left(-E_a/RT\right) \tag{7}$$

$$\eta \propto \exp\left(-E_b/RT\right) \tag{8}$$

where E_a and E_b are the activation energy of electrons and lattice, respectively. The dependence of real permittivity of Si_3N_4 ceramics on temperature at three representative frequencies and the best fitting diagram according to Eqs. (6)–(8) are shown by solid lines in **Figure 9**.

It can be clearly seen that the dielectric constant gradually increases as tempera-ture increased, starting from room temperature to 800°C, and results show that the real permittivity is well distributed on the predicted curves with coefficient of determination (R^2) ranging from 0.91 to 0.93. The characteristic parameters fitted from the temperature dependence of permittivity at three representative frequencies by the Trust-Region algorithm are also listed in **Figure 8**. The activation energy of electrons E_a is distributed between 15.46 and 17.49 KJ/mol, while the activation energy of

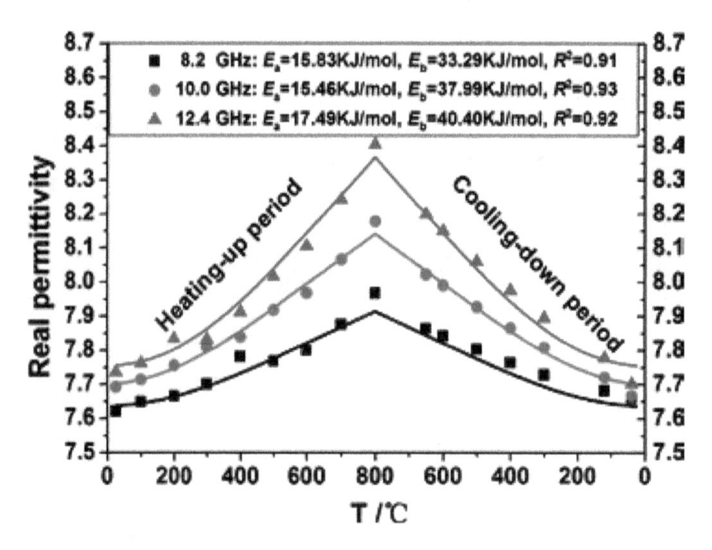

Figure 9.
Dependences of real permittivity of Si_3N_4 ceramics on temperature (reprinted with permission from Ref. [48]).

lattice E_b is distributed between 33.29 and 40.40 KJ/mol. The activation energy of electronic E_a is less than that of lattice E_b, which is mainly attributed to the binding force between the electrons and nucleus being lower than the covalent bonding force of lattice. Another important feature to be noticed is that the real permittivity of Si_3N_4 ceramics shows symmetrical features between the heating-up and cooling-down periods. The excellent thermo-stability of dielectric properties of Si_3N_4 ceramics has established the foundation for high-temperature radar absorbing materials.

3.4 High-temperature dielectric behaviors of multilayer C_f/Si_3N_4 composites

The evolution of ε' and ε'' for multilayer C_f/Si_3N_4 composites with temperature and frequency is illustrated in **Figure 10(a)** and, **(b)** respectively. It can be clearly seen that both ε' and ε'' of C_f/Si_3N_4 composites are enhanced with temperature. Similar phenomenon was also observed in other microwave-absorbing/shielding materials [24–26, 61–63]. Besides, as shown in **Figure 10(c)**, the loss tangent of C_f/Si_3N_4 composites remains higher than 0.6 over X-band and increases with increase in temperature. After comparison with low-loss β-SiC [64] as well as Si_3N_4, it is fairly clear that short carbon fibers are dominantly responsible for the frequency and temperature-dependent permittivity of C_f/Si_3N_4 composites. This is attributed to the unique microstructure (named skin-core structure), which was confirmed from transmission electron microscope analysis and selected-area electron-diffraction patterns previously [65, 66]. As illustrated in **Figure 11**, the core region consists of graphitic basal planes with random orientation, whereas graphitic basal planes are parallel to the fiber axis in the skin region [66, 67]. Therefore, electron migration behaviors both within and between graphitic basal planes inside carbon fibers would lead to electric charge accumulation at interfaces between carbon fibers and Si_3N_4 matrix (usually referred to as space charge polarization) when exposed to electromagnetic field. Furthermore, this space charge polarization is supposed to be

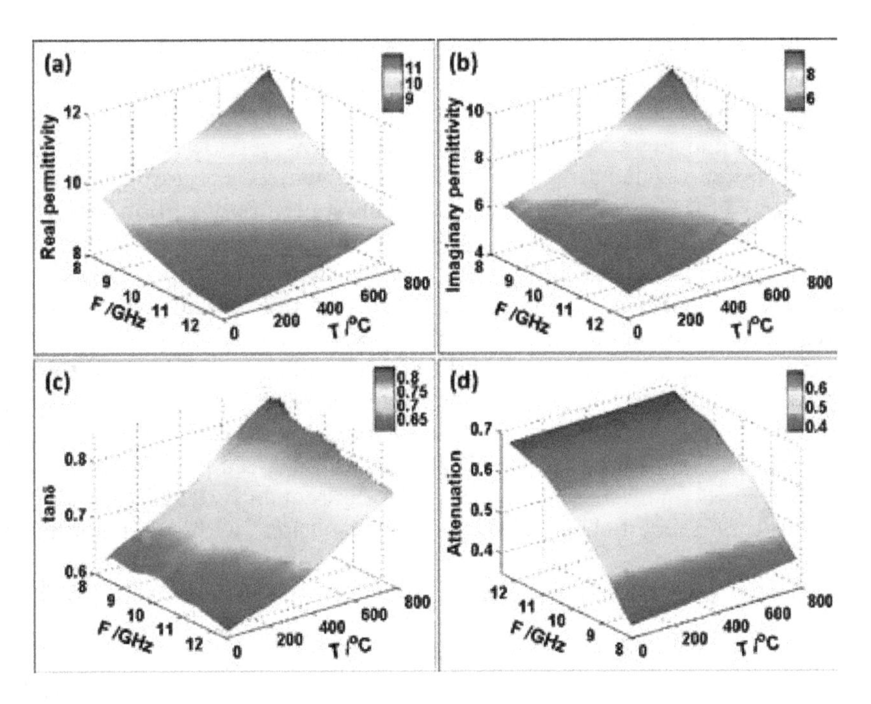

Figure 10.
Three-dimensional plots of dielectric properties of C_f/Si_3N_4 composites versus frequency and temperature: (a) real part, (b) imaginary part, (c) loss tangent, and (d) attenuation coefficient (reprinted with permission from Ref. [39]).

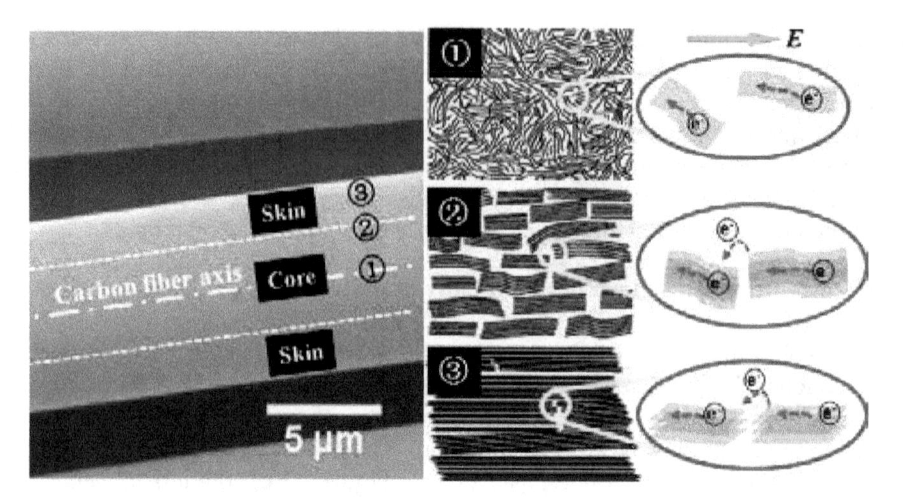

Figure 11.
Sketches of microstructure arrangement and electronic motion in carbon fiber (reprinted with permission from Ref. [39]).

enhanced with temperature rise since more and more electrons would be excited, which is in accordance with results in **Figure 10(a)** and **(b)**.

It should also be noted here that the lattice vibration also enhances with temperature rise, corresponding to the enhancement of scattering effect on migration of electrons. Fortunately, the energy of lattice vibration, which is described by phonons, is so small [68] that the scattering effect on electron migration could be neglected reasonably.

Attenuation coefficient is a quantity that characterizes how easily a material can be penetrated by incident microwave. A large attenuation coefficient means that the beam is quickly "attenuated" (weakened) as it passes through the material. The intrinsic attenuation coefficient of multilayer C_f/Si_3N_4 composites along with temperature and frequency derived from S parameters (i.e. $A = 1\text{-}S_{11}^2\text{-}S_{21}^2$) is illustrated in **Figure 10(d)**. It can be clearly observed that attenuation coefficient of C_f/Si_3N_4 composites enhances continuously with increase in frequency, and reaches a maximum value around 0.7 at 12.4 GHz, which is almost twice as much as that reported for C_f/SiO_2 composites [25]. This superior absorption performance suggests the multilayer C_f/Si_3N_4 composites to be competitive high-temperature microwave-absorbing materials. Another noteworthy phenomenon in **Figure 10(d)** is that the absorption coefficient remains steady in the investigated temperature range regardless of the loss tangent. As discussed previously, the overall absorbing performance results from reflection on the surface and attenuation inside the materials. However, the increasing permittivity means more severe impedance mismatch between air and the absorbing material, corresponding to reduction of penetrated electromagnetic energy. It could be explained by the reduced part of energy, which compensates the enhanced loss tangent, and consequently leads to a relatively steady absorption coefficient of C_f/Si_3N_4 composites with temperature rise.

3.5 Modeling of temperature-dependent dielectric responses for C_f/Si_3N_4 composites

As explicated in Section 3.2, the dielectric responses for multilayer C_f/Si_3N_4 composites have been evaluated experimentally over X-band. Additionally, the corresponding theoretical relationship of complex permittivity versus frequency at room temperature has been successfully established, which could be expressed as:

$$\varepsilon'(\omega) = \frac{c_1}{\omega^2} + c_2 \tag{9}$$

$$\varepsilon''(\omega) = \frac{\omega^2 + c_3}{c_4 \cdot \omega^3 + c_5 \cdot \omega} \tag{10}$$

where ω refers to the angular frequency (i.e., $\omega = 2\pi f$), and c_1, c_2, c_3, c_4, and c_5 are all pre-experimental parameters that are mainly associated with the surface density of short carbon fiber layers and thickness of Si_3N_4 layers. Herein, we first demonstrate this "room-temperature model" is still available at each evaluated temperature coverage up to 800°C or even higher with the help of nonlinear fitting technology. The best fitting curves of experimental data based on Eqs. (9) and (10) are depicted in **Figure 12**.

As expected, the measured results at each evaluated temperature agree quite well with the theoretical curve with coefficient of determination (R^2) above 0.98. These observed results suggest that both ε' and $\omega\varepsilon''$ of C_f/Si_3N_4 composites are still inversely proportional to the frequency square ω^2 within the temperature range of 25–800°C. However, a universal model coupled with frequency as well as temperature is urgently needed. Actually, a great deal of effort has been made to model frequency dispersive behaviors of permittivity for dielectrics [45]. It is well established that the development of all dielectric relaxation models that came after classical Debye's could be explicated as:

$$\varepsilon = \varepsilon_\infty + \frac{\varepsilon_s - \varepsilon_\infty}{1 + j\omega\tau} \tag{11}$$

where ε_s and ε_∞ are "static" and "infinite frequency" dielectric constants, respectively; j is the imaginary unit (i.e. equals to $\sqrt{-1}$); ω is the angular frequency; and τ is called the relaxation time. However, it is well known that the Debye model is originally developed for spherical polarizable molecules with a single relaxation time and without interaction between them [69]. Nevertheless, the short carbon fibers act as dipoles when exposed to altering electromagnetic fields oscillating in the microwave band, and electromagnetic field around each dipole is supposed to be coupled with that of neighboring dipoles. Furthermore, the relaxation time of chopped carbon fibers is supposed to be distributed over an interval rather than taking a constant value due to the variation of the fibers' length and the

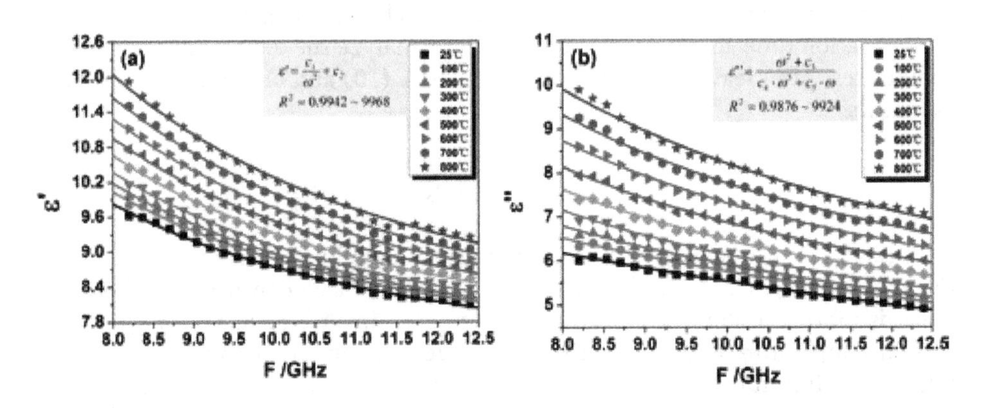

Figure 12.
(a) The real and (b) imaginary parts of permittivity for C_f/Si_3N_4 composites in X-band at the evaluated temperature (reprinted with permission from Ref. [39]).

heterogeneity of fibers' internal structure. This distribution property brings the Debye model (Eq. (11)) into more empirical and given by [25, 26, 70]:

$$\varepsilon = \varepsilon_\infty + \frac{\varepsilon_s - \varepsilon_\infty}{1 + (j\omega\tau)^{1-\alpha}} \tag{12}$$

where α is a parameter which determines the width of the distribution of relaxation time. All parameters in Eq. (12) should be a function of temperature. Substitution of Euler's formula $e^{jx} = \cos(x) + j\sin(x)$ in Eq. (12) yields

$$\varepsilon' = \varepsilon_\infty + \frac{(\varepsilon_s - \varepsilon_\infty)\cdot\left[1 + (\omega\tau)^{1-\alpha}\cdot\sin\frac{\alpha\pi}{2}\right]}{1 + 2(\omega\tau)^{1-\alpha}\cdot\sin\frac{\alpha\pi}{2} + (\omega\tau)^{2(1-\alpha)}} \tag{13}$$

$$\varepsilon'' = \frac{(\varepsilon_s - \varepsilon_\infty)\cdot(\omega\tau)^{1-\alpha}\cdot\cos\frac{\alpha\pi}{2}}{1 + 2(\omega\tau)^{1-\alpha}\cdot\sin\frac{\alpha\pi}{2} + (\omega\tau)^{2(1-\alpha)}} \tag{14}$$

We finally obtain the relationship between ε' and ε'', which could be expressed as

$$\left(\varepsilon' - \frac{\varepsilon_s + \varepsilon_\infty}{2}\right)^2 + \left(\varepsilon'' + \frac{\varepsilon_s + \varepsilon_\infty}{2}\tan\frac{\alpha\pi}{2}\right)^2 = \frac{(\varepsilon_s - \varepsilon_\infty)^2}{4\cos^2\frac{\alpha\pi}{2}} \tag{15}$$

Eq. (15) suggests that the locus of the permittivity in the (ε',ε'') complex plane should still be a circular arc with different radius. To put this into perspective, the experimental data are re-plotted in **Figure 13**, in which the best fitted circles based on Eq. (15) are marked as solid curves. As seen, high agreement between the proposed model with experimental data is observed. Besides, the center tends to shift toward greater values of the ε' axis with increase in temperature, suggesting enhanced dielectric strength with temperature according to the Eq. (15). This may be attributed to the Enhanced electron concentration which participated in the polarization process occurs with temperature.

In addition, it is important to highlight some additional details. Firstly, the y-coordinates of fitted circular centers remain so small (~10^{-3}) that Eq. (12) would be reduced to the classical Debye expression. From this point of view, the electronic polarization of short carbon fibers still follows the classic Debye relaxation process. Besides, the effects of temperature and electromagnetic interaction between neighboring short carbon fibers on relaxation time could be neglected reasonably.

Relaxation time also is one of key factors to analyze the dielectric behaviors for composites. After leaving out the effect of α, Eqs. (10) and (11) could be rewritten as follows:

$$\varepsilon' = \frac{1}{\tau}\cdot\frac{\varepsilon''}{\omega} + \varepsilon_\infty \tag{16}$$

It can be clearly seen from Eq. (16) that the real and imaginary parts of permittivity in (ε', ε''/ω) coordinate should be linear, which is also confirmed in **Figure 13**. What is more, the slope of each line is exactly the inverse of relaxation time τ at a certain temperature evaluated. In this case, we can come to a conclusion that the relaxation time for multilayer C_f/Si_3N_4 composites is weakly dependent on the temperature since the differences between fitted lines are quite modest. For clarity, a detailed plot of τ as a function of temperature is also illustrated as an inset in **Figure 14**. An increase from 216.1 to 250.2 ps has been derived when samples are

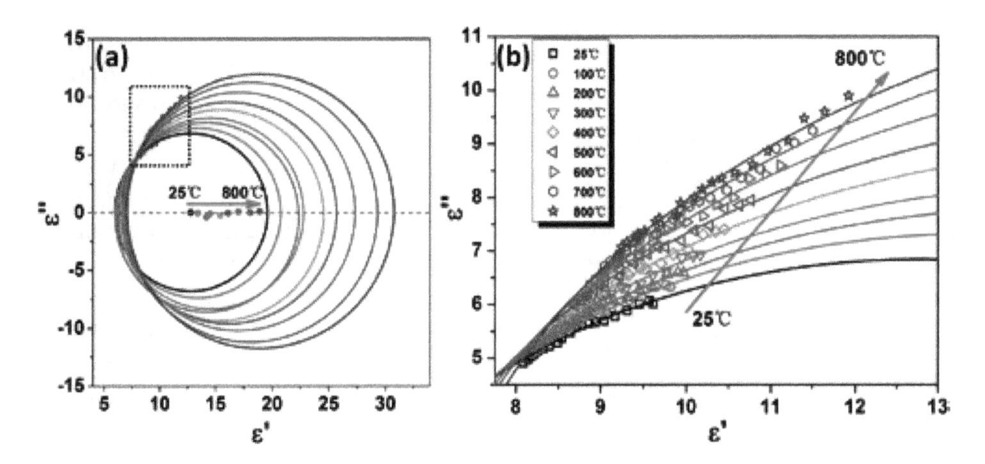

Figure 13.
(a) Argand diagram of C_f/Si_3N_4 composites at different temperature, (b) detailed view for the region marked by black box in (a) (reprinted with permission from Ref. [39]).

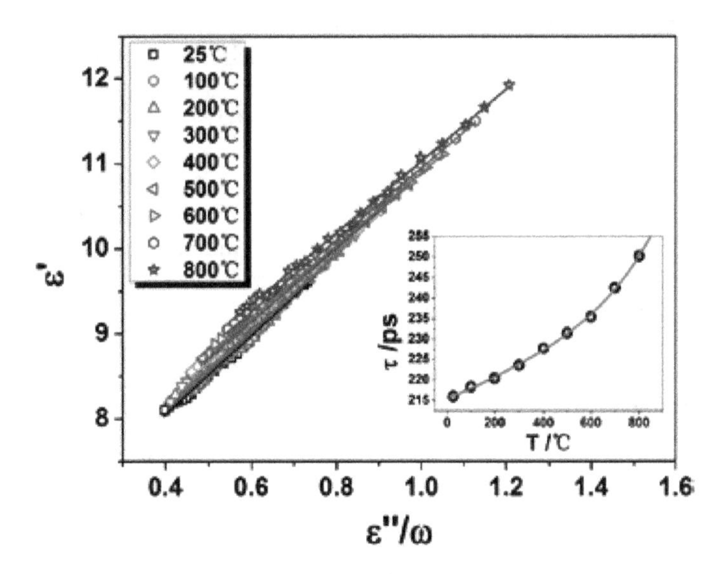

Figure 14.
Plot of ε' as a function of ε"/ω with an inset of temperature-dependent relaxation time (reprinted with permission from Ref. [39]).

heated from room temperature to 800°C. This gradually increasing trend may be ascribed to the enhancement of scattering effect between electrons for increase both the number and energy of electrons. Furthermore, the relaxation time τ (216.1–250.2 ps) is almost twice as much as a single cycle for time-harmonic electromagnetic wave in X-band ($t_0 = 1/f =$ 80.65–121.95 ps). As a result, electronic migration could not keep up with the pace of external alternating electronic field, which leads to continuous decrease of permittivity with frequency increase.

4. Conclusion

In this chapter, microwave dielectric properties of multilayer C_f/Si_3N_4 composites fabricated via gelcasting and pressureless sintering were intensively studied in X-band. Firstly, a strong frequency dependence of the real and imaginary parts of permittivity at room temperature was observed at the X-band. Particularly, an equivalent RC circuit model concerning the frequency-dependent permittivity of multilayer C_f/Si_3N_4 composites has been established. The predicted results reveal that both ε' and $(\omega \cdot \varepsilon'')$ are inversely proportional to the frequency square, and agree quite well with the measured data. Secondly, high-temperature dielectric behaviors of Si_3N_4 ceramic show that both permittivity and loss tangent exhibit excellent thermo-stability with temperature coefficient lower than 10^{-3}°C^{-1} within the temperature range of 25–800°C. A revised dielectric relaxation model with Lorentz correction for as-prepared Si_3N_4 ceramics was established and validated by experi-mental data. The activation energy of electrons (15.46–17.49 KJ/mol) was demon-strated to be slightly smaller than that of lattice (33.29–40.40 KJ/mol). Finally, the microwave attenuation coefficient of multilayer C_f/Si_3N_4 composites was inclined to be independent of temperature, and a maximum value of 0.7 could be achieved. The obvious positive temperature coefficient characteristic for permittivity is mainly attributed to the enhancement of electric polarization and relaxation of electron migration for graphitic basal planes of short carbon fibers. In addition to the fact that the room-temperature model concerning frequency-dependent permittivity is still available when extended into the full range of temperature coverage up to 800°C, an empirical equation with respect to temperature-dependent permittivity of multilayer C_f/Si_3N_4 composites has been established. It is concluded that the mea-sured complex permittivity of multilayer C_f/Si_3N_4 composites is well distributed on circular arcs with centers actually kept around the real (ε') axis in (ε', ε'') complex plane. Furthermore, the relaxation time as a function of temperature also has been derived. Results suggest that the relaxation time for multilayer C_f/Si_3N_4 composites increases from 216.1 to 250.2 ps when heated from room temperature to 800°C, and is almost twice as much as a single cycle for electromagnetic wave in X-band which leads to continuous decrease in permittivity with frequency increase. These findings point to important guidelines for analyzing high-temperature dielectric behaviors and revealing fundamental mechanisms for carbon fiber functionalized composites including but not limited to C_f/Si_3N_4 composites.

Acknowledgements

The authors would like to acknowledge the generous funding from the National Key Research and Development Program of China (Grant No. 2017YFA0204600), the State Key Development Program for Basic Research of China (Grant No. 2011CB605804), and the National Natural Science Foundation of China (Grant No. 51802352).

Conflict of interest

The authors declared that they have no conflicts of interest to this work.

Author details

Heng Luo[1,2*], Lianwen Deng[1] and Peng Xiao[2]

1 School of Physics and Electronics, Central South University, Changsha, China

2 State Key Laboratory of Powder Metallurgy, Central South University, Changsha, China

*Address all correspondence to: luohengcsu@csu.edu.cn

References

[1] Mohan RR, Varma SJ, Sankaran J. Impressive electromagnetic shielding effects exhibited by highly ordered, micrometer thick polyaniline films. Applied Physics Letters. 2016;**108**(15):154101. DOI: 10.1063/1.4945791

[2] Zhao Y-T et al. Transparent electromagnetic shielding enclosure with CVD graphene. Applied Physics Letters. 2016;**109**(10):103507. DOI: 10.1063/1.4962474

[3] Zeng Z et al. Thin and flexible multi-walled carbon nanotube/waterborne polyurethane composites with high-performance electromagnetic interference shielding. Carbon. 2016;**96**:768-777. DOI: 10.1016/j.carbon.2015.10.004

[4] Wen B et al. Reduced graphene oxides: Light-weight and high-efficiency electromagnetic interference shielding at elevated temperatures (Adv. Mater. 21/2014). Advanced Materials. 2014;**26**(21):3357-3357. DOI: 10.1002/adma.201470138

[5] Shahzad F et al. Electromagnetic interference shielding with 2D transition metal carbides (MXenes). Science. 2016;**353**(6304):1137-1140. DOI: 10.1126/science.aag2421

[6] Zhang Y et al. Additive manufacturing of carbon nanotube-photopolymer composite radar absorbing materials. Polymer Composites. 2016. DOI: 10.1002/pc.24117

[7] Hayashida K, Matsuoka Y. Electromagnetic interference shielding properties of polymer-grafted carbon nanotube composites with high electrical resistance. Carbon. 2015;**85**:363-371. DOI: 10.1016/j.carbon.2015.01.006

[8] Li N et al. Electromagnetic interference (EMI) shielding of single-walled carbon nanotube epoxy composites. Nano Letters. 2006;**6**(6):1141-1145. DOI: 10.1021/nl0602589

[9] Micheli D et al. X-Band microwave characterization of carbon-based nanocomposite material, absorption capability comparison and RAS design simulation. Composites Science and Technology. 2010;**70**(2):400-409. DOI: 10.1016/j.compscitech.2009.11.015

[10] Al-Saleh MH, Sundararaj U. A review of vapor grown carbon nanofiber/polymer conductive composites. Carbon. 2009;**47**(1):2-22. DOI: 10.1016/j.carbon.2008.09.039

[11] Han M et al. Hierarchical graphene/SiC nanowire networks in polymer-derived ceramics with enhanced electromagnetic wave absorbing capability. Journal of the European Ceramic Society. 2016;**36**(11):2695-2703. DOI: 10.1016/j.jeurceramsoc.2016.04.003

[12] Huang X et al. A three-dimensional graphene/Fe3O4/carbon microtube of sandwich-type architecture with improved wave absorbing performance. Scripta Materialia. 2016;**120**:107-111. DOI: 10.1016/j.scriptamat.2016.04.025

[13] Olmedo L, Hourquebie P, Jousse F. Microwave absorbing materials based on conducting polymers. Advanced Materials. 1993;**5**(5):373-377. DOI: 10.1002/adma.19930050509

[14] Huang Y et al. The microwave absorption properties of carbon-encapsulated nickel nanoparticles/silicone resin flexible absorbing material. Journal of Alloys and Compounds. 2016;**682**:138-143. DOI: 10.1016/j.jallcom.2016.04.289

[15] Chen Y et al. Enhanced electromagnetic interference shielding efficiency of polystyrene/graphene composites with magnetic Fe_3O_4 nanoparticles. Carbon. 2015;**82**:67-76. DOI: 10.1016/j.carbon.2014.10.031

[16] Al-Saleh MH, Saadeh WH, Sundararaj U. EMI shielding effectiveness of carbon based nanostructured polymeric materials: A comparative study. Carbon. 2013;**60**:146-156. DOI: 10.1016/j.carbon.2013.04.008

[17] Qin F, Brosseau C. A review and analysis of microwave absorption in polymer composites filled with carbonaceous particles. Journal of Applied Physics. 2012;**111**(6):061301. DOI: 10.1063/1.3688435

[18] Chung DDL. Carbon materials for structural self-sensing, electromagnetic shielding and thermal interfacing. Carbon. 2012;**50**(9):3342-3353. DOI: 10.1016/j.carbon.2012.01.031

[19] Liu L, Das A, Megaridis CM. Terahertz shielding of carbon nanomaterials and their composites – A review and applications. Carbon. 2014;**69**:1-16. DOI: 10.1016/j.carbon.2013.12.021

[20] Qin F, Peng H-X. Ferromagnetic microwires enabled multifunctional composite materials. Progress in Materials Science. 2013;**58**(2):183-259. DOI: 10.1016/j.pmatsci.2012.06.001

[21] Yin X et al. Electromagnetic properties of Si–C–N based ceramics and composites. International Materials Reviews. 2014;**59**(6):326-355. DOI: 10.1179/1743280414Y.0000000037

[22] Duan W et al. A review of absorption properties in silicon-based polymer derived ceramics. Journal of the European Ceramic Society. 2016;**36**(15):3681-3689. DOI: 10.1016/j. jeurceramsoc.2016.02.002

[23] Yin X et al. Dielectric, electromagnetic absorption and interference shielding properties of porous yttria-stabilized zirconia/silicon carbide composites. Ceramics International. 2012;**38**(3):2421-2427. DOI: 10.1016/j.ceramint.2011.11.008

[24] Wen B et al. Reduced graphene oxides: Light-weight and high-efficiency electromagnetic interference shielding at elevated temperatures. Advanced Materials. 2014;**26**(21):3484-3489. DOI: 10.1002/adma.201400108

[25] Cao M-S et al. The effects of temperature and frequency on the dielectric properties, electromagnetic interference shielding and microwave-absorption of short carbon fiber/silica composites. Carbon. 2010;**48**(3):788-796. DOI: 10.1016/j.carbon.2009.10.028

[26] Wen B et al. Temperature dependent microwave attenuation behavior for carbon-nanotube/silica composites. Carbon. 2013;**65**:124-139. DOI: 10.1016/j.carbon.2013.07.110

[27] Kong L et al. High-temperature electromagnetic wave absorption properties of $ZnO/ZrSiO_4$ composite ceramics. Journal of the American Ceramic Society. 2013;**96**(7):2211-2217. DOI: 10.1111/jace.12321

[28] Iveković A et al. Current status and prospects of SiCf/SiC for fusion structural applications. Journal of the European Ceramic Society. 2013;**33**(10):1577-1589. DOI: 10.1016/j.jeurceramsoc.2013.02.013

[29] Liu H, Tian H, Cheng H. Dielectric properties of SiC fiber-reinforced SiC matrix composites in the temperature range from 25 to 700°C at frequencies between 8.2 and 18GHz. Journal of Nuclear Materials. 2013;**432**(1-3):57-60. DOI: 10.1016/j.jnucmat.2012.08.026

[30] Tian H, Liu H-T, Cheng H-F. A high-temperature radar absorbing

structure: Design, fabrication, and characterization. Composites Science and Technology. 2014;**90**:202-208. DOI: 10.1016/j.compscitech.2013.11.013

[31] Azarhoushang B, Soltani B, Zahedi A. Laser-assisted grinding of silicon nitride by picosecond laser. The International Journal of Advanced Manufacturing Technology. 2017;**93**(5):2517-2529. DOI: 10.1007/s00170-017-0440-9

[32] Xu Y et al. Shear-thinning behavior of the $CaO\text{-}SiO_2\text{-}CaF_2\text{-}Si_3N_4$ system mold flux and its practical application. International Journal of Minerals, Metallurgy, and Materials. 2017;**24**(10):1096-1103. DOI: 10.1007/s12613-017-1500-8

[33] Zuo K-H, Zeng Y-P, Jiang D-L. The mechanical and dielectric properties of Si_3N_4-based sandwich ceramics. Materials & Design. 2012;**35**:770-773. DOI: 10.1016/j.matdes.2011.09.019

[34] Yang J, Yu J, Huang Y. Recent developments in gelcasting of ceramics. Journal of the European Ceramic Society. 2011;**31**(14):2569-2591. DOI: 10.1016/j.jeurceramsoc.2010.12.035

[35] Luo H et al. Dielectric properties of Cf–Si3N4 sandwich composites prepared by gelcasting. Ceramics International. 2014;**40**(6):8253-8259. DOI: 10.1016/j.ceramint.2014.01.023

[36] Bocanegra-Bernal MH, Matovic B. Mechanical properties of silicon nitride-based ceramics and its use in structural applications at high temperatures. Materials Science and Engineering A. 2010;**527**(6):1314-1338. DOI: 10.1016/j.msea.2009.09.064

[37] Dambatta YS et al. Ultrasonic assisted grinding of advanced materials for biomedical and aerospace applications—a review. The International Journal of Advanced Manufacturing Technology. 2017;**92**(9):3825-3858. DOI: 10.1007/s00170-017-0316-z

[38] Tatli Z, Thompson DP. Low temperature densification of silicon nitride materials. Journal of the European Ceramic Society. 2007;**27**(2-3):791-795. DOI: 10.1016/j.jeurceramsoc.2006.04.010

[39] Luo H et al. Modeling for high-temperature dielectric behavior of multilayer Cf/Si3N4 composites in X-band. Journal of the European Ceramic Society. 2017;**37**(5):1961-1968. DOI: 10.1016/j.jeurceramsoc.2016.12.028

[40] Peng C-H, Chen PS, Chang C-C. High-temperature microwave bilayer absorber based on lithium aluminum silicate/lithium aluminum silicate-SiC composite. Ceramics International. 2014;**40**(1 Part A):47-55. DOI: 10.1016/j. ceramint.2013.05.101

[41] Yuan J et al. High dielectric loss and microwave absorption behavior of multiferroic $BiFeO_3$ ceramic. Ceramics International. 2013;**39**(6):7241-7246. DOI: 10.1016/j.ceramint.2013.01.068

[42] Qing YC et al. Microwave absorbing ceramic coatings with multi-walled carbon nanotubes and ceramic powder by polymer pyrolysis route. Composites Science and Technology. 2013;**89**:10-14. DOI: 10.1016/j.compscitech.2013.09.007

[43] Liu H, Cheng H, Tian H. Design, preparation and microwave absorbing properties of resin matrix composites reinforced by SiC fibers with different electrical properties. Materials Science and Engineering B. 2014;**179**:17-24. DOI: 10.1016/j.mseb.2013.09.019

[44] Chen M et al. Gradient multilayer structural design of CNTs/SiO_2 composites for improving microwave absorbing properties. Materials & Design. 2011;**32**(5):3013-3016. DOI: 10.1016/j.matdes.2010.12.043

[45] Luo H, Xiao P, Hong W. Dielectric behavior of laminate-structure Cf/Si_3N_4 composites in X-band. Applied Physics Letters. 2014;**105**(17):172903. DOI: 10.1063/1.4900932

[46] Cui TJ, Smith D, Liu R. Metamaterials: Theory, Design, and Applications. 233 Spring Street, New York, USA Springer Publishing Company, Incorporated; 2009. p. 368

[47] Liu Y, Zhang X. Metamaterials: A new frontier of science and technology. Chemical Society Reviews. 2011;**40**(5):2494-2507. DOI: 10.1039/C0CS00184H

[48] Shao S et al. Effect of temperature on dielectric response in X-band of silicon nitride ceramics prepared by gelcasting. AIP Advances. 2018;**8**(7):075127. DOI: 10.1063/1.5033965

[49] Amirat Y, Shelukhin V. Homogenization of time harmonic Maxwell equations and the frequency dispersion effect. Journal de Mathématiques Pures et Appliquées. 2011;**95**(4):420-443. DOI: 10.1016/j.matpur.2010.10.007

[50] Haijun Z et al. Complex permittivity, permeability, and microwave absorption of Zn- and Ti-substituted barium ferrite by citrate sol–gel process. Materials Science and Engineering B. 2002;**96**(3):289-295. DOI: 10.1016/S0921-5107(02)00381-1

[51] Joshi A, Kumar S, Verma NK. Study of dispersion, absorption and permittivity of an synthetic insulation paper—with change in frequency and thermal aging. NDT & E International. 2006;**39**(1):19-21. DOI: 10.1016/j.ndteint.2005.05.004

[52] Kamenetsky FM. Frequency dispersion of rock properties in equations of electromagnetics. Journal of Applied Geophysics. 2011;**74**(4):185-193. DOI: 10.1016/j.jappgeo.2011.04.004

[53] Razzitte AC, Fano WG, Jacobo SE. Electrical permittivity of Ni and NiZn ferrite–polymer composites. Physica B: Condensed Matter. 2004;**354**(1-4):228-231. DOI: 10.1016/j.physb.2004.09.054

[54] Rica RA, Jiménez ML, Delgado AV. Electric permittivity of concentrated suspensions of elongated goethite particles. Journal of Colloid and Interface Science. 2010;**343**(2):564-573. DOI: 10.1016/j.jcis.2009.11.063

[55] Tan YQ et al. Giant dielectric-permittivity property and relevant mechanism of $Bi_{2/3}Cu_3Ti_4O_{12}$ ceramics. Materials Chemistry and Physics. 2010;**124**(2-3):1100-1104. DOI: 10.1016/j.matchemphys.2010.08.041

[56] Moysés Araújo C et al. Electrical resistivity, MNM transition and band-gap narrowing of cubic GaN: Si. Microelectronics Journal. 2002;**33**(4):365-369. DOI: 10.1016/S0026-2692(01)00133-1

[57] Fernandez JRL et al. Electrical resistivity and band-gap shift of Si-doped GaN and metal-nonmetal transition in cubic GaN, InN and AlN systems. Journal of Crystal Growth. 2001;**231**(3):420-427. DOI: 10.1016/S0022-0248(01)01473-7

[58] Schuurmans FJP, Vries Pd, Lagendijk A. Local-field effects on spontaneous emission of impurity atoms in homogeneous dielectrics. Physics Letters A. 2000;**264**(6):472-477. DOI: 10.1016/S0375-9601(99)00855-5

[59] Hao Wang YC, Kaneta Y, Iwata S. First-principles investigation of the structural, electronic and optical properties of olivine-Si3N4 and olivine-Ge3N4. Journal of Physics: Condensed Matter. 2006;**18**(47):10663-10676. DOI: 10.1088/0953-8984/18/47/012

[60] Xu M et al. Theoretical prediction of electronic structures and optical properties of Y-doped γ-Si3N4. Physica B: Condensed Matter. 2008;**403** (13-16):2515-2520. DOI: 10.1016/j.physb.2008.01.042

[61] Yang H-J et al. Silicon carbide powders: Temperature-dependent dielectric properties and enhanced microwave absorption at gigahertz range. Solid State Communications. 2013;**163**:1-6. DOI: 10.1016/j.ssc.2013.03.004

[62] Wang H et al. High temperature electromagnetic and microwave absorbing properties of polyimide/multi-walled carbon nanotubes nancomposites. Chemical Physics Letters. 2015;**633**:223-228. DOI: 10.1016/j.cplett.2015.05.048

[63] Dou Y-K et al. The enhanced polarization relaxation and excellent high-temperature dielectric properties of N-doped SiC. Applied Physics Letters. 2014;**104**(5):052102. DOI: 10.1063/1.4864062

[64] Liu Y et al. Transmission electron microscopy study of the microstructure of unidirectional C/C composites fabricated by catalytic chemical vapor infiltration. Carbon. 2013;**51**:381-389. DOI: 10.1016/j.carbon.2012.08.070

[65] Koziel S, Bekasiewicz A. Multi-objective optimization of expensive electromagnetic simulation models. Applied Soft Computing. 2016;**47**: 332-342. DOI: 10.1016/j.asoc.2016.05.033

[66] Zhou G et al. Microstructure difference between core and skin of T700 carbon fibers in heat-treated carbon/carbon composites. Carbon. 2011;**49**(9):2883-2892. DOI: 10.1016/j. carbon.2011.02.025

[67] Qin X et al. A comparison of the effect of graphitization on microstructures and properties of polyacrylonitrile and mesophase pitch-based carbon fibers. Carbon. 2012;**50**(12):4459-4469. DOI: 10.1016/j.carbon.2012.05.024

[68] Cai Y et al. First-principles study of vibrational and dielectric properties of β-Si3N4. Physical Review B. 2006;**74**(17):174301. DOI: 10.1103/PhysRevB.74.174301

[69] Kirkwood JG, Fuoss RM. Anomalous dispersion and dielectric loss in polar polymers. The Journal of Chemical Physics. 1941;**9**(4):329-340. DOI: 10.1063/1.1750905

[70] Cole KS, Cole RH. Dispersion and absorption in dielectrics I. Alternating current characteristics. The Journal of Chemical Physics. 1941;**9**(4):341-351. DOI: 10.1063/1.1750906

8

High-Frequency Permeability of Fe-Based Nanostructured Materials

Mangui Han

Abstract

Frequency dependence of permeability of magnetic materials is crucial for their various electromagnetic applications. In this chapter, the approaches to tailor the high-frequency permeability of Fe-based nanostructured materials are discussed: shape controlling, particle size distribution, coating treatments, phase transitions and external excitations. Special attention is paid on the electromagnetic wave absorbing using these Fe-based magnetic materials (Fe-Si-Al alloys, Fe-Cu-Nb-Si-B nanocrystalline alloys and Fe nanowires). Micromagnetic simulations also have been presented to discuss the intrinsic high-frequency permeability. Negative imaginary part of permeability ($\mu'' < 0$) is proved possible under the spin transfer torque effect.

Keywords: complex permeability, electromagnetic wave absorbing, negative imaginary part of permeability, micromagnetic simulation, natural resonance

1. Introduction

The high-frequency permeability values are crucial for many applications of electromagnetic (EM) materials. At low frequency region, a typical application is magnetic cores for transformers or inductors; these devices demand the larger real parts of permeability ($\mu' > 0$) and lower imaginary parts of permeability ($\mu'' > 0$ and $\mu'' \rightarrow 0$) for magnetic materials. With the working frequency increases, many electronic devices work above gigahertz (GHz). The unwanted electromagnetic wave pollution is problematic, which can be overcome by using the electromagnetic wave attenuation composites with the absorbing frequency band falling into the GHz zone. Therefore, the electromagnetic wave absorbing materials require their working frequencies to be above 1 GHz and larger μ'' values for efficient absorption via the typical magnetic loss mechanism (natural resonance). Due to the small magnetocrystalline anisotropy constants of spinel ferrites, the magnetic loss peaks of spinel ferrites are below 1 GHz. Although the natural resonance frequencies of hexagonal ferrites can be a few GHz, the shortcoming of both ferrites (spinel and hexagonal) is the poor temperature stability; in other words, the permeability decreases faster with increasing environmental temperature, making these ferrites not suitable candidates for EM attenuation applications. The physical law governing the relationship of initial permeability, working frequency and magnetization is the well-known "Snoek's law" [1]. This law tells us that if we want to increase the working frequency of a specific ferrite (*Ms* is constant), we have to sacrifice their permeability and vice versa. People often tailor the $\mu \sim f$ spectrum of a ferrite by cation substitution or microstructures by changing the sintering conditions. We think this technique is time-consuming and inefficient for mass production. In this chapter, we propose Fe-based nanostructured materials (Fe-Cu-Nb-Si-B alloys, also called "FINEMET") as ideal EM attenuation materials, which offer us greater freedom to tailor their high-frequency permeability spectra and possess good temperature stabil-ity. Firstly, we will discuss how to increase the permeability by controlling the shape of particles, and then we move to discuss the impacts of size distribution, phase transitions and coating on the permeability. Next, we will discuss the cause of fre-quently observed broad permeability spectra using the micromagnetic simulations. Finally, we will prove how we can obtain negative imaginary parts of permeability via spin transfer torque effect. Our discussions mainly focus on the electromagnetic wave absorbing (attenuation) applications, which require large imaginary parts of perme-ability and working frequency above 1 GHz.

2. Shape controlling

Soft magnetic materials can give large initial permeability, but their working frequency is far below 1 GHz due to their small magnetocrystalline anisotropic constants. As per Snoek's law, if the working frequency is shifted to GHz range, the permeability has to be compromised. To increase the working frequency, we have to look at other anisotropy fields, such as shape anisotropy and stress (or external field)-induced anisotropy. Shape anisotropy is most commonly employed. Materials in forms of thin films, microwires, nanowires and flakes can possess a strong shape anisotropy. Here, we present how to enhance the permeability of a soft ferromag-netic material (Fe-Si-Al alloys) above 1 GHz via shape controlling. The Fe-Si-Al alloy ingot (composition: $Fe_{84.94}Si_{9.68}Al_{5.38}$) was prepared using the hydrogen reduction method in a furnace with the starting materials (Si, Al and Fe with high purity). Subsequently, the Fe-Si-Al ingot was first pulverized into particles with irregular shapes; later these particles were further milled into flakes under different milling times (10, 20 and 30 hours, respectively). Our traditional ball milling pro-cess description can be found in our published paper [2]. The scanning electron microscopy images of Fe-Si-Al particles are shown in **Figure 1**. Clearly, the preliminarily pulverized particles have irregular shapes. The traditional ball milling process can transform them into flakes after 10, 20 or 30 hours of milling. The typical milling result is illustrated in **Figure 1b** showing flaky particles milled for 30 hours. The high-frequency complex permeability values were measured within 0.5–10 GHz using a network analyzer (Agilent 8720ET). The measured samples were prepared by mixing the Fe-Si-Al particles and wax homogeneously (weight ratio: alloy particles/wax = 4:1). The measured sample has an annular shape with inner and outer diameters of 3 and 7 mm, respectively.

The dependence of high-frequency complex permeability on particle shape is shown in **Figure 2**. Obviously, flaky particles have much larger values in both real and imaginary parts of permeability compared to the irregularly shaped particles. This finding is named "enhanced permeability." The enhanced permeability is more apparent at the lower frequency range. Besides, with increasing the milling hours, the enhanced permeability is stronger. For example, the μ' value of flakes after being milled for 30 hours is found to be 4.4 at 0.5 GHz. However, it is only 1.3 for the irregular shaped particle. Within 0.5–7 GHz, the μ' values of flakes after being milled for 30 hours are evidently larger than those of particles with irregular shapes. With regard to the imaginary part values (μ''), the flakes of Fe-Si-Al exhibit larger

Figure 1.
SEM pictures: (a) preliminarily pulverized particles and (b) particles with flaky shapes prepared by 30 hours of ball milling.

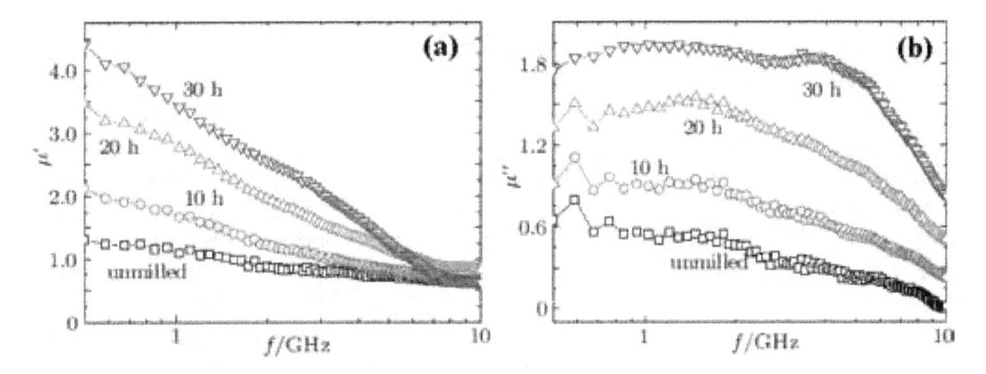

Figure 2.
Dependence of high-frequency permeability on particle shapes.

values within the whole frequency range studied than those of particles with irregular shapes, as shown in **Figure 2b**. Our explanations for the enhanced permeability are as follows: Fe-Si-Al alloy is a metallic alloy with a conducting feature. The eddy current effect is stronger for irregular shape but can be greatly suppressed by milling the irregular particles into flakes that have lower thickness. Besides, Snoek's law governing the relationship between the permeability and shape factor is given as [3]

$$(\mu_s - 1)f_r^2 = \left(\frac{\gamma}{2\pi}\right)^2 \times 4\pi M_s \times (H_k + 4\pi M_s D_z) \tag{1}$$

where D_z is a factor related to the shape, also known as the demagnetization factor for the normal direction of the particle plane. D_z is about $4\pi/3$ for a sphere and is about 4π for a flake. Accordingly, the enhanced permeability value can be observed by increasing D_z from $(4\pi/3)$ to (4π) by controlling particle shapes. That is also the reason why people often fabricate ferromagnetic thin films to obtain enhanced high-frequency permeability. Thin films can be viewed as an extreme case of "all flakes well aligned," which therefore are found to have much larger permeability values resulting from the "flakes" and in-plane induced uniaxial anisotropy. Furthermore, the well-known Snoek's law for bulk materials describes

$$(\mu_s - 1)f_r = \frac{2}{3}\gamma' 4\pi M_s \tag{2}$$

No shape-related demagnetization factor is found in this equation.

3. Particle size distribution

As discussed before, the high-frequency permeability can be greatly enhanced by controlling the shapes of particles and alignment of flakes (i.e., a thin film can be treated as a large quantity of well-aligned flakes in a plane). In this section, we will focus on the effect of size distribution of flakes on the dynamic permeability. We choose the composites containing Fe-Cu-Nb-Si-B alloy particles (a.k.a. "FINEMET" alloy) for this purpose. The reason to choose this Fe-based nanostructured material for its special nanoscale structure is the nano-grains are well dispersed in an amorphous matrix (it can be thought as a core-shell structure). The amorphous matrix has the larger resistivity which can reduce the detrimental effects of eddy current on high-frequency permeability. Besides, both amorphous phase and nano-grains are ferromagnetic, which avoids the effect non-magnetic coatings have on ferromagnetic particles (it will be discussed in the next section), and as a result both phases contribute to the permeability.

The phase transition temperatures of a "FINEMET" alloy can be identified by taking a differential scanning calorimeter (DSC) curve. As shown in **Figure 3**, two exothermal peaks are found in the DSC curve: one at 530°C (called T_{x1}) and the other at 672°C (called T_{x2}). T_{x1} is known as the primary crystallization temperature, and T_{x2} is the secondary crystallization temperature. In order to form nanoscale grains, as-prepared ribbons may be annealed above T_{x1} (see zone I) or T_{x2} (see zone II). It was previously reported that nanoscale Fe(Si) grains can be formed when the alloys are annealed above T_{x1}. Annealing at different temperatures and with different time will give rise to different volume fractions of ferromagnetic phases (amorphous phase and nanocrystallized phase), size of nano-grains and different kinds of magnetic phases ($T > T_{x2}$, zone II), such as Fe-B phases. More details on the phase transitions can be found in literature [4]. We believe that controlling the annealing (in other words, phase transition) will give us abundant ways to tailor the high frequency of Fe-based nanostructured materials to meet the specific requirement of an EM wave absorbing application. Here, we propose that without preparing materials with new composi-tions, just annealing the nanocrystalline soft magnetic alloys (such as FINEMET, NANOPERM or HITPERM) to tailor the high-frequency permeability is an economic approach from the perspective of mass production.

Preparation of FINEMENT amorphous ribbon, annealing treatments and milling processes can be found in our published paper [5]. Nanostructures of sample heat

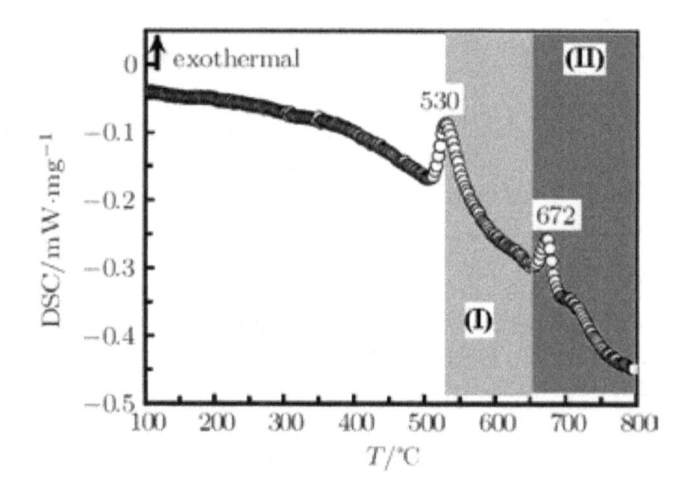

Figure 3.
DSC curve of FINEMET alloy.

treated at 540°C are characterized by a transmission electron microscopy (TEM) and shown in **Figure 4**. **Figure 4a** is the bright-field TEM image, whereas **Figure 4b** shows the typical features of Fe-based nanostructured alloys, where the amorphous matrix is surrounding the nanocrystalline grains. The average grain size is about 14.38 nm. XRD results show the nanocrystalline grains to be α-Fe(Si). It should be noted that the effect of annealing and ball milling on phase transitions can be investigated using Mössbauer technique and XRD measurements. Especially, Mössbauer technique can detect fine distinctions in the transformations of crystal structures due to the high-energy ball milling [5]. Using the milling procedure described in the previous section, we can effectively transform the annealed ribbons into flakes, which are shown in **Figure 5**. These flakes are divided into two categories by a shaking sieve: the large flakes (**Figure 5a**) and the small flakes (**Figure 5b**). The typical thickness of flakes of both categories is also shown in **Figure 5c** and **d**. For large flakes, the width of flakes is about 23–111 μm, average width is estimated to be about 81.1 μm and average thickness was found to be 4.5 μm. For small flakes, the width of flakes is about 3–21 μm, average width is estimated about 9.4 μm and thickness is averaged to be 1.3 μm.

Frequency dependence of permeability of these two kinds of flakes is shown in **Figure 6**. At the beginning of measurement frequency (i.e., at 0.5 GHz), the real part of permeability, which can be named "initial permeability" (μ_s'), is found to be about 4.6 for large flakes and about 3.9 for small flakes. Initial imaginary part of permeability (μ_s'') is about 1.3 for large flakes and about 0.9 for small flakes. Interestingly, it should be pointed out that wide magnetic loss spectra ($\mu \sim f$) can be found for both kinds of flakes, as indicated in **Figure 6b**. Wide magnetic loss peak is advantageous for electromagnetic noise attenuation composites with a wide working frequency band. The loss peak (μ'' is maximum) is found to be 3.0 GHz for large flakes and 5.5 GHz for small flakes, whereas μ'' (max) is 2.1 for the composite with large flakes, and for composites with small flakes, μ'' (max) is about 1.6.

For materials with high resistivity, such as ferrites magnetic loss above 1 GHz is often ascribed to the natural resonance mechanism. The frequency of natural resonance is closely associated with the magnetocrystalline anisotropy as per Snoek's law. One of typical magnetic materials in this case is M-type hexagonal ferrites. In our case, however, the sources of the observed broad magnetic loss peaks are believed to be due to the distribution of localized magnetic anisotropy field, which is the resultant of distribution of shapes (shape anisotropy fields), and distribution of interaction fields among particles. Moreover, eddy current effect is another cause

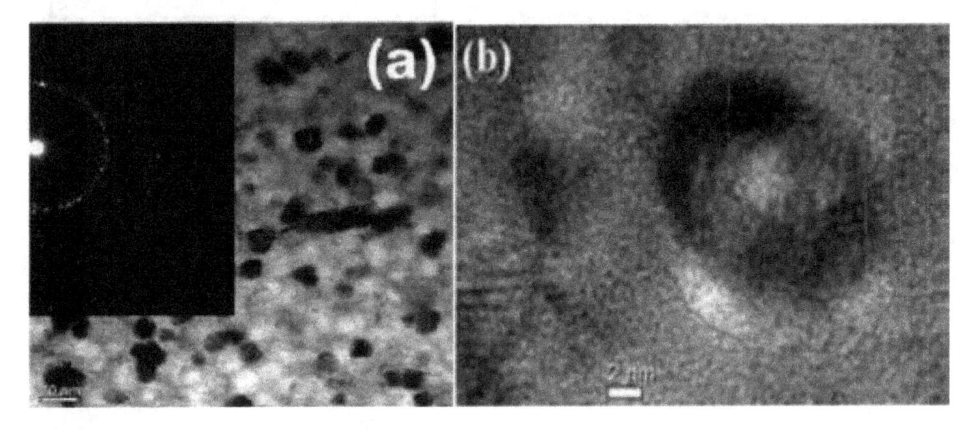

Figure 4.
TEM photographs of Fe-Cu-Nb-Si-B nanostructured alloys. (a) Image showing nano-grains. The inset showing the select area electron diffraction. (b) Nano-grains circled in the high-resolution image.

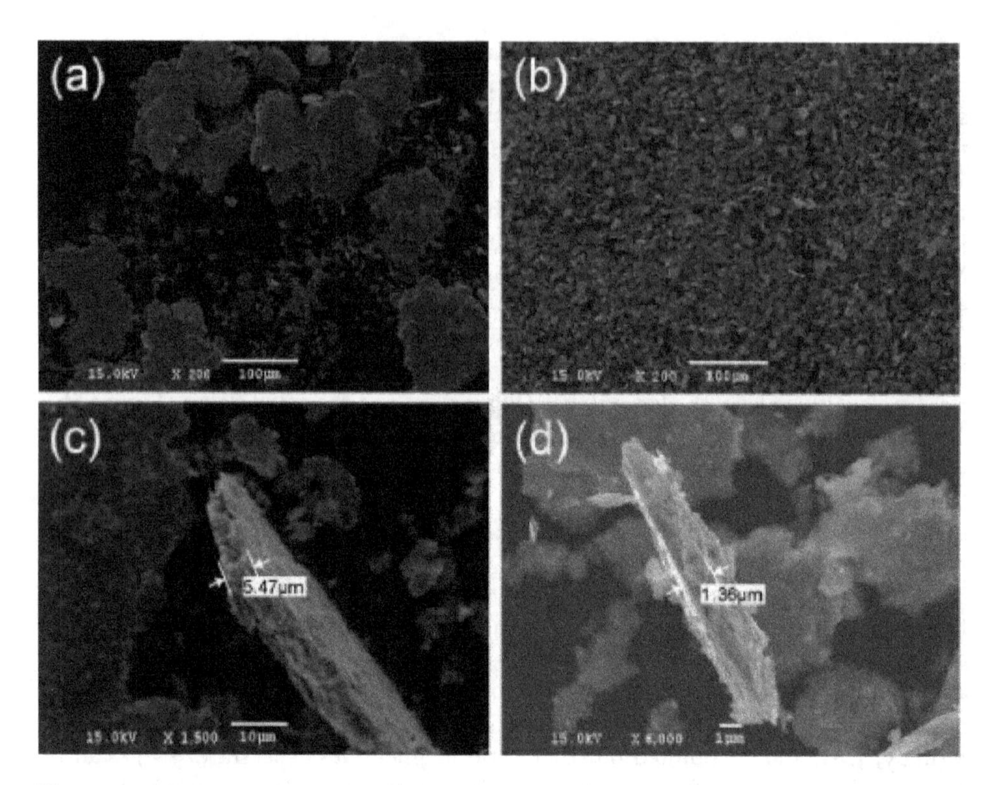

Figure 5.
SEM images of two categories of flakes: (a) large flakes; (b) small flakes; (c) typical thickness of large flakes; (d) typical thickness of small flakes.

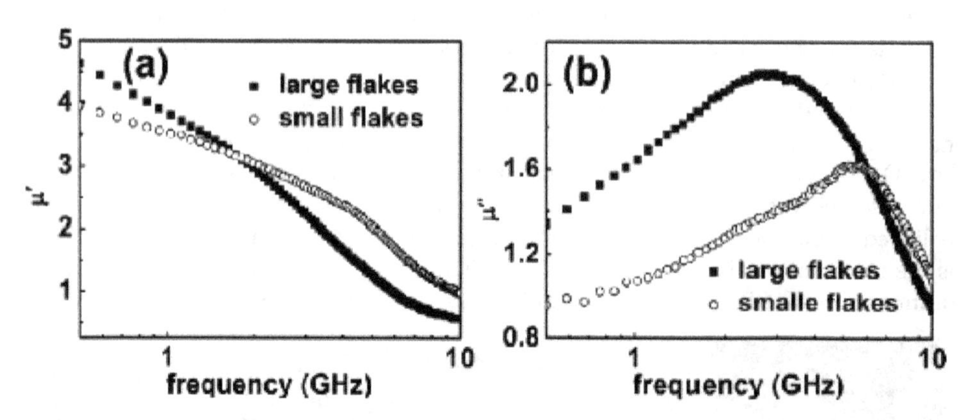

Figure 6.
High-frequency permeability of composites contained different sizes of Fe-Cu-Nb-Si-B flakes. (a) $\mu' \sim f$ spectra and (b) $\mu \sim f$ spectra.

for broadening the spectra of permeability. In order to interpret the observed dissimilarities in the high-frequency permeability of composites with these two categories of flakes, Snoek's law with shape factors included is employed and given as below:

$$\mu_s = 1 + \frac{4\pi M_s}{(H_k + 4\pi M_s N_h)} \tag{3}$$

$$f_r = \frac{\gamma}{2\pi}\sqrt{H_k^2 + 4\pi M_s H_k (N_\perp + N_h) + (4\pi M_s)^2 N_\perp N_h} \tag{4}$$

$$N_{\perp} = \frac{\alpha_r^2}{\alpha_r^2 - 1} \times \left(1 - \sqrt{\frac{1}{\alpha_r^2 - 1}} \times ars \sin \frac{\sqrt{\alpha_r^2 - 1}}{\alpha_r}\right) \quad (5)$$

$$N_h = \frac{1 - N_{\perp}}{2} \quad (6)$$

where α_r is the width/thickness ratio (often called "aspect ratio") of a flake. The demagnetization factor along the direction of thickness and width is $N_{\perp}$ and N_h, respectively. The saturation magnetization of material under studied is denoted as M_s. Magnetocrystalline anisotropy field is denoted as H_k. The total magnetic anisotropy field is given in the denominator of Eq. (3). For our samples, these two kinds of flakes are obtained under the same milling process and made from the same material. Therefore, large flakes and small flakes have same M_s and H_k values. The main differences among them are size distribution, aspect ratio and thickness. The aspect ratio is calculated as 7.23 and 18.02, respectively, for small flakes and large flakes according to the measured geometrical parameters. Subsequently, demagnetization factors ($N_{\perp}$ and N_h) have been calculated as per Eqs. (5) and (6) and are shown in **Table 1**. It can be seen that larger flakes having the larger initial permeability values are because they have smaller N_h values, as indicated by Eq. (3). Moreover, the finding that their magnetic loss peaks are found at lower frequencies can be understood according to Snoek's law: the inversely proportional relationship between the initial permeability and the loss peak. The $\mu' \sim f$ spectrum of large flakes drops more rapidly than small flakes. When f > 1.8 GHz, the real part (μ'(z, the real partflakes drops more rapidly ty and the loss peak. Theower frequencies, with increasing frequency, eddy current becomes a more serious issue in the large flakes. As we pointed out, FINEMET alloys are metallic and well-conducting materials. When they work above 1 GHz, the eddy current effect will be unavoidable. The electromagnetic wave will interact with the part of magnetic materials which are within the so-called "skin depth." As a result, high-frequency permeability spectrum depends on this eddy current effect. Stronger eddy current effect will give rise to the smaller volume fraction of magnetic materials interacting with the EM wave. Consequently, a faster dropping of $\mu' \sim f$ spectrum is observed.

For electromagnetic wave absorbing application, the simplest example is that composites containing the flakes work as single layer on a perfectly conducting substrate (such as the surface of aircrafts). The absorbing properties of a normal incident EM wave can be assessed by the reflection loss (RL, in dB) based on the equations as follows:

$$Z_{in} = Z_0 \sqrt{\frac{\mu}{\varepsilon}} \tanh\left(j \frac{2\pi f d}{c} \sqrt{\mu\varepsilon}\right) \quad (7)$$

$$R.L. = 20 \log \left|\frac{Z_{in} - Z_0}{Z_{in} + Z_0}\right| \quad (8)$$

where "*d*" is the thickness of composite layer, "*c*" is the velocity of light, Z_0 is the impedance of free space and Z_{in} is the characteristic impedance at the free

	$N_{\perp}$	N_h	$(N_{\perp} + N_h)$	$N_{\perp}N_h$
Small flakes	0.816	0.092	0.908	0.075
Large flakes	0.919	0.041	0.960	0.038

Table 1.
Demagnetization factors of two kinds of flakes.

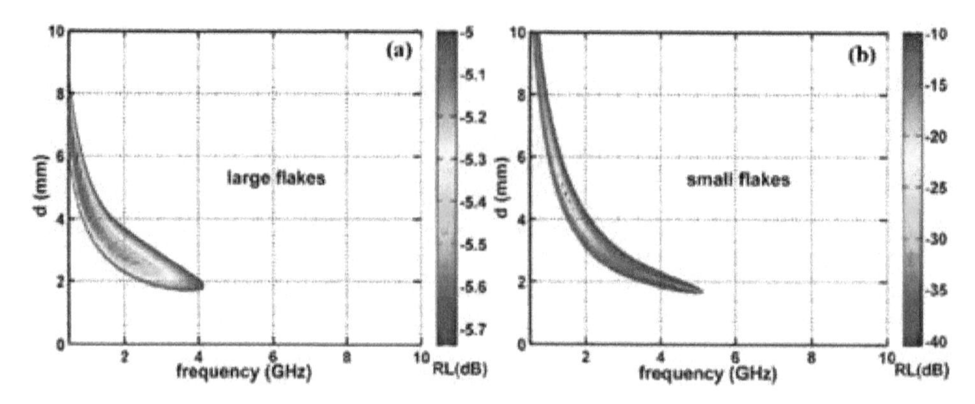

Figure 7.
Contour maps showing the absorbing properties of single layer composites with different flakes: (a) large flakes and (b) small flakes.

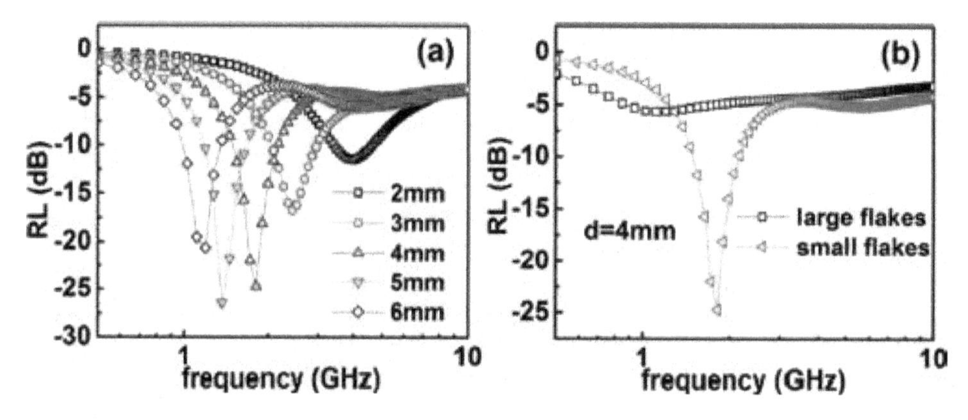

Figure 8.
(a) Composites filled with smaller flakes and with different thickness and (b) composite filled with different flakes but with same thickness (4 mm).

space/absorber interface. In this chapter, all "t" values are in mm unit. "μ" and "ε" are the measured relative complex permeability and permittivity, respectively. The measured permittivity values can be found and have been studied in our paper [6]. The potential absorbing performances of composites with different thickness values are illustrated in **Figure 7**. Clearly, composites containing smaller flakes will have excellent absorbing performances in terms of RL as well as reduced absorber thickness. The superior absorbing properties are also shown in **Figure 8** by selecting a few thicknesses of single layer of composites filled with smaller flakes.

4. Coating treatments

Since high-frequency permittivity of metallic flakes is much larger than their permeability, this difference will cause a serious mismatch of impedance ($Z_{in} >> Z_0$), which will deteriorate the absorbing properties of flakes. Common means of reducing impedance mismatch is to coat the metallic particles with a layer of oxides with high resistivity (such as SiO_2, TiO_2, Al_2O_3, etc.). Although these layers are effective in decreasing the permittivity, they are not ferro- (or ferri-) magnetic and cannot take part in the absorbing of EM wave via magnetic losses. Therefore, we propose to coat the FINEMET flakes with ferrimagnetic layer (such as $NiFe_2O_4$ ferrite) so as to realize two objectives: to reduce the permittivity and to absorb the EM wave by the high-resistivity layer.

$NiFe_2O_4$ layers were fabricated on the Fe-Cu-Nb-Si-B flakes using a low-temperature chemical plating route. In a simplified description, Fe-Cu-Nb-Si-B nanocrystalline flakes were added into a flask containing a bath solution at 333 K. As for the bath solution, the well-designed molar ratios of $NiCl_2$, $FeCl_2$ and KOH solutions were prepared in the flask. Meanwhile, oxygen gas was introduced into the solution until the chemical reactions were completed. Subsequently, the flask was heated at 333 K for 40 min. When the chemical reaction was finished, the deionized flakes were collected and dried at 333 K for 12 h. Elaborative experimental descriptions can be found in our published paper [7]. The morphologies of coated flakes are presented in **Figure 9a** and **b**, respectively. TEM images in **Figure 9c** show that the thickness of coating layer is estimated to be about 17.73–55.61 nm. The energy dispersion spectrum (EDS) and XRD measurements of uncoated and coated flakes confirm the formation of $NiFe_2O_4$ spinel ferrite. XRD patterns are given in **Figure 10**, which indicate the formation of spinel ferrite phase. The magnetic hysteresis loops of coated and uncoated flakes are given in Ref. [7] and show that the saturation magnetization of coated flakes drops from 129.33 to 96.54 emu/g.

The impacts of spinel ferrite coating layer on the high-frequency permittivity and permeability are shown in **Figure 11**. When the flakes are coated, ε' drops from 61.49 to 33.02 at 0.5 GHz, whereas ε'' also drops from 21.39 to 1.16 at 0.5 GHz. As shown, the complex permittivity values are significantly decreased within the lower frequency band and are believed to result from increased resistivity. The permeability values are also found to be a little reduced, due to the fact that Ms value of

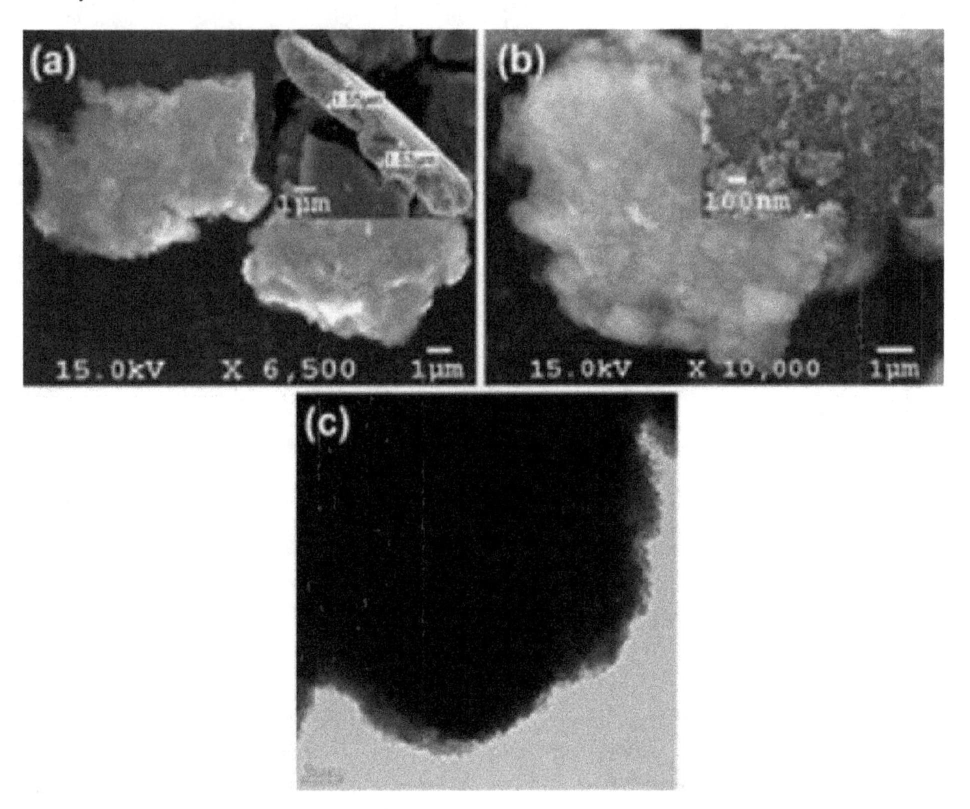

Figure 9.
SEM images of (a) uncoated flakes and (b) coated flakes. (c) TEM image of a coated flake. Inset in (a): typical thicknesses of the sample. Inset in (b): surface roughness of a coated flake.

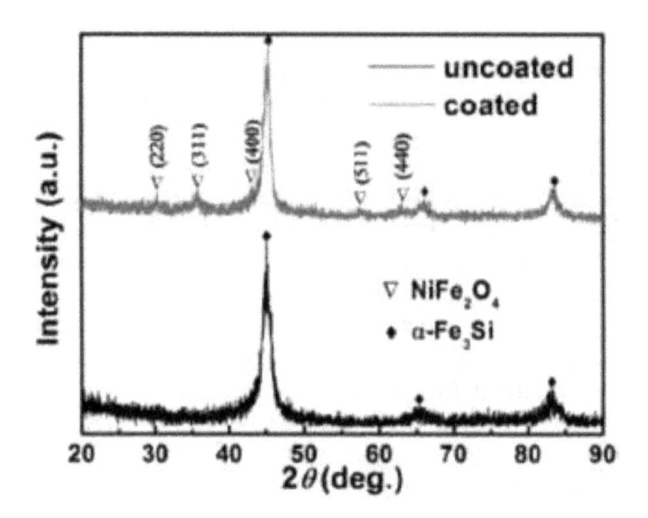

Figure 10.
XRD patterns of uncoated and coated flakes.

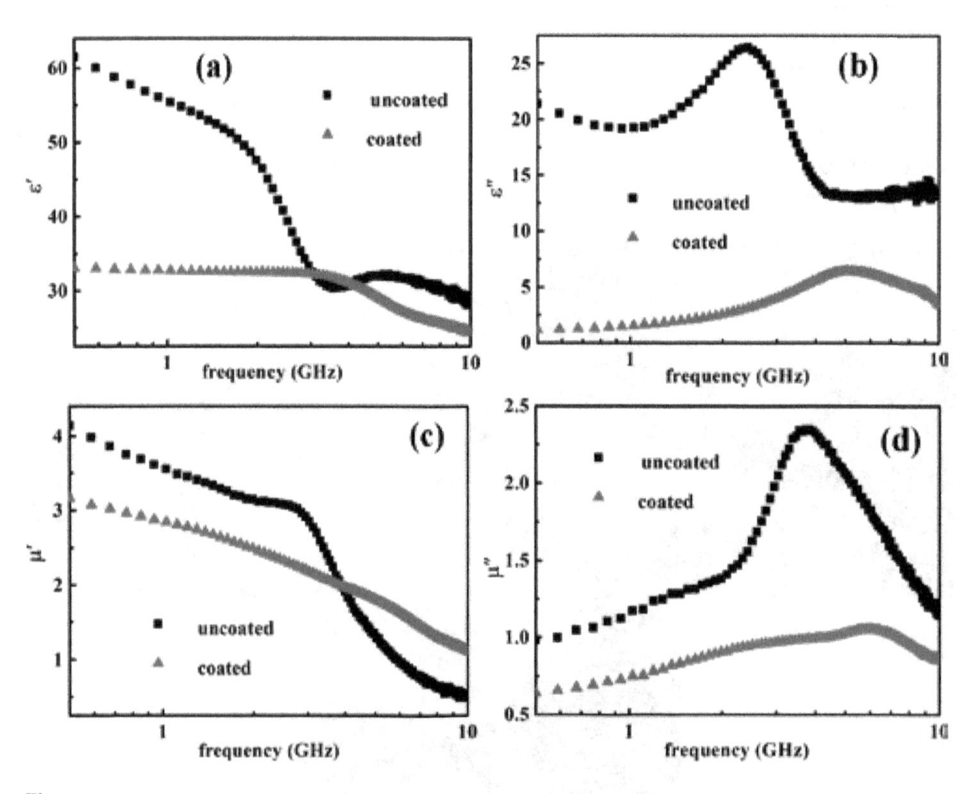

Figure 11.
The impact of ferrite coating on the permittivity and permeability of flakes. (a) $\varepsilon' \sim f$ and (b) $\varepsilon \sim f$. (c) $\mu' \sim f$ and (d) $\mu \sim f$.

coated flakes is less than that of uncoated flakes; see **Figure 11c** and **d**. Since the spinel ferrite layer has a smaller Ms value than the FINEMET alloy, the reduced permeability can be understood as per Snoek's law. It is interesting to point out that $\mu \sim f$ spectra of coated flakes are not significantly fluctuated as much as the uncoated flakes. The previous section showed that eddy current effect in uncoated flakes is strong and results in a large reduction of μ values. Such a large reduction is

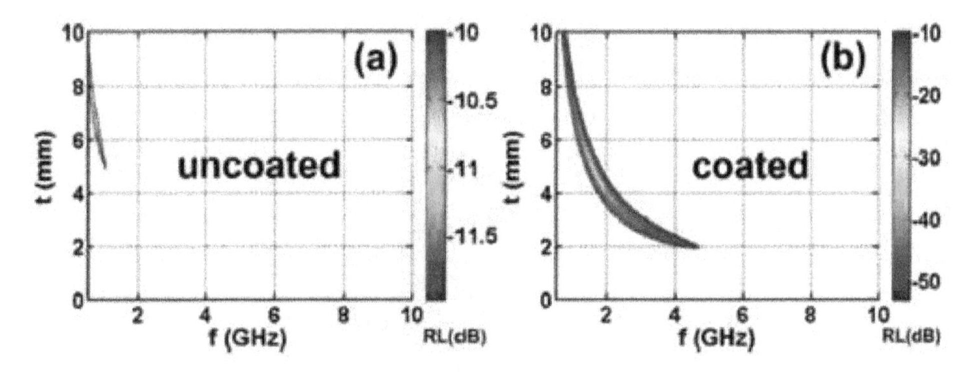

Figure 12.
Contour maps showing the absorbing properties (in terms of RL) of composites with different thicknesses. (a) Uncoated flakes and (b) coated flakes.

greatly suppressed in coated particles, which means that the high-resistivity coating ferrite effectively reduces the impact of eddy currents on μ values.

The impact of coating with a high-resistivity ferrite on EM absorbing performances is shown in **Figure 12**. We use a contour map to illustrate the reflection loss of absorbers filled with flakes which are coated and uncoated. For composites containing uncoated flakes, the complex ε values are much larger than the complex μ values. Due to this large difference in ε and μ, the impedance mismatch (Z_{in} and Z_0) according to Eq. 7 is significant, and consequently it results in the worsening of absorbing properties compared to absorbers filled with coated flakes. Obviously, the high-resistivity coating can effectively lessen the impedance mismatch and improve absorbing performance. Moreover, the thickness of absorbing composite is greatly decreased if coated flakes are used as absorbent composite.

5. Origins of multi-peaks in the intrinsic permeability spectra

As discussed before, broad permeability spectra are commonly observed in the composites filled with magnetic particles. The debate on the causes is intense. We believe that a broad intrinsic permeability spectrum results from superposition of many natural resonance peaks. Intrinsic permeability can be retrieved from measured permeability using one of the mixing rules [8]. However, origins of multi-peaks in the intrinsic permeability spectra have not been well answered, and it is essential for the designing of electromagnetic devices or materials. In order to exclude impact of non-intrinsic factors (e.g., inhomogeneous microstructure, eddy current, size distribution and particle shape) on the understanding of the origin of multi-peaks, we design several "1 × 3" iron nanowire (Fe-NW) arrays. Each array has a different interwire spacing. Each nanowire in the array is identical in geometry. Each one has a cuboid shape: the length is 100 nm (set as the "z"axis). The cross section is 10 nm ("y" axis) × 20 nm ("x" axis).

In this study, we will discuss the impact of interwire distance on the intrinsic permeability for only two interwire distances: 2.5 and 60 nm, as depicted in **Figure 13**. The static magnetic properties and dynamic responses of Fe nanowire arrays are simulated using micromagnetic simulation. First-order-reversal-curve (FORC) technique is used to simulate and analyse the impact of interwire distance on the static magnetic properties. The dynamic response of magnetization under excitation of pulse magnetic field can be described by the Landau-Lifshitz-Gilbert equation:

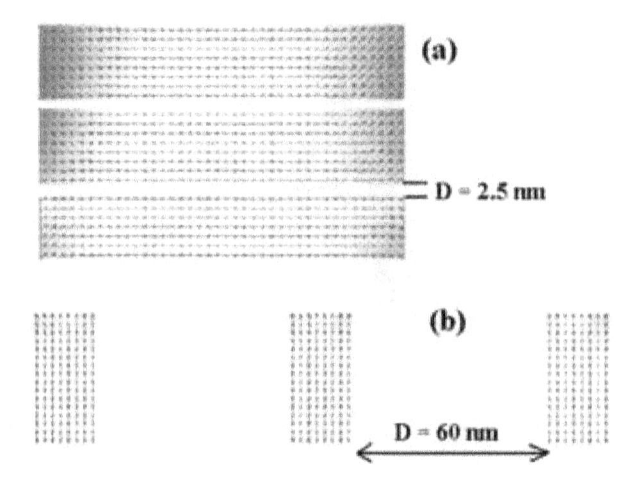

Figure 13.
The equilibrium states of magnetization for two kinds of interwire distance: (a) D = 2.5 nm and (b) D = 60 nm (only partial sections of nanowire are presented for saving space).

$$\frac{\partial \vec{M}\,(r_i,t)}{\partial t} = -\gamma\,\vec{M}\,(r_i,t) \times \vec{H}_{eff}(r_i,t) - \frac{\gamma\alpha}{M_s}\vec{M}\,(r_i,t) \times \left[\vec{M}\,(r_i,t) \times \vec{H}_{eff}(r_i,t)\right] \quad (9)$$

More details on simulation procedures and setting parameters can be found in our published paper [9]. FORC approach is based on the Preisach hysteresis theory and is helpful in investigating factors that determine local magnetic properties of materials. According to FORC measurement procedure, the array is at first saturated at a field value (H_s) in one direction; next, the magnetic field is decreased to a field (called "reversal field," H_a); and then the array is again saturated from H_a to H_s, which will trace out one partial magnetization curve (i.e., a FORC curve). Following the same procedure, a series of FORC curves can be obtained starting from different H_a values to the H_s value, which will fill the interior of the major hysteresis loop. Each data point of a FORC curve is denoted as M(H_a, H_b), where H_b is the applied field. According to the Preisach model, a major hysteresis loop is consisted to be a set of hysterons, and the probability density function ρ (H_a, H_b) of hysteron ensemble can be calculated by a mixed second derivative as follows, where ρ wh$_a$, Hb) is often called the FORC distribution, and is expressed as

$$\rho(H_a, H_b) = -\frac{1}{2}\frac{\partial^2 M}{\partial H_a \partial H_b} \quad (10)$$

Usually, the FORC distribution is illustrated in the diagram of (H_u vs. H_c), as shown in **Figure 14a**. H_u is the local interaction field and H_c the local coercivity. Relationship between (H_a, H_b) coordinate system and (H_u, H_c) coordinate system is given as $H_c = (H_b - H_a)/2$ and $H_u = (H_a + H_b)/2$.

Clearly, there are no domain walls existing in our simulations, as shown in **Figure 13**. Therefore, the impact of domain wall on the permeability can be excluded. In addition, coercivity is therefore only decided by localized effective magnetic field (no domain wall movements involved). Localized magnetic fields decide the precession of local magnetizations, which will precess around these effective local fields. Each precession has an eigenfrequency, which is also called " natural resonance frequency. " Under the excitation of a pulse magnetic field ((h(t) = 100 exp(-10^9t), t in second, hour in A/m) is) perpendicular to "z" axis, the simulated intrinsic permeability spectra are shown in **Figures 14b** and **15b**.

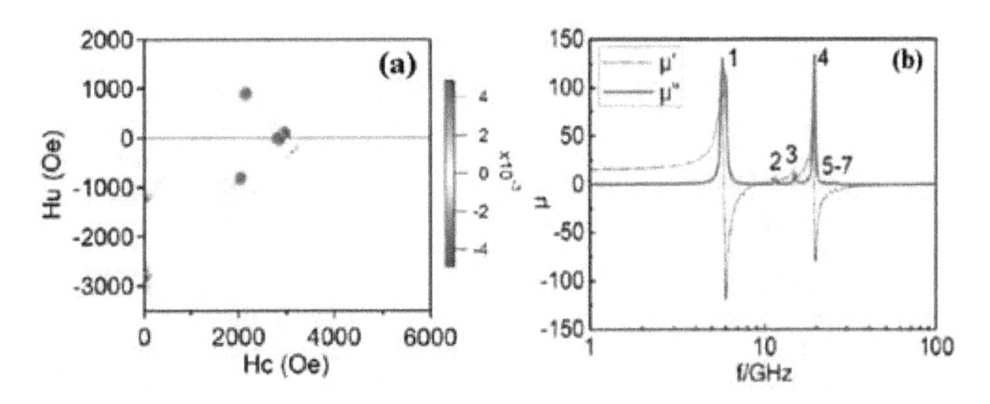

Figure 14.
FORC distribution and intrinsic permeability of nanowire array when interwire distance (D) is 2.5 nm.

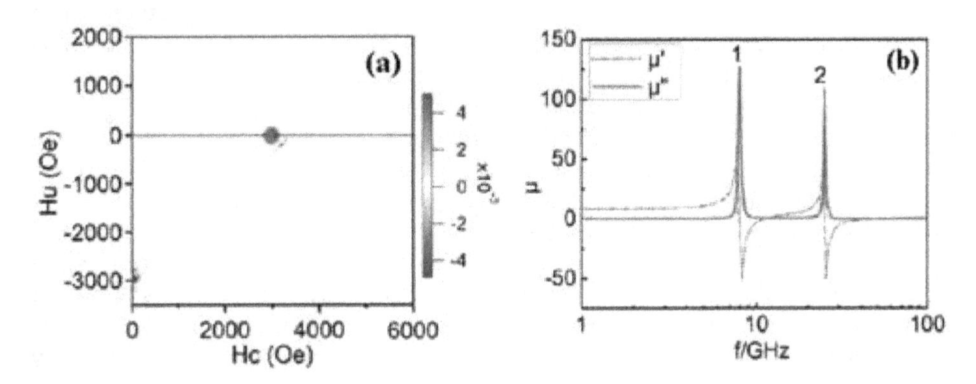

Figure 15.
FORC distribution and intrinsic permeability of nanowire array when D = 60 nm.

Corresponding FORC distributions are shown in **Figures 14a** and **15a**. Obviously as seen in **Figure 14**, two stronger resonance peaks at f = 5.75 and 19.5 GHz and several weak resonance peaks at f = 11.75, 15, 22.25, 24.25 and 26.5 GHz are found for the nanowire array with D as 2.5 nm. The previous studies of both ours and by others show that the resonance peaks found at lower frequency of 5.75 GHz are often named "edge mode" [10], which are resulting from precessions at the ends of nanowires. The frequencies of "edge mode" found by us are very close to those reported by others [10, 11]. The peak found at 19.5 GHz can be identified as "bulk mode." The eigenfrequency of bulk mode for an isolated nanowire can be calculated by the following equation:

$$f = \gamma' \sqrt{\{H + (Nx - Nz)M_s\} \times \{H + (Ny - Nz)M_s\}} \quad (11)$$

where H (the external DC magnetic field) is along the easy axis, demagnetization factor is N and γ' is the gyromagnetic ratio (for Fe: 2.8×10^6 Hz/Oe). For our nanowire arrays, Nx + Ny + Nz = 1, Nz is 0, Nx is 1/3 and Ny is 2/3. Then, the eigenfrequency frequency (natural resonance frequency) of bulk mode is calculated as 28.2 GHz. This calculated value is larger than those for bulk mode simulated in the arrays with different D values. This difference is a result from the fact that the calculated value is based on Eq. (11) which is for uniform precession without considering interactions among nanowires. However, the simulated values are obtained under the circumstance of interaction among NW. It was found that "the calculated f_r" and "the simulated f_r" are in good agreement in an isolated nanowire [10]. Our simulations show that when the interwire distance increases, FORC diagrams and the intrinsic $\mu \sim f$ spectra vary differently. In addition, the reversal process of nanowire array is found to be different. Please refer to our published paper for more details [9]. When interwire distance increases, magnetization reversal behaviour progressively changes from "the sequential mode" to "the independent mode." Moreover, the finding that the amount of weak resonance peaks decreases with increasing D is worth noting. With regard to the origin of weak resonance peaks in **Figure 14b**, we presume that it is the consequence of superposition of interaction field, magnetocrystalline field and shape anisotropy field. As depicted in FORC diagrams (see **Figures 14** and **15**), the interaction field can be either negative or positive. Effective field acting on some magnetization can be smaller (or larger) than the effective field related to "bulk mode" in an isolated nanowire. Accordingly, some resonance frequency is smaller (or larger) than that of bulk mode, which can be clearly observed in **Figure 14**. With increasing interwire distance, interaction among nanowires gradually disappears; therefore, these smaller peaks gradually vanish. From the stand point of electromagnetic (EM) attenuation application, it means that volume fraction of magnetic particles in a composite should not be diluted in order to have many resonance peaks for expanding the attenuation bandwidth. As shown in **Figure 15**, there are only two strong resonance peaks found when D is 60 nm, which may be explained using the same physical mechanism. The "calculated f_r" (bulk mode) is about 28.2 GHz and is close to the "simulated f_r" (about 25 GHz).

Impact of interwire distance on interaction among NWs is shown in **Figures 14a** and **15a**. Obviously, with increasing interwire distance (D), data points approach together (results of other D values were shown in Ref. 9), and H_u becomes "zero" when D is equal or greater than 60 nm. Here, when reversal resistance comes only from localized effective field (H_{eff}), then effective field can be approximated by coercivity field (i.e., $H_c \approx H_{eff}$). Hence, when D increases, localized H_c is always larger than zero. The "scattered" characteristic of H_c value vanishes when D is 60 nm, which means that reversal behaviour of M changes into the "independent mode," as shown in Ref. 9. In addition, peak magnitude of "edge mode" in an isolated "cylindrical" nanowire is usually smaller than "bulk mode" [10, 11]. However, all peak magnitudes of "edge mode" in our "cuboid" shape nanowires are comparable with bulk mode, which suggests that high-frequency magnetic loss due to the "edge mode" cannot be neglected in our case. Moreover, smaller fr value of "edge mode" means that localized effective magnetic field governing its precession is weaker. It means that reversal of magnetization commonly starts from the ends of nanowires, which has been observed in our study.

To better understand the changes of intrinsic $\mu \sim f$ spectra, studies on difference in magnetic moment orientations in equilibrium states are necessary, as illustrated in **Figure 13**. Apparently, magnetic moment orientations around the ends of nanowires at equilibrium states are distinct. Such orientation pattern of magnetic moments strongly relies on the local effective magnetic field. The first-order--reversal-curve diagram is useful to know the distribution of local effective magnetic fields. Hence, we believe any factor affecting the equilibrium state of magnetic moments will change the intrinsic $\mu \sim f$ spectra. If the equilibrium state is rebuilt, then the permeability spectra under same excitation can be recovered too (we name it as "***memory effect***"). Finally, we want to point out that such traits of "scattered" values of H_c and H_u in these simulations differ from the continuous distribution measured in most nanowires deposited in nanoporous templates (such as AAO templates). The reason is that continuous distribution of H_c and H_u in a FORC diagram results from inhomogeneity in nanopore sizes of template, nanoscale electrochemically deposited polycrystals and multimagnetic domains, which will give rise to plentiful irreversible magnetization processes.

6. Negative imaginary parts of permeability ($\mu'' < 0$)

As stated before, high-frequency permeability of Fe-based conducting nanostructured materials can be tailored by shapes, phase transitions, coating and size distributions. Most importantly, all imaginary parts of permeability are positive ($\mu'' > 0$), which are accepted as an unalterable natural principle. This is correct when there are no other excitations except AC magnetic field based on the fact that positive μ'' manifests energy dissipation of magnetization precession. What happens to the imaginary parts of permeability if there is another excitation, which can compensate the energy loss during the precession? Furthermore, is it possible to tailor the high-frequency permeability by electric current? If possible, it will be possible for us to design intelligent electromagnetic devices. In this section, we propose for the first time an approach to accomplish such a goal via spin transfer torque (STT) effect. According to STT effect, when polarized electrons flow through zone of metallic ferromagnetic material with nonuniformly oriented magnetic moments (e.g., magnetic domain walls), the spins of electrons will exert torques on them, which will change the dynamic responses of precession of magnetic moments [13]. When STT is strong enough, it can even switch the direction of magnetic moments. Here we only demonstrate an example using micromagnetic simulation; our detailed results can be found in Ref. [12]. The object of micromagnetic simulations is an isolated Fe nanowire with a length of 400 nm (set as "x" axis) and the diameter of 10 nm, as shown in **Figure 16**. The dynamic response of magnetization (a group of magnetic moments in a direction, also can be called "macro spin") can be simulated under two external excitations: AC magnetic field and electrically polarized current. The AC magnetic field is applied along the "z" axis, and polarized electrons are flowing along the "x" axis. The software of micromagnetic simulation is a three-dimensional object-oriented micromagnetic framework (OOMMF). The dynamic responses of precession are simulated by solving the Landau-Lifshitz-Gilbert (LLG) equation as a function of time. When the STT effect is incorporated, the modified LLG equation is expressed as [13, 14]:

$$\frac{d\vec{M}}{dt} = \gamma \vec{H}_{eff} \times \vec{M} + \frac{\alpha}{Ms}\left(\vec{M}_s \times \frac{d\vec{M}}{dt}\right) - \left(\vec{u} \cdot \vec{\nabla}\right)\vec{M} + \frac{\beta}{Ms}\vec{M} \times \left[\left(\vec{u} \cdot \vec{\nabla}\right)\vec{M}\right] \quad (12)$$

where α (damping constant) is set as 0.01 for the simulations; γ is the Gilbert gyromagnetic ratio (2.21×10^5 mA^{-1} S^{-1}); $\boldsymbol{M}$ is the magnetization; $\boldsymbol{H_{eff}}$ is the

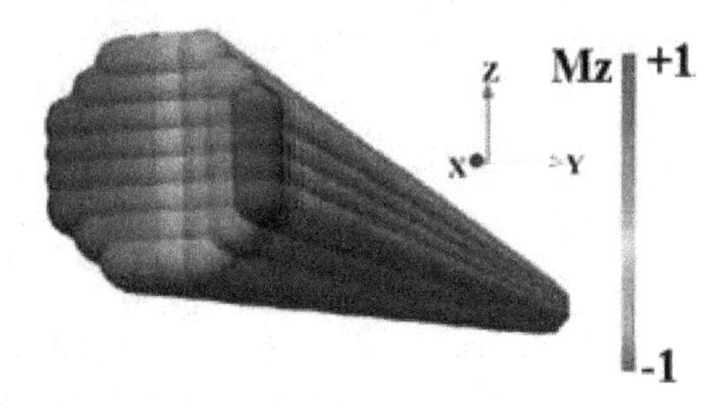

Figure 16.
Inhomogeneous orientation of magnetic moments of an isolated nanowire.

effective magnetic field which consists of demagnetization field, exchange interac-tion, anisotropic field and external applied field; and finally, β (nonadiabatic spin transfer parameter) is set as 0.02 and is used to consider the impact of temperature on the dynamics of precession. The vector $\boldsymbol{u}$ is defined as

$$\vec{u} = -\frac{g\mu_B P}{2eM_s}\vec{J} \tag{13}$$

where $\boldsymbol{g}$ is the Landé factor, $\boldsymbol{e}$ is the electron charge, μ_B is the Bohr magneton, $\boldsymbol{J}$ is the current density and $\boldsymbol{P}$ is the polarization ratio of current and set as 0.5. The simulation parameters for iron (Fe) nanowire are $\boldsymbol{Ms} = 17 \times 10^5$ A/m, magnetocrystalline anisotropy constant, $\boldsymbol{K_1} = 4.8 \times 10^4$ J/m^3, exchange stiffness constant and $\boldsymbol{A} = 21 \times 10^{-12}$ J/m. Also, the nanowire is discretized into many tetrahedron cells (cell size: $5 \times 1.25 \times 1.25$ nm). The cell is smaller than the critical exchange length. To obtain high-frequency permeability spectra, these steps are followed: Firstly, the remanent magnetization state should be acquired after the external field is removed. Secondly, a pulse magnetic field is applied perpendicular to the long axis (x-axis) of the wire. The pulse magnetic field has the form $\boldsymbol{h(t)} = 1000 \exp(-10^9 t)$ ($\boldsymbol{t}$ in second, ***hour*** in A/m). A polarized current (density J is 3.0×10^{12} A/m^2) is flowing along x-axis. Thirdly, the dynamic responses of magnetization in time-domain are recorded under these two excitations. Using the Fast Fourier Transform (FFT) technique, the dynamic responses in frequency domain are obtained. The high-frequency permeability spectra are calculated based on the definition of permeability:

$$\mu(f) = 1 + \frac{m(f)}{h(f)} = 1 + \chi'(f) - i\chi''(f) \tag{14}$$

where $\boldsymbol{m(f)}$ is magnetization under both excitations and the pulse magnetic field ($\boldsymbol{h}$) works as the perturbing field. The following relations exist: $\mu' = \boldsymbol{1} + \chi'(f)$, $\mu'' = \chi(f)$. The differences in the high-frequency permeability under different excitations have been investigated.

When single nanowire is under the excitation of only an AC magnetic field (h), two Lorentzian-type resonance peaks are found in the permeability spectrum, as shown in **Figure 17a**. One is located approximately at 18 GHz, and the other is located approximately at 31.5 GHz. According to our previous studies, the major resonance is called "bulk mode," which has larger magnetic loss and is manifested by the larger μ'' value, which is ascribed to the precession of perfectly aligned magnetic moments within the nanowire body excluding the misaligned magnetic moments at both ends of nanowire. The minor resonance peak located at 18 GHz is often called "edge mode," whose smaller magnitudes of real and imaginary parts are due to smaller volume fractions of magnetic moments at the ends of nanowire. The edge mode is due to the precession and resonance of misaligned magnetic moments. Physically, if external field is not applied, the magnetizations are aligned along the local effective field ($\boldsymbol{H_{eff}}$). Once the perturbation field ($\boldsymbol{h}$) is applied to excite the precession, $\boldsymbol{M}$ will move away from their equilibrium positions with small angles, and $\boldsymbol{M}$ will then precess around H_{eff}. When precession goes on, the y-component of M ($\boldsymbol{M_y}$) will have nonzero values, as shown in **Figure 17b**. During the period of precession, the energy absorbed from the AC magnetic field will be gradually dissipated via damping mechanisms. M will restore to its equilibrium positions, and M_y will be zero; see **Figure 17b**. The energy loss is manifested by a positive imaginary part of permeability ($\mu'' > 0$). When the nanowire is excited under both the pulse magnetic field and polarized current, the high-frequency permeability

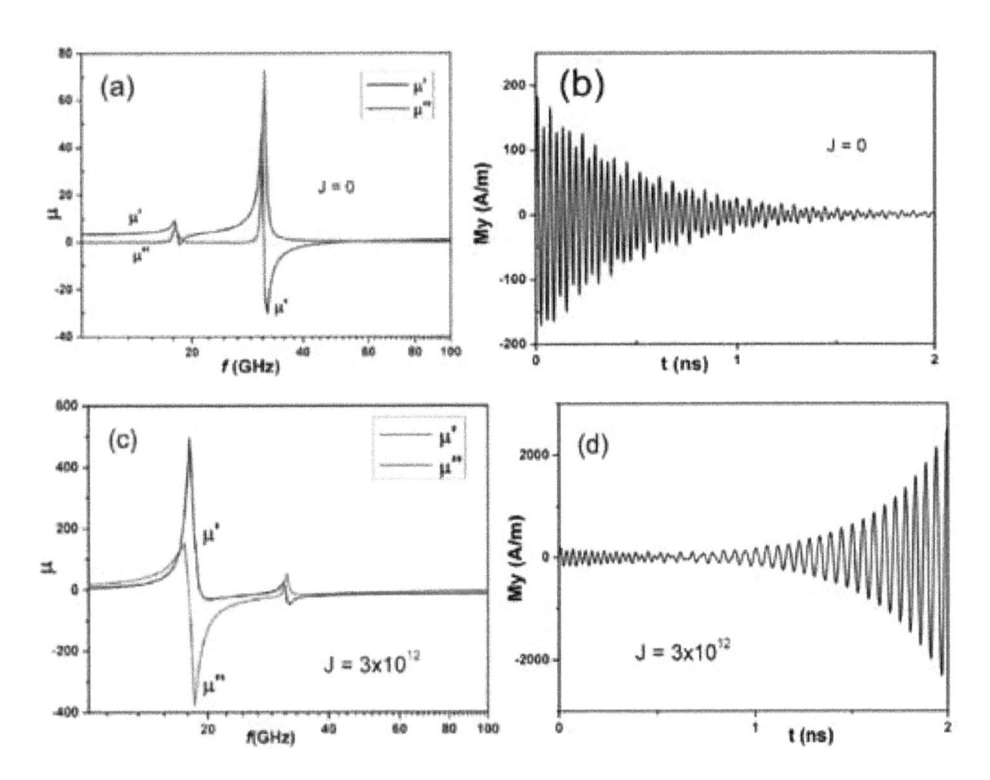

Figure 17.
Permeability spectra and responses of $\mathbf{M}_y$ component without and with STT excitation ($J \neq 0$).

spectrum is naturally found to be different and is illustrated in **Figure 17c**. The previous minor resonance becomes the major resonance with negative μ'' values. The "M_y" component does not vanish gradually; see **Figure 17d**. According to the STT effect, a spin-polarized current flows through the nanowire; STTs only act on the magnetic moments at the ends of nanowire. These STTs will counteract the torques due to effective magnetic field, which will then bring the magnetizations back to their equilibrium positions. When the STT is strong enough, it can maintain the precession angle ($\mu'' = 0$) and even enlarge the precession angle ($\mu'' < 0$) or switch the direction of magnetization. Additionally, negative μ'' values are found only for "edge mode" at lower frequency. This is because the STT effect only acts on magnetic moments with a spatially nonuniform orientation, which are only found at both ends of nanowires. The cause for minor peak becoming major peak under STT effect is due to the fact that the spin transfer torques are consistently acting on

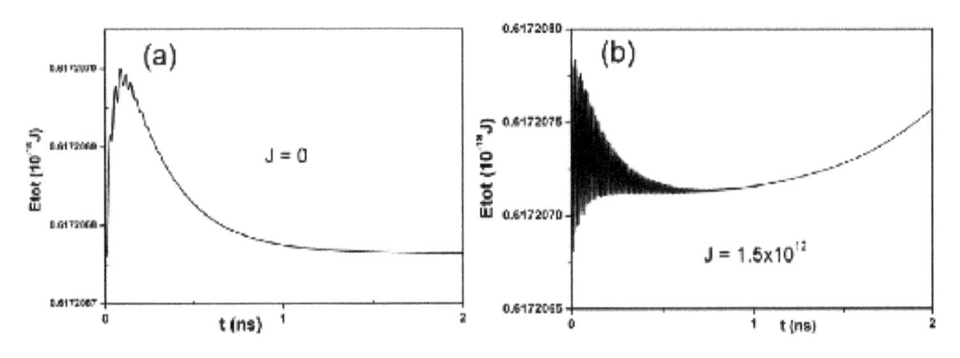

Figure 18.
The variation of total energy without and with STT excitation.

the magnetic moments; therefore, the precession angle is enlarging, and as a result, M_y component is also increased. From the perspective of total energy (E_{total}) of magnetic moment ensemble, when no STT effect is involved, E_{total} first increases due to the excitation of pulse magnetic field and then gradually attenuates due to energy loss via damping mechanisms. However, when STT effect exists, E_{total} is not attenuated to a constant; see **Figure 18**.

Finally, it should be pointed out that although others have also reported negative μ'' in other materials, there are no convincing physical mechanisms that have been provided to support their results ($\mu'' < 0$), and negative μ'' is mostly likely due to measurement errors.

7. Conclusions

In this chapter, we have shown several approaches to tailor the high-frequency permeability of ferromagnetic materials: shapes, particle size distribution, heat treatments, coating and spin transfer torque effect. Micromagnetic simulations are used to explain the origins of multi-peaks in permeability spectra of nanowire array. In addition, just like the simultaneous excitation method (pulse ***H*** field + ***STT*** effect) shown here, we want to propose the other possibilities (pulse ***H*** field + *the other excitation*) to tailor $\mu \sim f$ spectra and to realize negative imaginary parts of permeability ($\mu'' < 0$), such as femtosecond laser pulses or photomagnetic pulses.

Acknowledgements

This work is supported by the National Science Foundation of China (Grant No. 61271039) and the International Scientific Collaboration of Sichuan Province (Grant No. 2015HH0016).

Author details

Mangui Han
School of Materials and Energy, University of Electronic Science and Technology of China, Chengdu, China

*Address all correspondence to: han_mangui@yahoo.com

References

[1] Snoek JL. Dispersion and absorption in magnetic ferrites at frequencies above one mc/s. Physica. 1948;**14**:207. DOI: 10.1016/0031-8914(48)90038-X

[2] Han M, Deng L. Mössbauer studies on the shape effect of $Fe_{84.94}Si_{9.68}Al_{5.38}$ particles on their microwave permeability. Chinese Physics B. 2013; **22**(8):083303. DOI: 10.1088/1674-1056/22/8/083303

[3] Acher O, Adenot AL. Bounds on the dynamic properties of magnetic materials. Physical Review B. 2000;**62**: 11324. DOI: 10.1103/PhysRevB.62.11324

[4] McHenry ME, Willard MA, Laughlin DE. Amorphous and nanocrystalline materials for applications as soft magnets. Progress in Materials Science. 1999;**44**:291. DOI: 10.1016/S0079-6425 (99)00002-X

[5] Han M, Guo W, Wu Y, et al. Electromagnetic wave absorbing properties and hyperfine interactions of Fe–Cu–Nb–Si–B nanocomposites. Chinese Physics B. 2014;**23**(8):083301. DOI: 10.1088/1674-1056/23/8/083301

[6] Wu Y, Han M, Liu T, Deng L. Studies on the microwave permittivity and electromagnetic wave absorption properties of Fe-based nano-composite flakes in different sizes. Journal of Applied Physics. 2015;**118**:023902. DOI: 10.1063/1.4926553

[7] Wu Y, Han M, Deng L. Enhancing the microwave absorption properties of Fe-Cu-Nb-Si-B nanocomposite flakes by coating with spinel ferrite $NiFe_2O_4$. IEEE Transactions on Magnetics. 2015; **51**(11):2802204. DOI: 10.1109/TMAG.2015.2444656

[8] Lagarkov AN, Rozanov KN. High-frequency behavior of magnetic composites. Journal of Magnetism and Magnetic Materials. 2009;**321**:2082. DOI: 10.1016/j.jmmm.2008.08.099

[9] Deng D, Han M. Discussing the high frequency intrinsic permeability of nanostructures using first order reversal curves. Nanotechnology. 2018;**29**: 445705. DOI: 10.1088/1361-6528/aad9c2

[10] Gerardin O, Le Gall H, Donahue MJ, Vukadinovic N. Micromagnetic calculation of the high frequency dynamics of nano-size rectangular ferromagnetic stripes. Journal of Applied Physics. 2001;**89**:7012. DOI: 10.1063/1.1360390

[11] Han XH, Liu RL, Liu QF, Wang JB, Wang T, Li FS. Micromagnetic simulation of the magnetic spectrum of two magnetostatic coupled ferromagnetic stripes. Physica B. 2010; **405**:1172-1175. DOI: 10.1016/j.physb.2009.11.030

[12] Han M, Zhou W. Unusual negative permeability of single magnetic nanowire excited by the spin transfer torque effect. Journal of Magnetism and Magnetic Materials. 2018;**457**:52-56. DOI: 10.1016/j.jmmm.2018.02.031

[13] Thiaville A, Nakatani Y, Miltat J, Suzuki Y. Micromagnetic understanding of current-driven domain wall motion in patterned nanowires. Europhysics Letters. 2005;**69**:990. DOI: 10.1209/epl/i2004-10452-6

[14] Yan M, Kakay A, Gliga S, Hertel R. Beating the walker limit with massless domain walls in cylindrical nanowires. Physical Review Letters. 2010;**104**: 057201. DOI: 10.1103/PhysRevLett.104.057201

9

Electromagnetic Wave Absorption Properties of Core-Shell Ni-Based Composites

Biao Zhao and Rui Zhang

Abstract

Currently, high efficiency electromagnetic wave absorption plays an important role to keep away from the detection of aircraft by radar and reduce information leakage in various electronic equipment. Among the candidates of electromagnetic absorbers, ferromagnetic Ni materials possess high saturation magnetization and high permeability at high frequency (1–18 GHz), which is widely used to prepare thinner absorbing materials along with strong electromagnetic absorption properties. However, the metallic materials usually have relatively high electrical conductivity, and their permeability decreases rapidly at high frequency thanks to the eddy current losses, which is generally named as skin-depth effect. To address this issue, one effective way is to design core-shell structured Ni based composites combining magnetic cores with dielectric shells. This chapter focuses on the state-of-the-art of the microwave absorption properties of Ni-based core-shell composites, and the related electromagnetic attenuation theory about how to enhance absorption properties is also discussed in detail.

Keywords: ferromagnetic Ni, core-shell structure, dielectric loss, magnetic loss, impedance match

1. Introduction

The discovery of electromagnetic (EM) waves boosted the development of transmission technology. It is well accepted that the various electromagnetic waves are widely applied in numerous areas and make our daily life convenient [1]. Extensive electronics devices, such as mobile phones, WiFi, Near Field Communication (NFC), and wireless charging, are developing and play an important role in modern life [2, 3]. With the extensive practical applications of electronic devices and densely packed systems, electromagnetic interference (EMI) becomes a more and more serious problem, which would lead to pernicious impacts on equipment performance, human health, as well as surrounding environment [4]. Furthermore, our individual mini device produces unwanted EM waves, which would influence other nearby devices [5]. Moreover, the global need for some EM waves, such as for military radar stealth, is also boosting, which produces plentiful concerns for human health. Therefore, the protection of electromagnetic radiation has been widely concerned by the whole society. Recently, development of high performance microwave absorbing materials (MAMs) have attracted great interests to eliminate the electromagnetic pollution [6, 7].

Recently, considerable research attention has been focused on core-shell structure for innovative electromagnetic absorption due to the potential to combine the individual properties of each component or achieve enhanced performances through cooperation between the components [8–10]. Liu and co-workers have fabricated core-shell structured Fe_3O_4@TiO_2 with Fe_3O_4 as cores and hierarchical TiO_2 as shells and the Fe_3O_4@TiO_2 core-shell composites displayed the enhanced microwave absorption properties than pure Fe_3O_4 [11]. Chen and co-workers have successfully synthesized core/shell Fe_3O_4/TiO_2 composite nanotubes with superior microwave attenuation properties [12]. Combining Fe_3O_4 and TiO_2 can take advantage of both the unique magnetic properties of Fe_3O_4 and strong dielectric characteristics of TiO_2 as well as core-shell structure, and therefore offer an avenue to achieve excellent microwave-absorption performance. In this kind of core/shell configurations, the magnetic materials regarding as cores, which could improve the permeability of the composites, is conductive to the enhancement of the magnetic loss. The dielectric materials considering as shells, which are supposed to play the roles not only as a center of polarization but also as an insulating medium between the magnetic particles, lead to the increased dielectric loss and good impedance match. The high-efficiency microwave absorption properties resulted from the enhanced magnetic loss, dielectric loss, reduced eddy current loss and impedance match [8]. Thus, the traditional microwave absorbing materials holding a core-shell structure may improve their microwave absorption capabilities.

It is well known Ni is regarded as a typical magnetic metal material, which are supposed to have numerous applications in many fields such as magnetic recording devices, clinical medicine, catalysis and so on [13, 14]. It is worth pointing that Ni was also proved to be as a competitive candidate for high-efficiency electromagnetic absorption materials to address the electromagnetic interference and pollution problems because Ni can provide more beneficial features, such as high saturation magnetization, distinguishable permeability, and compatible dielectric loss ability in the gigahertz range when compared with those nonmagnetic EM absorption materials. However, single-component electromagnetic absorption materials easily suffer from mismatched characteristic impedance and poor microwave absorption performance. Moreover, Ni would generate eddy current induced by microwave in GHz range because of high conductivity. The eddy current effect may cause impedance mismatching between the absorbing materials and air space, which would make microwave reflection rather than absorption. This issue is a challenge to handle for scientists. Thus, for the sake of getting superior microwave absorption ability, a promising pathway is to compound the Ni products with an inorganic or nonmagnetic constituent to produce a core@shell configuration. Numerous literatures have been carried out to cover the magnetic Ni with inorganic or nonmagnetic shells. For example, Ni/SnO_2 core-shell composite [15], carbon-coated Ni [16], Ni/ZnO [17], Al/AlOx-coated Ni [18], Ni@Ni_2O_3 core-shell particles [19], Ni/polyaniline [20], and CuO/Cu_2O-coated Ni [21] show the better microwave absorption performance than the pure core or shell materials. Thus, the EM wave absorption abilities of Ni particles can be obviously enhanced after coating inor-ganic and nonmagnetic shells.

Herein, we report the microwave absorption properties of core-shell structured Ni based composited and discuss how does core-shell ameliorate the electromagnetic wave absorption properties and also investigate the related electromagnetic attenuation theory in detail.

2. Core-shell Ni@oxide composite as microwave absorbers

2.1 Core-shell Ni@ZnO composites as microwave absorbers

For the ZnO nanostructural materials, due to the features of lightweight semi-conductive properties and its easily mass synthesis, they are expected to be the potential applications in EM wave absorbing materials [22]. Therefore, when the Ni particles were covered by ZnO shell, the electromagnetic properties of Ni would be boosted, correspondingly. Moreover, it is well accepted that the EM absorp-tion properties are closely related with their morphologies. Herein, the various morphologies of Ni/ZnO composites were synthesized by control of the amounts of $NH_3{\cdot}H_2O$, and the microwave absorption properties of these Ni/ZnO composites have been investigated based on the complex permittivity and permeability.

2.1.1 Preparation of core-shell Ni@ZnO composites

Ni microspheres were prepared based on our previous publication [15]. Ni/ZnO composites were synthesized through a facile hydrothermal method [23]. Typically, 0.05 g of the as-obtained Ni microspheres was distributed in 60 mL distilled water. Then 0.45 g of $Zn(CH_3COO)_2{\cdot}2H_2O$ and a certain amounts of ammonia solution were added into the mixture solution. The mixture was transferred into a Teflon-lined stainless steel autoclave, and maintained at 120°C for 15 h. The precipitates were collected by centrifugation, washed several times with distilled water and absolute ethanol, respectively. For the convenience of discussion, the Ni/ZnO prepared at 1 mL $NH_3{\cdot}H_2O$, 2 mL $NH_3{\cdot}H_2O$ and 3 mL $NH_3{\cdot}H_2O$ were denoted as SA, SB and SC, respectively.

Figure 1a displays the representative FESEM image of the Ni particles, which possesses uniformly spherical shape and the diameter is about 1.0–1.2 μm. The SEM images of the Ni/ZnO obtained at different contents of $NH_3{\cdot}H_2O$ are displayed in **Figure 1b–d**. **Figure 1b** exhibits that the as-prepared Ni/ZnO product is composed of plentiful ZnO polyhedrons with the diameter of 0.2–0.5 μm covered on the surface of Ni particles to generate special core-shell structure if small amount of $NH_3{\cdot}H_2O$ (1 mL) was added. If the content of $NH_3{\cdot}H_2O$ is lifted to 2.0 mL, the football-like Ni/ZnO samples with the size of 4–5 μm could be observed (**Figure 1c**). One can infer that the thickness of ZnO polyhedron is about 2-3 μm, which is larger than that of Ni microsphere. Therefore, the Ni microspheres are completely coated by ZnO, thus, we could not see the existence if individual Ni microspheres. When the content of $NH_3{\cdot}H_2O$ is further improved to 3 mL, ZnO rods and Ni microsphere coexist in the final products (**Figure 1d**), separately. These results indicate that the morphology of Ni/ZnO composite can be effectively adjusted by controlling the $NH_3{\cdot}H_2O$ content.

Figure 2 depicts the schematic diagram of the generation for various shapes of Ni/ZnO composite. First, the distributed Ni microspheres are fabricated through a chemical reduction method. Following, the different shapes of Ni/ZnO compos-ites are fabricated by the addition of various $NH_3{\cdot}H_2O$ contents. The ZnO nuclei is prone to plant along special crystal planes and finally generate polyhedron-like or rod-like ZnO products. It is accepted that ZnO is a polar crystal with a polar c-axis ([0001] direction) [24]. In the solution system, the $NH_3{\cdot}H_2O$ consists of the positive hydrophilic group (NH_4^+) and negative hydrophobic group (OH^-). The positive hydrophilic groups would link with the basic cells of crystalline growth $[Zn(OH)_4]^{2-}$ easily by Coulomb force, which means that the positive hydrophilic groups turn into the carriers of $[Zn(OH)_4]^{2-}$; on the other hand, due to existence

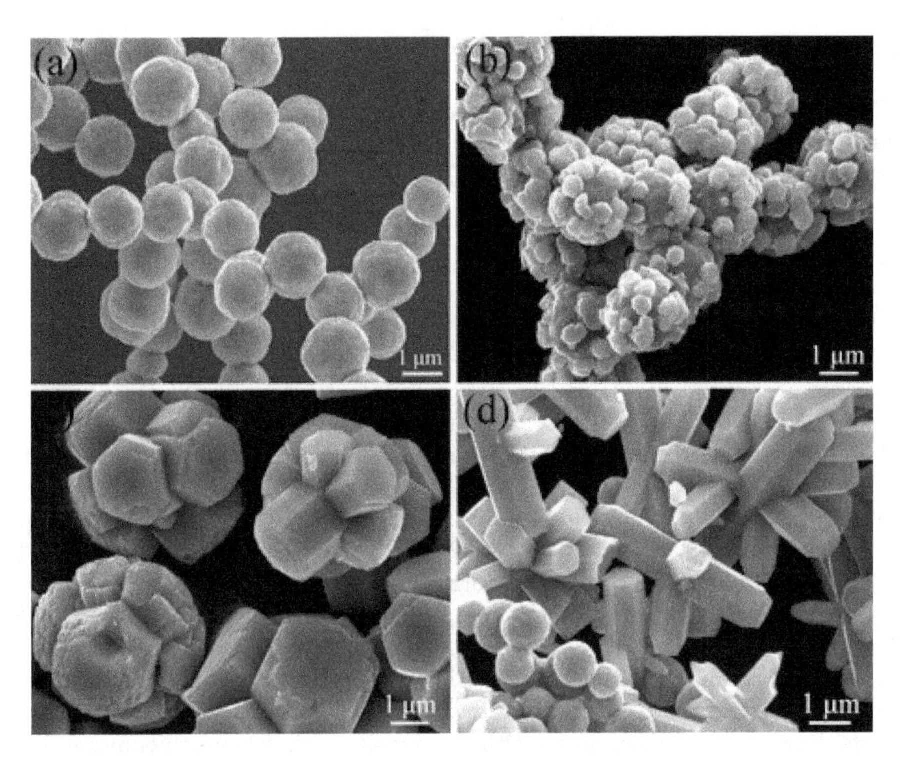

Figure 1.
(a) SEM image of pure Ni microspheres and (b–d) SEM images of as-prepared Ni/ZnO samples prepared at various concentration of $NH_3 \cdot H_2O$: (b) 1 mL, (c) 2 mL, and (d) 3 mL [23] (permission from Elsevier).

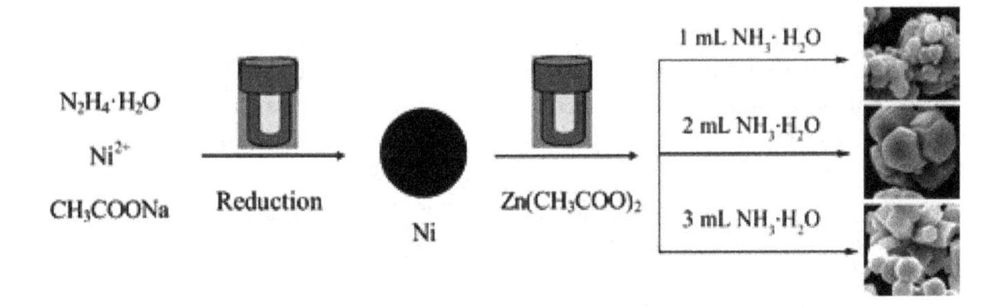

Figure 2.
Schematic illustration of formation of various morphologies of Ni/ZnO composites [23] (permission from Elsevier).

of van der Waals force, the connection between the negative hydrophobic groups and the non-polar lateral surfaces of ZnO would take place, indicating that the occurrence of hydrophobic films thanks to the negative hydrophobic groups on the non-polar lateral surfaces [25]. The basic cells of growing $[Zn(OH)_4]^{2-}$ attracted by the positive hydrophilic groups would move to the polar axial surface (0001) easily to integrate together and remain at the suitable lattice locations but difficultly reach the non-polar lateral remains due to the presence of the hydro-phobic films. It indicates that the positive polar surface (0001) grows quicker than that of non-polar lateral surfaces in a fixed content of $NH_3 \cdot H_2O$. As a result, more OH^- ions assimilate and hamper the growth on the positively charged (0001) surface, forcing a shape transition [26]. At the high content of $NH_3 \cdot H_2O$, long ZnO nanorods could be produced because of the fast growth rate along the [0001] direction [27].

2.1.2 EM properties of core-shell Ni@ZnO composites

To reveal the electromagnetic wave absorption properties of SA, SB and SC paraffin composites, the reflection loss (RL) values of the Ni/ZnO samples are calculated based on following equations [28]:

$$RL = 20\log_{10}|(Z_{in} - Z_0)/(Z_{in} + Z_0)| \quad (1)$$

$$Z_{in} = Z_0\sqrt{\frac{\mu_r}{\varepsilon_r}}\tanh\left(j\frac{2\pi fd\sqrt{\mu_r\varepsilon_r}}{c}\right) \quad (2)$$

Herein Z_0 is the impedance of free space, Z_{in} is the input impedance of the material, f is the frequency of the microwave, c is the velocity of microwave in free space, μ_r and ε_r are, respectively, the relative complex permeability and permittivity, and d is the thickness of the absorber. The RL values of the three samples with a thickness of 2.0 mm are displayed in **Figure 3a**. The SA sample holds the outstanding EM wave absorption performances. A strong peak (−48.6 dB) could be seen at 13.4 GHz. The RL less than −10 dB (90% absorption) reaches 6.0 GHz (10.5–16.5 GHz). Furthermore, the RL less than −20 dB (99% microwave dissipation) is also obtained in the range of 11.5–14.2 GHz. But, for the SB and SC samples, they present inferior microwave dissipation capabilities. **Figure 3b** depicts the simulated RL of SA paraffin-composite with various thicknesses in the frequency of 1–18 GHz. Clearly, one can notice that the optimal RL shifts into lower fre-quency range along with an increased thickness, indicating that we could adjust the absorption bandwidth by tuning absorber thickness. From above analysis, one can note that the minimal RL of −48.6 dB could be observed at 13.4 GHz with a layer thickness of 2.0 mm. The effective absorption (below −10 dB) bandwidth could be monitored in the frequency of 9.0–18.0 GHz by control of the absorber thickness between 1.5 mm and 2.5 mm. Furthermore, the frequency with RL below −20 dB could be observed at 11.1–16.2 GHz with thickness of 1.8–2.2 mm. For the SA sample, the enhanced microwave absorption properties are stemmed from the good impedance match, synergistic effect between dielectric loss and magnetic loss, and special core-shell microstructures, which could induce the interference of microwave multiple reflection [29]. In addition, the compact polyhedron ZnO coat-ing brings the metal/dielectric interfaces, in which the interface polarization boosts the microwave dissipation. For the football-like Ni/ZnO (SB), the size of ZnO is so big that the Ni microspheres could not interact with incident microwave, which

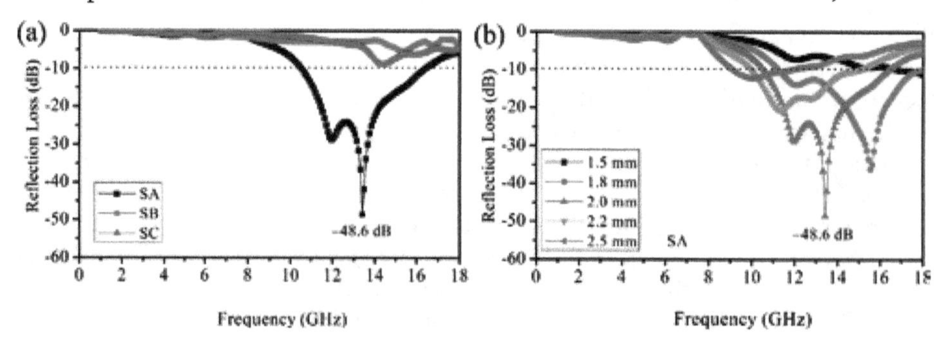

Figure 3.
(a) Frequency dependences of RL with the thickness of 2.0 mm for the three samples SA, SB, SC; (b) RL of Ni/ZnO (SA) paraffin composite of varying thicknesses [23] (permission from Elsevier).

gives rise to the mismatch between the magnetic loss and dielectric loss, leading to inferior microwave absorption. For the SC, due to the presence of uncover Ni, these uncoated Ni microspheres play a negative in the wave-absorption of materials thanks to the occurrence of a significant skin effect when its surface is irradiated by microwaves [30].

On the basis of transmission line theory, the suitable microwave absorption properties are determined by two key factors. One factor is the impedance match, which need the complex permittivity is close to the complex permeability, and the other one is the EM attenuation ability, which dissipates the microwave energy through dielectric loss or magnetic loss. The EM attenuation was determined by the attenuation constant α, which can be expressed as [31, 32]:

$$\alpha = \frac{\sqrt{2}\pi f}{c} \times \sqrt{(\mu''\varepsilon'' - \mu'\varepsilon') + \sqrt{(\mu''\varepsilon'' - \mu'\varepsilon')^2 + (\mu'\varepsilon'' + \mu''\varepsilon')^2}} \quad (3)$$

where f is the frequency of the microwave and c is the velocity of light. **Figure 4** displays the frequency dependence of the attenuation constant. The SA possesses the biggest α in all measured frequency ranges, meaning the outstanding attenuation. Moreover, based on the above equation, one can notice that the attenuation constant is closely related to the values of ε″ and μ″. The highest ε″ and μ″ values are the responsible for the highest α in Ni@ZnO (SA) core/shell structures, which is related to interface polarization and relaxation. As a result, the enhancement of the microwave absorption properties for the dielectric coating originates from the improvement of dielectric loss and magnetic loss.

2.2 Core-shell Ni@CuO composites as microwave absorbers

Nowadays, CuO is well accepted as an important p-type semiconductor, which holds the unique features of narrow band gap (E_g = 1.2 eV), and has captured more and more interests. This material has proved to exhibit widespread potential applications in optical switches, anode materials, field emitters, catalyst, gas sensors, photoelectrode and high-temperature micro-conductors [33, 34]. Recently, CuO has been realized as an efficient material for the preparation of microwave absorbing materials [21, 35, 36]. Herein, we fabricated the core-shell structural composites with Ni cores and rice-like CuO shells via a facile method. The microwave absorp-tion properties of Ni,CuO and Ni/CuO composites are studied in term of complex

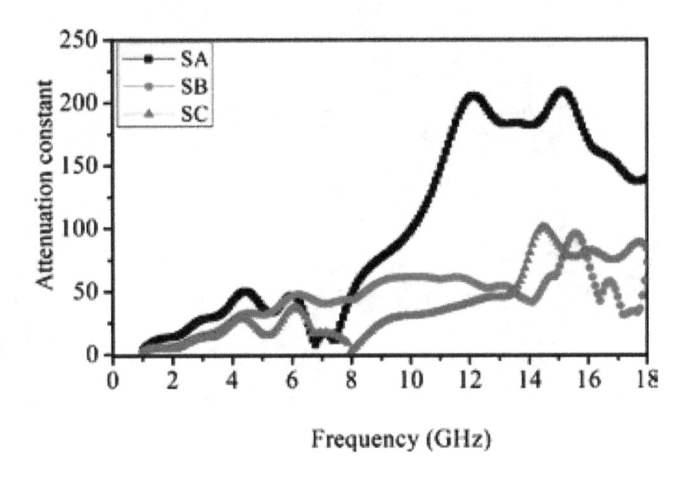

Figure 4.
Attenuation constant of Ni/ZnO samples-paraffin composites versus frequency [23] (permission from Elsevier).

permittivity and permeability. In comparison with pristine Ni and CuO, rice-like CuO-coated Ni composites displayed the enhanced microwave absorption properties. Furthermore, we also studied the effects of CuO amounts on microstructures and microwave absorption properties of Ni/CuO composites in detail.

2.2.1 Preparation of core-shell Ni@CuO composites

The Ni microspheres were prepared by a solvothermal method, which was described in our pervious literature [15]. Synthesis of CuO nanoflakes: $CuCl_2{\cdot}2H_2O$ (0.36 g) was dissolved in a mixture of distilled water (60 mL) and ammonia (2 mL) under continuously stirring (30 min); The final mixture was transferred into a Teflon-lined autoclave and heated hydrothermally at 150°C for 15 h.

Synthesis of CuO rice-coated Ni core/shell composites [37]: the as-prepared Ni microspheres (0.05 g) and $CuCl_2{\cdot}2H_2O$ (0.36 g) were both added in distilled water (60 mL). Then, the ammonia (2 mL) was introduced into the mixture. Finally, the prepared mixture was moved into a Teflon-lined autoclave. The Teflon-lined autoclave was sealed and kept at 150°C for 15 h. The Ni/CuO composites prepared at 0.18 g $CuCl_2{\cdot}2H_2O$, 0.36 g $CuCl_2{\cdot}2H_2O$ and 0.54 g $CuCl_2{\cdot}2H_2O$ were denoted as S-1, S-2 and S-2, respectively.

Figure 5c, d exhibits FESEM micrographs of Ni@CuO composites with different magnification after hydrothermal treatment at 150°C for 15 h. It can be observed that the products are composed of CuO rices-coated smooth Ni microspheres heterostructures with the diameter of 1.0–1.2 μm. One can notice that rice-like CuO/Ni composites hold rough surfaces, which results from compactly aggregated panicle-shape CuO nanostructures. In order to get more information about microstructure of Ni/CuO composite, TEM and HR-TEM images of Ni/CuO composites are carried out. The core-shell structure of Ni/CuO composite can be

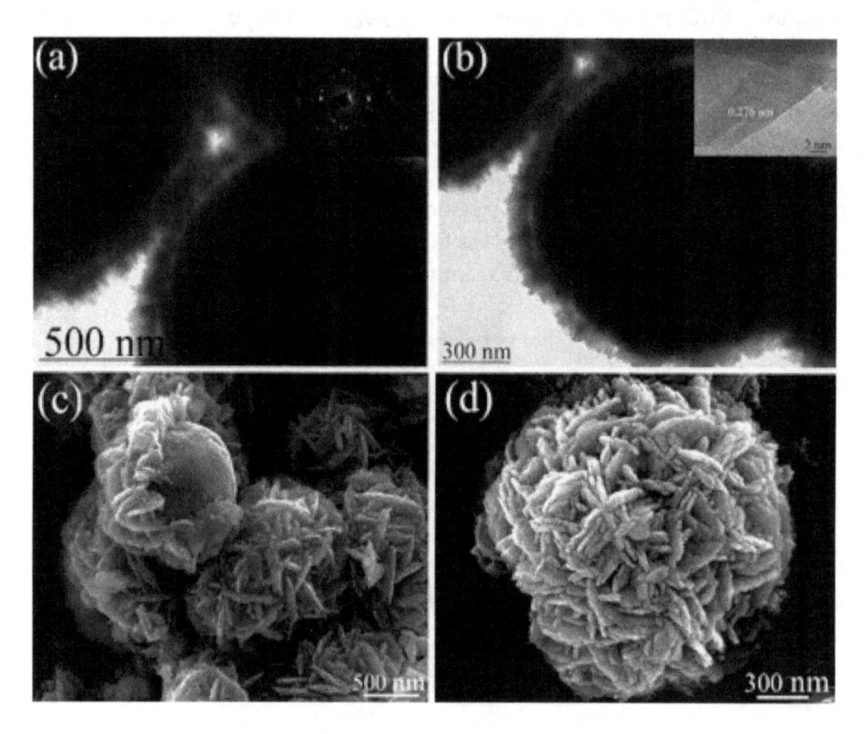

Figure 5.
Typical TEM and HRTEM images of the as-prepared Ni/CuO structures: (a) low magnification, inset of (a) shows the SAED pattern; (b) high magnification; inset of (b) shows the HRTEM. (c, d) FESEM images of the Ni microsphere-CuO rice core-shell structures [37] (permission from RSC).

clearly observed from **Figure 5a, b**. The inset SAED pattern of the CuO particles indicated that CuO particles are polycrystalline (**Figure 5a**). The HRTEM image (inset of **Figure 5b**) displays that the lattice spacing is 0.276 nm, which is in good agreement with the (110) lattice spacing of CuO. Based on the SEM and TEM results, it can be concluded that the CuO is deposited on the surface of Ni, the core-shell composites are obtained under this procedure.

Figure 6 exhibits the morphologies of the obtained products with different molar ratio of the $CuCl_2{\cdot}2H_2O$ to Ni microspheres. Noticeably, the surfaces of all samples turns coarser in comparison with the pure Ni microspheres, which indicates the successful coating of the CuO nanoparticles on the pristine Ni surfaces. Furthermore, the shape and coverage density of CuO materials could be controlled by tuning the content of precursor (Cu^{2+}). When the molar ratio of the $CuCl_2{\cdot}2H_2O$ to Ni microspheres in the precursor solution is 1: 0.85 (S-1), one can find (**Figure 6a, b**) that the Ni microspheres are coated by a large number of CuO nanorices. But, due to the low content of precursor (Cu^{2+}), we just could obtain thin CuO shell. If the

Figure 6.
FESEM images of hierarchical Ni/CuO core-shell heterostructures with different molar ratio: (a, b) S-1; (c, d) S-2; and (e, f) S-3 [37] (permission from RSC).

molar ratio of precursor is enhanced to 2: 0.85 (S-2), one can see that the aggregation occurs and CuO nanorices are compactly covered on the smooth surfaces of Ni microspheres to produce coarser thick CuO shells (**Figure 6c, d**). If the molar ratio is improved continuously to 3:0.85 (S-3), a thick layer of compact CuO nanoflakes coated on Ni microspheres could be observed (**Figure 6e, f**). Based on above results, the microstructures and coverage density of CuO shells can be effectively monitored by selecting a suitable content of Cu^{2+}.

2.2.2 EM properties of core-shell Ni@CuO composites

To compare and assess the EM wave absorption properties of Ni, Ni/CuO core-shell composites, and CuO nanoflakes, the paraffin (30 wt%, which is transparent to microwave) are mixed with as-obtained products, and pressed into a ring shape with an outer diameter of 7.00 mm and an inner diameter of 3.04 mm. The microwave absorption abilities of these as-fabricated products could be evaluated by the RL val-ues, which could be simulated on the basis of the complex permeability and permit-tivity with the measured frequency and given layer thickness [38, 39]. As presented in **Figure 7a**, the three Ni/CuO composites show the superior microwave-absorption properties to those of the pure Ni microspheres and the CuO nanoflakes. Taking an example, when the thickness is 2 mm, the S-1 sample exhibits the enhanced EM-wave absorption with the minimal RL value of −15.6 dB at 11.9 GHz among the five samples. From Eqs. (1) and (2), one can find that the thickness of the absorber is one important factor, which would affect the position of minimal RL value and the absorption bandwidth. Therefore, the RL values of Ni/CuO samples with different thicknesses are also calculated. Compared with S-2 (**Figure 7c**) and S-3 (**Figure 7d**) samples, the S-1 (**Figure 7b**) displays the outstanding microwave absorption per-formances. The lowest RL of the S-1 sample is −62.2 dB at 13.8 GHz with the only thickness of 1.7 mm. The effective absorption (below −10 dB) bandwidth can be tuned between 6.4 GHz and 18.0 GHz by adjusting thickness in 1.3–3.0 mm.

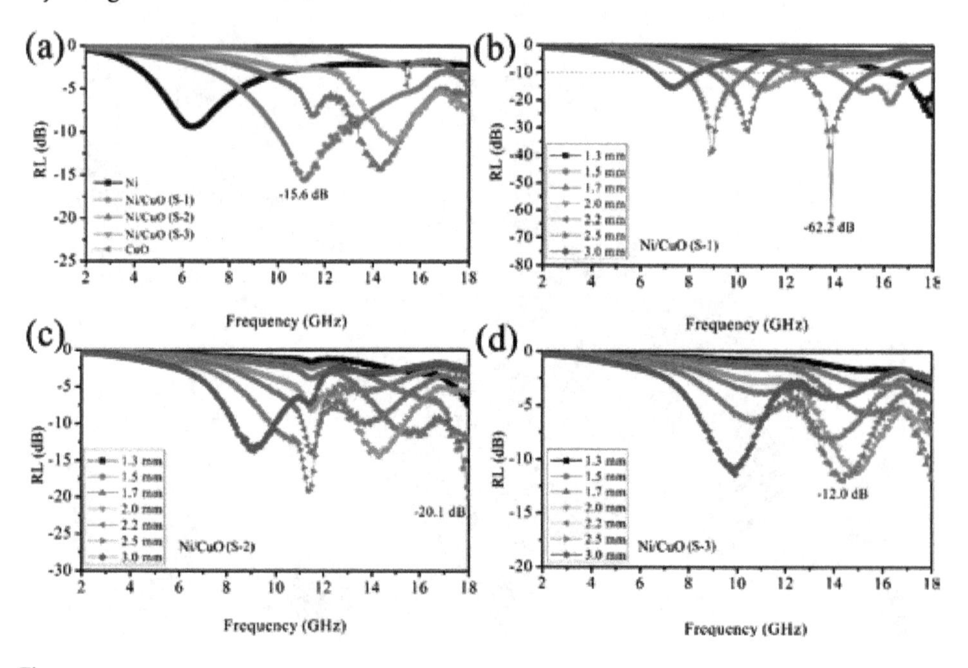

Figure 7.
(a) Comparison of RL of the five as-obtained samples with a thickness of 2.0 mm. The RL values of (b) S-1, (c) S-2, and (d) S-3 samples with various thicknesses [37] (permission from RSC).

Notably, the optimal RL peaks gradually shift toward lower frequencies with an increased absorber thickness, which can be described by quarter-wavelength cancelation model that the incident and reflected waves in the absorber are out of phase 180° and causing the reflected waves in the air-absorber interface are totally cancelled [40]. The enhanced microwave absorption property of core-shell Ni/CuO composites can be obtained by tuning the content of CuO. The rice-like CuO shell is expected to be helpful for the dissipation proprieties of the core/shell compos-ites. The CuO shells are covered on the surface of Ni microspheres to produce the special core-shell structure, which brings metal-dielectric hetero-interfaces to cause interfacial polarization. It is supposed that the interfacial polarization taken place in heterostructures consisting of at least two constituents [17, 41, 42]. This type of polarization occurring at the interfaces results from the movement of charge carri-ers between different compositions, which accumulate the moving charge at these interfaces. When irradiated by alternating EM fields, the accumulated charge would redistribute periodically between Ni cores and CuO shells, which are favorable for the microwave dissipation. However, for the S-2 and S-3 samples, thanks to the high content of CuO, we cannot observe the synergistic effect between Ni cores and CuO shells, which gives rise to inferior microwave absorption.

3. Core-shell Ni@non-oxide composite as microwave absorbers

3.1 Ni@ZnS composites as microwave absorbers

It is well known that ZnS, a wide band-gap semiconductor with the band-gap energy (Eg) of 3.6 eV, has been widely applied in displays, sensors, and lasers for many years [43, 44]. As far as I am concerning, the publications about the microwave absorption properties of ZnS and core/shell structured Ni/ZnS are not reported. Herein, we synthesized the core-shell structured composites with Ni cores and ZnS nanowall shells through a facile method. The microwave absorption properties of Ni, ZnS and Ni@ZnS composites are detailedly investigated in the frequency of 2–18 GHz.

3.1.1 Preparation of core-shell Ni@ZnS composites

ZnS nanowall-coated Ni composites were fabricated via a two-stage method [45]. First, Ni microspheres were synthesized based on our previous paper [15]. Second, Ni microspheres are coated by ZnS nanowalls to generate the core-shell structural composites. In the modified procedure, the as-obtained Ni microspheres (0.05 g) and $Zn(CH_3COO)_2{\cdot}2H_2O$ (0.45 g) were added in a mixture solution of etha-nol (30 mL) and distilled water (30 mL). Then, $Na_2S{\cdot}9H_2O$ (0.50 g) and ammonia solution (4 mL) were added into the mixture solution with intensely stirring for 20 min. Finally, the mixture was moved into a Teflon-lined stainless steel autoclave, and kept at 100°C for 15 h. In order to study the effect of core-shell structure on the microwave absorption properties of the Ni/ZnS composite, the pure ZnS particles were also prepared according to the above procedure without addition of Ni microspheres.

Inset of **Figure 8a** presents the XRD profiles of Ni microspheres, ZnS particles and Ni@ZnS composites. For the Ni microspheres, all the diffraction peaks can be well indexed to the face-centered cubic (fcc) structure of nickel (JCPDS No. 04-0850). For the ZnS particles, all diffraction peaks can be indexed to a typical zinc blende structured ZnS, which is consistent with the standard value for bulk ZnS (JCPDS Card No. 05-0566). The crystal structure of core/shell structured Ni/ZnS

Figure 8.
SEM images of (a) Ni microspheres, (b) ZnS particles, and (c, d) the as-prepared Ni/ZnS composites. The inset in (b) is the magnified SEM image of ZnS particles. Inset in Figure 8a is the XRD patterns of Ni microspheres, ZnS particles and Ni/ZnS composites [45] (permission from RSC).

composites is also investigated by XRD measurements. Noticeably, we expectantly observed the diffraction peaks, which are in good accordance with Ni and ZnS, respectively. One can conclude that the as-obtained core/shell structural composites are made up of crystalline Ni and ZnS. **Figure 8a** presents the SEM image of the Ni microspheres. One can notice that the products have a relatively uniform spherical shape with the diameter of 0.7–1.0 μm. The pure ZnS products appear to have irreg-ular shapes (**Figure 8b**). In raw ZnS particles, the formation of ZnS is via a two-step pathway. The fresh nanoparticles incline to aggregate for the sake of decreasing the surface energy. Therefore, we could obtain the irregular gathering ZnS particles. Whereas, as for the Ni/ZnS system, ZnS particles were generated via the template way (heterogeneous nucleation, raw Ni as template). Therefore, the variation of the shape and dimensions of ZnS particles in pure ZnS and core-shell Ni/ZnS compos-ites could be seen. **Figure 8c, d** present the SEM images of core-shell Ni/ZnS. In comparison with pure Ni (**Figure 8a**), one significant distinction is clearly observed between the Ni/ZnS composites and the pure Ni particles. The distinction is that the Ni particles are absolutely wrapped by the ZnS nanowalls. The large-scale SEM image in **Figure 8d** suggests that the as-prepared Ni/ZnS composites show crinkled and rough textures, which are similar with the reduced graphene oxide sheets [46]. The thickness of ZnS nanowall is about 10 nm.

3.1.2 EM properties of core-shell Ni@ZnS composites

The relative complex permittivity (ε' and ε'') and permeability (μ' and μ'') of the Ni/paraffin, ZnS/paraffin and Ni@ZnS/paraffin composite samples are measured over a frequency of 2–18 GHz. **Figure 9a–c** manifest the real part (ε') and imaginary part (ε″) of the complex permittivity as a function of frequency. The ε′ of Ni/paraffin composite shows a gradual decrease with frequency (**Figure 9a**). However, the ε″ values are relative constant without significant change over the 2–18 GHz. The ε′ and ε″ of ZnS/paraffin composite presents constant value (4.5 and 0.5, respectively) in **Figure 9b**. The ε′ of the Ni/ZnS composite firstly reduces in the frequency of 2–15 GHz and then improves with increasing frequency (**Figure 9c**). Nevertheless, the ε″ exhibits the opposite tendency in the frequency of 2–18 GHz. One can note that the ε″ values of Ni/ZnS composite presents a peak in the 13–15 GHz range, which is originated from the natural resonance behavior of core-shell micro-structure [21, 47]. Furthermore, it can be found that the ε″ values of Ni/paraffin composite are larger than those of ZnS/paraffin composite and Ni@ZnS/paraffin composite. Based on the free electron theory [17], $\varepsilon'' \approx 1/2\ \pi\pi\pi\pi_0\ \rho\rho f$, where ρ is the resistiv-ity. The lower ε″ values of ZnS/paraffin composite and Ni@ZnS/paraffin composite indicate the higher electric resistivity. In general, a high electrical resistivity is favor-able for improving the microwave absorption abilities [48].

Figure 9 (d–f) present the curves of the real part (μ′) and imaginary part (μ″) of the complex permeability as a function of frequency for the Ni/paraffin, ZnS/paraffin and Ni@ZnS/paraffin composites. The μ′ and μ″ of Ni/paraffin composite are 0.81–1.59 and 0.05–0.51, respectively (**Figure 9d**). Compared with the complex permittivity (**Figure 9a**), the values of complex permeability is relatively small, which lead to mismatch impedance. The impedance match is required that complex

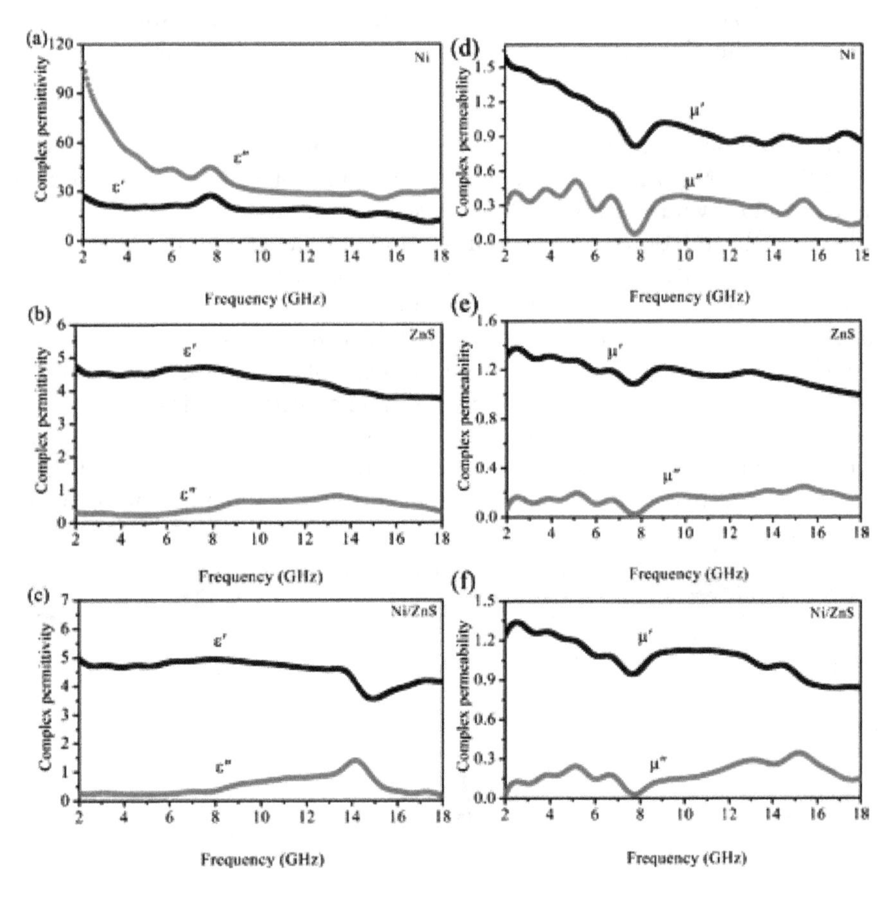

Figure 9.
Frequency dependence of the complex permittivity ($\varepsilon_r = \varepsilon' - j\varepsilon''$) of (a) Ni microspheres, (b) ZnS particles, and (c) Ni/ZnS composites; frequency dependence of the complex permeability ($\mu_r = \mu' - j\mu''$) of (d) Ni microspheres, (e) ZnS particles, and (f) Ni/ZnS composites [45] (permission from RSC).

permittivity is close to the permeability, which can make microwaves enter into the materials as much as possible [49]. The higher permittivity of the absorber plays a negative role in the impedance match [50], thus gives rise to inferior microwave absorption. From the **Figure 9e**, **f**, it can be found that the complex permeability of ZnS/paraffin and Ni@ZnS/paraffin composite exhibit the similar tendency with an increased frequency. The μ' values of 0.99–1.38 and 0.84–1.34 could be observed in ZnS/paraffin and Ni@ZnS/paraffin composites, respectively. The μ'' values are in the range of 0.02–0.24 and 0.03–0.34 for the ZnS/paraffin and Ni@ZnS/paraffin composites, respectively. Given the complex permittivity (**Figure 9b**, **c**), it can be found that the relation between permittivity and permeability is prone to be close (good impedance match). The good impedance match is beneficial for the microwave absorption. On the basis of the above results, one can deduce that the impedance match of ZnS/paraffin and Ni@ZnS/paraffin is superior to that of Ni/paraffin composite. Thus, the ZnS/paraffin and Ni@ZnS/paraffin composites may possess better dissipation abilities of microwave energy.

It is widely accepted that the RL values could be utilized to evaluate the microwave absorption abilities of EM materials. **Figure 10a** exhibits the calculated RL values of Ni, ZnS and Ni/ZnS paraffin composites with 70 wt% amounts at the thickness of 2.5 mm in the frequency range of 2–18 GHz. Because a paraffin matrix is transparent to microwaves, these results are generally considered as the wave-absorption abilities of the filler itself. It is noting that the microwave absorption properties of Ni particles are weak and the optimal RL value is only −3.04 dB at 5.28 GHz, which is due to the mismatch impedance. Another possible factor is that the skin effect could be observed in Ni microspheres, which is harmful for microwave absorption [30]. Compared with Ni particles, ZnS particles and ZnS nanowall-coated Ni composite presents the superior microwave absorption abilities, which stems from good impedance match. It is worth pointing that, for Ni@ZnS composite, the minimal RL of −20.16 dB is observed at 13.92 GHz and RL values below −10 dB are seen in the 12–16.48 GHz rang. **Figure 10b** displays the relationship between RL and frequency for the paraffin wax composites with 70 wt% Ni/ZnS in various thicknesses. The optimal RL is −25.78 dB at 14.24 GHz with the corresponding thickness of 2.7 mm. The effective absorption (less than −10 dB) bandwidth reaches 4.72 GHz (11.52–16.24 GHz). Interestingly, with increasing the sample thickness, the location of minimal absorption peaks almost keeps the same at various thicknesses without moving to lower frequency, which has also been recorded by other' groups [51]. The location of absorption peaks is in accordance with the natural resonance, which means the resonance behavior in permittivity influences the microwave absorption.

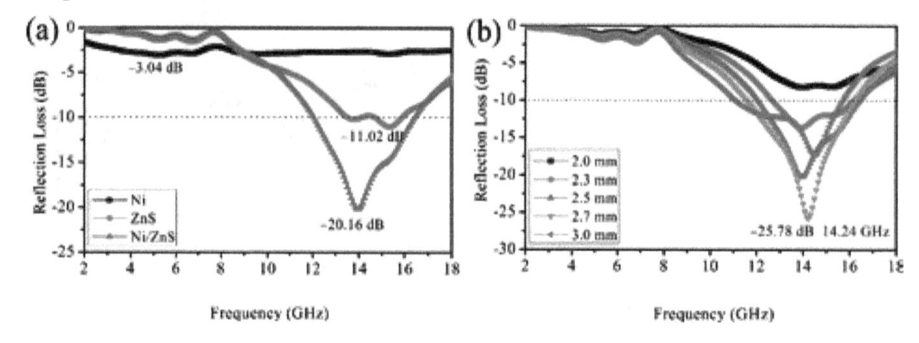

Figure 10.
(a) RL curves of Ni, ZnS and Ni/ZnS paraffin composite with 70 wt% loadings at the thickness of 2.5 mm; (b) RL curves of 70 wt% Ni/ZnS wax-composite at various thicknesses in the frequency of 2–18 GHz [45] (permission from RSC).

3.2 Urchin-like ZnS/Ni_3S_2@Ni composites as microwave absorbers

The EM absorption properties of nanomaterial are associated with the size, shape, and dimensionality. Metal sulfide nanomaterials have captured more and more attention due to their excellent properties and wide applications in electronic and optoelectronic devices. Nickel subsulfide and ZnS are the important categories in the metal sulfide family [52, 53] thanks to their various applications such as in lithium ion batteries, supercapacitors, dye-sensitized [54, 55] and charge transfer, anion exchange, electricity generation [56, 57], respectively. Due to the semiconductor properties of nickel subsulfide and ZnS, it can induce dipole and space charge polarizations when placed in the alternated electromagnetic field [16, 58]. Moreover, nickel subsulfide (Ni_3S_2) are highly metallic compared to insulating oxide compounds, which can cause conductive loss [59]. The core-shell Ni@ZnS com-posites with the improvement of electromagnetic properties were reported by our earlier literatures [45, 60]. To the best of our knowledge, the EM wave absorption of core-shell structured ternary ZnS/Ni_3S_2@Ni composite is hardly found in the pub-lished papers. Herein, a novel ZnS/Ni_3S_2@Ni composite with urchin-like core-shell structure is successfully synthesized, and it exhibits the excellent EM absorption and the absorption mechanism of such unique hierarchical microstructure is also discussed in detail.

3.2.1 Preparation of urchin-like ZnS/Ni_3S_2@Ni composites

The monodispersed Ni microspheres were prepared according to our previous literatures [37, 45, 62]. Synthesis of urchin-like core-shell structured ZnS/Ni_3S_2@Ni composite [61]: in brief, Ni (0.05 g) was dispersed in a mixture of aqueous solution of distilled water (30 mL) and ethanol (30 mL) containing 1.0 M NaOH. Then, 1 mmol $ZnCl_2$ and 2 mmol $Na_2S{\cdot}9H_2O$ were introduced into above mixture, respectively. The mixture was transferred to a Teflon-lined autoclave. The sealed autoclave was heated to 120°C for 15 h. To explore the possible generation mechanism of core-shell microstructure urchins and effects of temperatures on the shapes of target products, the temperature-control experiments (60°C, 80°C, 100° C and 120°C) were also conducted.

The phase constituent and structure of the as-prepared Ni microspheres and urchin-like ZnS/Ni_3S_2@Ni products are characterized by XRD. **Figure 11a** displays the XRD curve of uncoated Ni microspheres, which could be well assigned to face-centered cubic structure of Ni (JCPDS No. 040850). As presented in **Figure 11b**, except for the diffraction peaks of Ni, the other diffraction peaks could be attributed to the zinc-blende ZnS (JCPDS Card No. 05-0566) and Ni_3S_2 (JCPDS Card No. 76-1870). From these XRD patterns, it can be confirmed that the core-shell ZnS/Ni_3S_2@Ni composite is composed of nickel, nickel sulfide and zinc sulfide. It can be inferred that nickel has functioned as the template for in-situ generation of Ni_3S_2 and deposition of ZnS.

To investigate the morphology of the products, FESEM images are taken for Ni microspheres and ZnS/Ni_3S_2@Ni composite and the corresponding results were shown in **Figure 12**. From the **Figure 12a**, it can be seen that Ni products were com-posed of uniform distribution and smooth surface of microspheres. **Figure 12b, c** presents the different magnification FESEM images of core-shell ZnS/Ni_3S_2@Ni composite. Interestingly, from panoramic observation (**Figure 12b**), urchin-like products are optionally grown on the surfaces of Ni microspheres and the out-line of Ni spheres cannot be clearly observed due to the formation of ZnS/Ni_3S_2. Noticeably, the decease size of Ni particles could be observed, which suggests the depletion of Ni products. The thorns possess about 1 μ m and 50 nm in length and

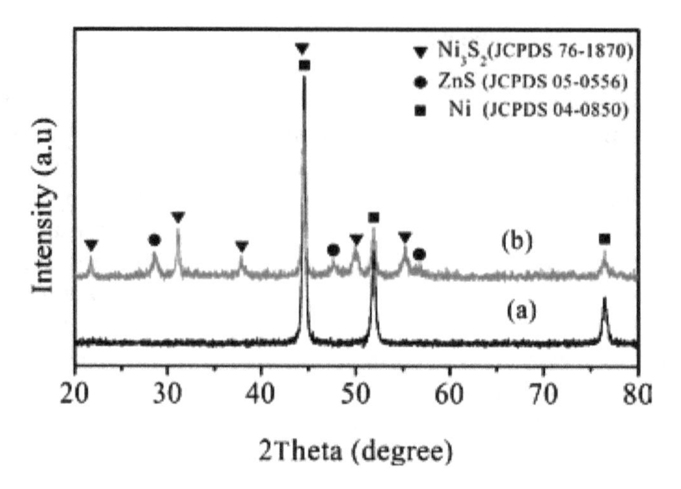

Figure 11.
XRD patterns of (a) pristine Ni microspheres and (b) as-prepared ZnS/Ni_3S_2@Ni [61] (permission from RSC).

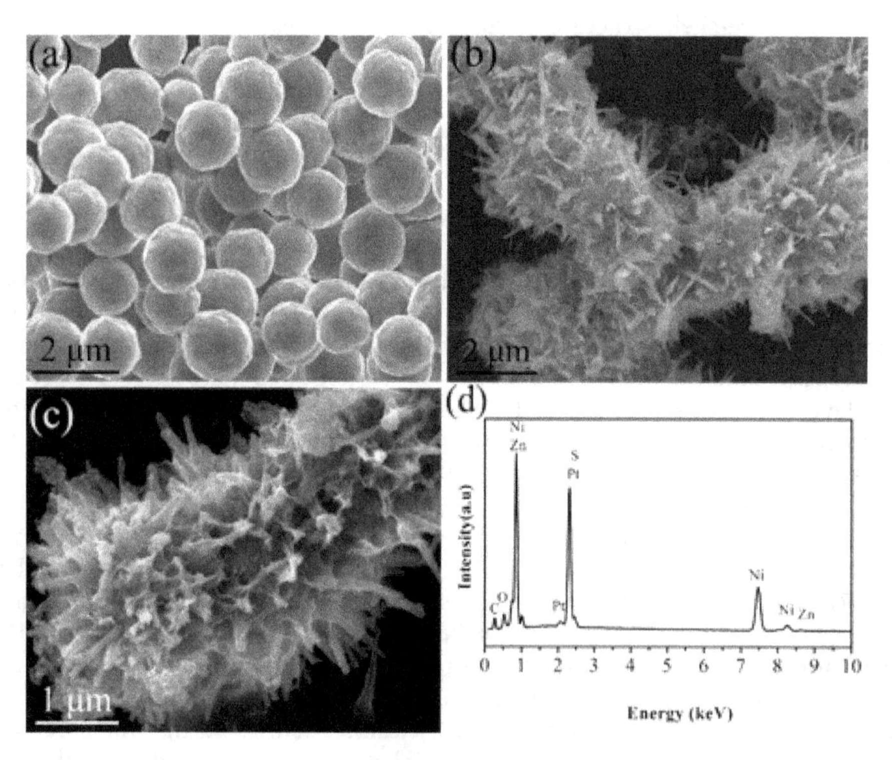

Figure 12.
FESEM images of (a) Ni microspheres, (b, c) urchin-like core-shell structured ZnS/Ni_3S_2@Ni composite, and (d) EDS profile of ZnS/Ni_3S_2@Ni composite [61] (permission from RSC).

diameter, respectively. Further observation from the high magnification FESEM image (**Figure 12c**), the existence of crumple products encircle Ni particles and the crinkled products are expected to be linked between Ni particles and thorns. **Figure 12d** presents the EDS of the ZnS/Ni_3S_2@Ni composite. The EDS reveals the presence of elements of S, Zn and Ni. Pt peaks are also seen in the EDS curve because the SEM sample is prepared by sputtering of platinum onto the sample.

It is assumed that the reaction temperature has an effect on the morphology of core-shell heterostructure. At low temperature (60°C), interestingly, there are plentiful waxberry-like products existed in **Figure 13a**. It is due to the fact that Ni

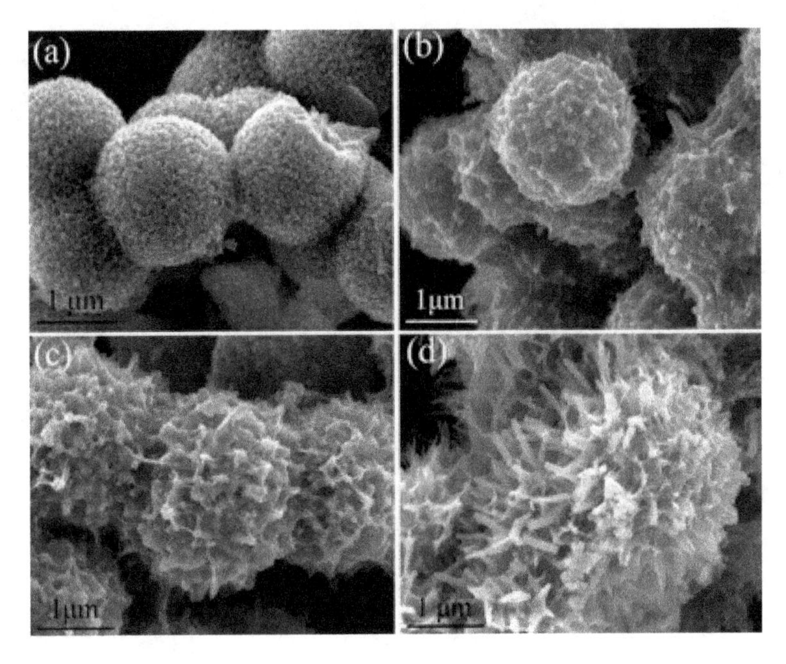

Figure 13.
FESEM images of core-shell ZnS/Ni_3S_2@Ni composites prepared at various reaction temperatures: (a) 60°C, (b) 80°C, (c) 100°C, and (d) 120°C [61] (permission from RSC).

microspheres were coated by wrinkle ZnS products. Ni_3S_2 nanoparticles are in-situ formed by depletion of Ni particles and then covered on the surface of left Ni products. With elevating the reaction temperatures to high temperatures (80°C), one can notice that some protuberant prickles were grown on the surfaces of core-shell composites (**Figure 13b**). When the reaction temperature is increased to 100° C, the presence of more and strong protuberant stabs on the crumple surfaces of composite can be obviously observed (**Figure 13c**). With further enhancing the temperature to 120°C, the target urchin-like core-shell structural ZnS/Ni_3S_2@Ni composites are formed with numerous of thorns or rods grown on the rugged surfaces (**Figure 13d**).

3.2.2 EM properties of urchin-like ZnS/Ni3S2@Ni composites

With the purpose of revealing the electromagnetic wave absorption properties, the representation RL values of the core-shell ZnS/Ni_3S_2@Ni paraffin-composites with different sample thicknesses are simulated. **Figure 14** reveals the RL values of the core-shell ZnS/Ni_3S_2@Ni composites prepared at different temperatures with thick-ness varies from 0.8 to 2.5 mm in the frequency range of 1–18 GHz. It can be found that urchin-like ZnS/Ni_3S_2@Ni composite exhibits outstanding electromagnetic absorption. The minimal RL is down to −27.6 dB at 5.2 GHz as the thickness is 2.5 mm. Notably, the reflection loss of urchin-like ZnS/Ni_3S_2@Ni is −21.6 dB at 13.3 GHz with the absorber thickness of 1.0 mm, and the valuable bandwidth (RL below −10 dB) could reach 2.5 GHz (12.2–14.7 GHz), which is better than those of the literatures about microwave absorption performances of dielectric/magnetic composites, such as Fe@SnO_2 (>−10 dB) [63], and Co/CoO(−14.5 dB) [64]. Meanwhile, the absorption peaks would move to lower frequencies and dual RL peaks could be obtained with an increased absorber thickness above 2.5 mm. This phenomenon could be described by the quarter-wavelength cancelation model [65].

According to above results, we propose a possible electromagnetic wave absorption mechanism for the core-shell ZnS/Ni_3S_2@Ni heterogeneous system When the

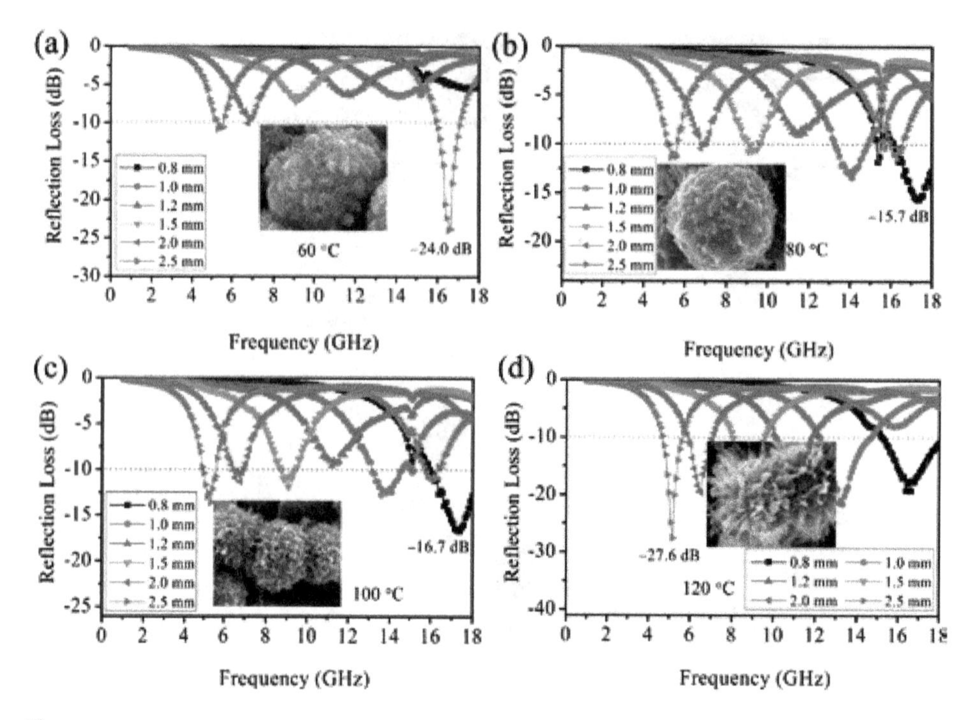

Figure 14.
The calculated RL of core-shell ZnS/Ni_3S_2@Ni composites prepared at various temperatures: (a) 60°C, (b) 80°C, (c) 100°C, and (d) 120°C with different thickness in the frequency range of 1–18 GHz [61] (permission from RSC).

ZnS/Ni_3S_2@Ni composite is subjected to EM wave radiation, the Ni_3S_2 thorns grown on the surfaces of Ni microspheres are expected as antenna receiver to allow electro-magnetic waves penetrate into interior of absorber as much as possible, namely called good impedance match [66, 67]. Moreover, EM absorption is also understood from the viewpoint of Ohmic heating induced by an alternating magnetic field [68], in which ZnS/Ni_3S_2@Ni composite with relatively high electric conductivity. Thirdly, due to the heterogeneous systems of core-shell ZnS/Ni_3S_2@Ni composites, the multi-interfaces between the ZnS, Ni_3S_2, and Ni are favorable for improvement of electromagnetic wave absorption thanks to the interaction between electro-magnetic radiation and charged multipoles at the interfaces [69]. The interfacial polarizations [42, 70] occurring at the interfaces are resulted from the migration of charge carriers through different conductivity properties of the composite material. During the irradiated by alternating electromagnetic field, an additional interfacial relaxation is produced, which is beneficial for the electromagnetic absorption [71]. On the other hand, the cooperation effect between dielectric loss at high frequency and magnetic loss at low frequency contribute to the improvement of electromag-netic absorption [72].

4. Conclusion

Various morphologies of Ni/ZnO composites are successfully fabricated by a hydrothermal method. The shapes of ZnO nanostructures could be effectively monitored by tuning the contents of $NH_3{\cdot}H_2O$. One can find that the morphology of ZnO plays an important role on the microwave absorption capabilities. The core-shell structural Ni/polyhedron ZnO presents relatively high dielectric loss, magnetic loss and microwave dissipation abilities compared with the other Ni/ZnO samples.

The minimal RL could reach −48.6 dB at 13.4 GHz in the absorber thickness of 2.0 mm. The absorption band with RL below −10 dB could reach 6.0 GHz between 10.5 and 16.5 GHz. The enhanced microwave dissipation abilities are attributed to the synergetic effect between dielectric loss and magnetic loss, strong dissipation ability, as well as the multiple polarization of the core/shell interfaces.

The hierarchical Ni-CuO heterostructures have been successfully synthesized by a two-step process. The as-prepared Ni-CuO products display a rice-like coating composite. Moreover, by tuning the molar ratio of $CuCl_2{\cdot}2H_2O$ to the Ni microspheres, different shapes and coverage densities of CuO coating are obtained. The effects of CuO amounts on the Ni microspheres for microwave absorption properties have been investigated. The thin CuO-coated Ni composites (S-1) exhibit the enhanced electromagnetic absorption properties. The optimal RL is −62.2 dB at 13.8 GHz with only thickness of 1.7 mm. The outstanding microwave absorption properties result from the strongest attenuation constant, interfacial polarization of and the synergetic effect between the dielectric loss and magnetic loss.

ZnS nanowall-covered Ni composite is fabricated via a hydrothermal template method. The as-obtained Ni/ZnS composites display the crumble and rough features and the thickness of ZnS nanowall is about 10 nm. In comparison with raw Ni and ZnS particles, the Ni@ZnS composites show superior microwave dissipa-tion abilities. The optimal RL of −25.78 dB could be obtained at 14.24 GHz and the valuable (less than −10 dB) band could reach 4.72 GHz (11.52–16.24 GHz) in the thickness of 2.7 mm. Moreover, the location of absorption peaks is almost similar at various thicknesses without moving to low frequency, which originates from natural resonance in permittivity.

Novel and interesting urchin-like ZnS/Ni_3S_2@Ni composites are synthesized through a two-step process including solution reduction and subsequently a template method. Crucially, the morphologies of the core-shell ZnS/Ni_3S_2@Ni composites are determined by the reaction temperature. Different ZnS/Ni_3S_2@Ni composites prepared at different temperatures show variable electromagnetic absorption responses, for which the urchin-like ZnS/Ni_3S_2@Ni obtained at 120°C show the enhanced EM absorption properties thanks to its promising dielectric loss and magnetic loss, good impedance match, as well as its unique urchin-like structure. Multiple dielectric resonances stemming from effective accumulation of different polarizations in the urchin-like structure are regarded to make a contribution to the enhancement of electromagnetic wave absorption. It is believed that *in situ* synthesis of core-shell ZnS/Ni_3S_2@Ni composites may open up a new avenue for the design and preparation of novel microwave absorbers with promising applica-tion potential.

In a word, the core-shell configuration is proved to be a promising pathway to design high-efficiency EM absorption properties of Ni based composites.

Author details

Biao Zhao[1,2*] and Rui Zhang[1,3]

1 Henan Key Laboratory of Aeronautical Materials and Application Technology, School of Material Science and Engineering, Zhengzhou University of Aeronautics, Zhengzhou, Henan, China

2 Department of Mechanical and Industrial Engineering, University of Toronto, Toronto, Ontario, Canada

3 School of Material Science and Engineering, Zhengzhou University, Zhengzhou, Henan, China

References

[1] Jian X, Wu B, Wei Y, Dou SX, Wang X, He W, et al. Facile synthesis of Fe_3O_4/GCs composites and their enhanced microwave absorption properties. ACS Applied Materials & Interfaces. 2016;**8**(9):6101-6109. DOI: 10.1021/acsami.6b00388

[2] Lv H, Guo Y, Yang Z, Cheng Y, Wang LP, Zhang B, et al. A brief introduction to the fabrication and synthesis of graphene based composites for the realization of electromagnetic absorbing materials. Journal of Materials Chemistry C. 2017;**5**(3):491-512. DOI: 10.1039/C6TC03026B

[3] Zhao B, Deng J, Zhang R, liang L, Fan B, Bai Z, et al. Recent advances on the electromagnetic wave absorption properties of Ni based materials. Engineered Science. 2018;**3**:5-40. DOI: 10.30919/es8d735

[4] Wang G, Gao Z, Tang S, Chen C, Duan F, Zhao S, et al. Microwave absorption properties of carbon nanocoils coated with highly controlled magnetic materials by atomic layer deposition. ACS Nano. 2012;**6**(12):11009-11017. DOI: 10.1021/nn304630h

[5] Ye F, Song Q , Zhang Z, Li W, Zhang S, Yin X, et al. Direct growth of edge-rich graphene with tunable dielectric properties in porous Si3N4 ceramic for broadband high-performance microwave absorption. Advanced Functional Materials. 2018;**28**(17):1707205. DOI: 10.1002/adfm.201707205

[6] Liu Q, Cao Q , Bi H, Liang C, Yuan K, She W, et al. CoNi@SiO2@TiO2 and CoNi@Air@TiO2 microspheres with strong wideband microwave absorption. Advanced Materials. 2016;**28**(3):486-490. DOI: 10.1002/adma.201503149

[7] Zhang Y, Huang Y, Zhang T, Chang H, Xiao P, Chen H, et al. Broadband and tunable high-performance microwave absorption of an ultralight and highly compressible graphene foam advanced materials. 2015;**27**(12):2049-2053. DOI: 10.1002/adma.201405788

[8] Xi L, Wang Z, Zuo Y, Shi X. The enhanced microwave absorption property of $CoFe_2O_4$ nanoparticles coated with a Co_3Fe_7–Co nanoshell by thermal reduction. Nanotechnology. 2011;**22**(4):045707. DOI: 10.1088/0957-4484/22/4/045707

[9] Zhou W, Hu X, Bai X, Zhou S, Sun C, Yan J, et al. Synthesis and electromagnetic, microwave absorbing properties of Core–Shell Fe_3O_4–Poly(3, 4-ethylenedioxythiophene) microspheres. ACS Applied Materials & Interfaces. 2011;**3**(10):3839-3845. DOI: 10.1021/am2004812

[10] Chen Y-J, Zhang F, G-g Z, X-y F, Jin H-B, Gao P, et al. Synthesis, multi-nonlinear dielectric resonance, and excellent electromagnetic absorption characteristics of Fe_3O_4/ZnO core/shell nanorods. The Journal of Physical Chemistry C. 2010;**114**(20):9239-9244. DOI: 10.1021/jp912178q

[11] Liu J, Che R, Chen H, Zhang F, Xia F, Wu Q , et al. Microwave absorption enhancement of multifunctional composite microspheres with spinel Fe_3O_4 cores and anatase TiO_2 shells. Small. 2012;**8**(8):1214-1221. DOI: 10.1002/smll.201102245

[12] Zhu C-L, Zhang M-L, Qiao Y-J, Xiao G, Zhang F, Chen Y-J. Fe_3O_4/TiO_2 core/shell nanotubes: synthesis and magnetic and electromagnetic wave absorption characteristics. The Journal of Physical Chemistry C. 2010;**114**(39):16229-16235. DOI: 10.1021/jp104445m

[13] Metin Ö, Mazumder V, Özkar S, Sun S. Monodisperse nickel nanoparticles and their catalysis in

hydrolytic dehydrogenation of ammonia borane. Journal of the American Chemical Society. 2010;**132**(5):1468-1469. DOI: 10.1021/ja909243z

[14] Xiong J, Shen H, Mao J, Qin X, Xiao P, Wang X, et al. Porous hierarchical nickel nanostructures and their application as a magnetically separable catalyst. Journal of Materials Chemistry. 2012;**22**(24):11927-11932. DOI: 10.1039/C2JM30361B

[15] Zhao B, Shao G, Fan B, Li W, Pian X, Zhang R. Enhanced electromagnetic wave absorption properties of Ni–SnO_2 core–shell composites synthesized by a simple hydrothermal method. Materials Letters. 2014;**121**:118-121. DOI: 10.1016/j.matlet.2014.01.081

[16] Wang H, Guo H, Dai Y, Geng D, Han Z, Li D, et al. Optimal electromagnetic-wave absorption by enhanced dipole polarization in Ni/C nanocapsules. Applied Physics Letters. 2012;**101**(8):083116. DOI: 10.1063/1.4747811

[17] Liu XG, Jiang JJ, Geng DY, Li BQ, Han Z, Liu W, et al. Dual nonlinear dielectric resonance and strong natural resonance in Ni/ZnO nanocapsules. Applied wPhysics Letters. 2009;**94**(5):053119. DOI: 10.1063/1.3079393

[18] Liu X, Feng C, Or SW, Jin C, Xiao F, Xia A, et al. Synthesis and electromagnetic properties of Al/AlOx-coated Ni nanocapsules. Materials Research Bulletin. 2013;**48**(10):3887-3891. DOI: 10.1016/j.materresbull.2013.05.110

[19] Wang B, Zhang J, Wang T, Qiao L, Li F. Synthesis and enhanced microwave absorption properties of Ni@Ni_2O_3 core–shell particles. Journal of Alloys and Compounds. 2013;**567**:21-25. DOI: 10.1016/j.jallcom.2013.03.028

[20] Dong XL, Zhang XF, Huang H, Zuo F. Enhanced microwave absorption in Ni/polyaniline nanocomposites by dual dielectric relaxations. Applied Physics Letters. 2008;**92**(1):013127. DOI: 10.1063/1.2830995

[21] Liu X, Feng C, Or SW, Sun Y, Jin C, Li W, et al. Investigation on microwave absorption properties of CuO/Cu_2O-coated Ni nanocapsules as wide-band microwave absorbers. RSC Advances. 2013;**3**(34):14590-14594. DOI: 10.1039/C3RA40937F

[22] Li H, Huang Y, Sun G, Yan X, Yang Y, Wang J, et al. Directed growth and microwave absorption property of crossed ZnO netlike micro-/nanostructures. The Journal of Physical Chemistry C. 2010;**114**(22):10088-10091. DOI: 10.1021/jp100341h

[23] Zhao B, Shao G, Fan B, Guo W, Xie Y, Zhang R. Facile synthesis of Ni/ZnO composite: Morphology control and microwave absorption properties. Journal of Magnetism and Magnetic Materials. 2015;**382**:78-83. DOI: 10.1016/j.jmmm.2015.01.053

[24] Wang ZL, Kong XY, Zuo JM. Induced growth of asymmetric nanocantilever arrays on polar surfaces. Physical Review Letters. 2003;**91**(18):185502. DOI: 10.1103/PhysRevLett.91.185502

[25] Yan J-F, Zhang Z-Y, You T-G, Zhao W, Yun J-N, Zhang F-C. Effect of polyacrylamide on morphology and electromagnetic properties of chrysanthemum-like ZnO particles. Chinese Physics B. 2009;**18**(10):4552. DOI: 10.1088/1674-1056/18/10/076

[26] Na J-S, Gong B, Scarel G, Parsons GN. Surface polarity shielding and hierarchical ZnO nano-architectures produced using sequential hydrothermal crystal synthesis and thin film atomic layer deposition. ACS Nano.

2009;**3**(10):3191-3199. DOI: 10.1021/nn900702e

[27] Jung S-H, Oh E, Lee K-H, Yang Y, Park CG, Park W, et al. Sonochemical preparation of shape-selective ZnO nanostructures. Crystal Growth & Design. 2007;**8**(1):265-269. DOI: 10.1021/cg070296l

[28] Flaifel MH, Ahmad SH, Abdullah MH, Rasid R, Shaari AH, El-Saleh AA, Appadu S. Preparation, thermal, magnetic and microwave absorption properties of thermoplastic natural rubber matrix impregnated with NiZn ferrite nanoparticles. Composites Science and Technology. 2014;**96**:103-108

[29] Li G, Xie T, Yang S, Jin J, Jiang J. Microwave absorption enhancement of porous carbon fibers compared with carbon nanofibers. The Journal of Physical Chemistry C. 2012;**116**(16):9196-9201. DOI: 10.1021/jp300050u

[30] Cooper ER, Andrews CD, Wheatley PS, Webb PB, Wormald P, Morris RE. Ionic liquids and eutectic mixtures as solvent and template in synthesis of zeolite analogues. Nature. 2004;**430**(7003):1012-1016. DOI: 10.1038/nature02860

[31] Zhang X, Dong X, Huang H, Lv B, Lei J, Choi C. Microstructure and microwave absorption properties of carbon-coated iron nanocapsules. Journal of Physics D: Applied Physics. 2007;**40**(17):5383. DOI: 10.1088/0022-3727/40/17/056

[32] Yan S, Zhen L, Xu C, Jiang J, Shao W.Microwave absorption properties of $FeNi_3$ submicrometre spheres and SiO_2@ $FeNi_3$ core–shell structures. Journal of Physics D: Applied Physics. 2010;**43**(24):245003. DOI: 10.1088/0022-3727/43/24/245003

[33] Pecquenard B, Le Cras F, Poinot D, Sicardy O, Manaud J-P. Thorough characterization of sputtered CuO thin films used as conversion material electrodes for lithium batteries. ACS Applied Materials & Interfaces. 2014;**6**(5):3413-3420. DOI: 10.1021/am4055386

[34] Kargar A, Jing Y, Kim SJ, Riley CT, Pan X, Wang D. ZnO/CuO Heterojunction branched nanowires for photoelectrochemical hydrogen generation. ACS Nano. 2013;**7**(12):11112-11120. DOI: 10.1021/nn404838n

[35] Zeng J, Xu J, Tao P, Hua W. Ferromagnetic and microwave absorption properties of copper oxide-carbon fiber composites. Journal of Alloys and Compounds. 2009;**487**(1-2):304-308. DOI: 10.1016/j.jallcom.2009.07.112

[36] Jun Z, Huiqing F, Yangli W, Shiquan Z, Jun X, Xinying C. Ferromagnetic and microwave absorption properties of copper oxide/cobalt/carbon fiber multilayer film composites. Thin Solid Films. 2012;**520**(15):5053-5059. DOI: 10.1016/j.tsf.2012.03.059

[37] Zhao B, Shao G, Fan B, Zhao W, Zhang R. Facile synthesis and enhanced microwave absorption properties of novel hierarchical heterostructures based on a Ni microsphere-CuO nano-rice core-shell composite. Physical Chemistry Chemical Physics. 2015;**17**(8):6044-6052. DOI: 10.1039/C4CP05229C

[38] Wang Z, Wu L, Zhou J, Cai W, Shen B, Jiang Z. Magnetite nanocrystals on multiwalled carbon nanotubes as a synergistic microwave absorber. The Journal of Physical Chemistry C. 2013;**117**(10):5446-5452. DOI: 10.1021/jp4000544

[39] Pan G, Zhu J, Ma S, Sun G, Yang X. Enhancing the electromagnetic performance of co through the phase-controlled synthesis of hexagonal

and cubic co nanocrystals grown on graphene. ACS Applied Materials & Interfaces. 2013;**5**(23):12716-12724. DOI: 10.1021/am404117v

[40] Wang C, Han X, Zhang X, Hu S, Zhang T, Wang J, et al. Controlled synthesis and morphology-dependent electromagnetic properties of hierarchical cobalt assemblies. The Journal of Physical Chemistry C. 2010;**114**(35):14826-14830. DOI: 10.1021/jp1050386

[41] Ortega N, Kumar A, Katiyar R, Rinaldi C. Dynamic magneto-electric multiferroics PZT/CFO multilayered nanostructure. Journal of Materials Science. 2009;**44**(19):5127-5142. DOI: 10.1007/s10853-009-3635-0

[42] Wen S, Liu Y, Zhao X, Cheng J, Li H. Synthesis, multi-nonlinear dielectric resonance and electromagnetic absorption properties of hcp-cobalt particles. Journal of Magnetism and Magnetic Materials. 2014;**354**:7-11. DOI: 10.1016/j.jmmm.2013.10.030

[43] Yan C, Xue D. Room temperature fabrication of hollow ZnS and ZnO architectures by a sacrificial template route. The Journal of Physical Chemistry B. 2006;**110**(114):7102-7106

[44] Gu F, Li CZ, Wang SF, Lü MK. Solution-phase synthesis of spherical zinc sulfide nanostructures. solution-phase synthesis of spherical zinc sulfide nanostructures. Langmuir. 2005;**22**(3):1329-1332

[45] Zhao B, Shao G, Fan B, Zhao W, Xie Y, Zhang R. ZnS nanowall coated Ni composites: facile preparation and enhanced electromagnetic wave absorption. RSC Advances. 2014;**4**(105):61219-61225. DOI: 10.1039/C4RA08095E

[46] Wang G-S, Wu Y, Wei Y-Z, Zhang X-J, Li Y, Li L-D, et al. Fabrication of Reduced Graphene Oxide (RGO)/Co3O4 nanohybrid particles and a RGO/Co3O4/Poly(vinylidene fluoride) composite with enhanced wave-absorption properties. ChemPlusChem. 2014;**79**(3):375-381. DOI: 10.1002/cplu.201300345

[47] Ren Y, Zhu C, Zhang S, Li C, Chen Y, Gao P, et al. Three-dimensional $SiO_2@Fe_3O_4$ core/shell nanorod array/graphene architecture: synthesis and electromagnetic absorption properties. Nanoscale. 2013;**5**(24):12296-12303. DOI: 10.1039/C3NR04058E

[48] Liu XG, Geng DY, Zhang ZD. Microwave-absorption properties of FeCo microspheres self-assembled by Al2O3-coated FeCo nanocapsules. Applied Physics Letters. 2008;**92**(24):243110. DOI: 10.1063/1.2945639

[49] Du Y, Liu T, Yu B, Gao H, Xu P, Wang J, et al. The electromagnetic properties and microwave absorption of mesoporous carbon. Materials Chemistry and Physics. 2012;**135**(2-3):884-891. DOI: 10.1016/j.matchemphys.2012.05.074

[50] He S, Wang G-S, Lu C, Liu J, Wen B, Liu H, et al. Enhanced wave absorption of nanocomposites based on the synthesized complex symmetrical CuS nanostructure and poly(vinylidene fluoride). Journal of Materials Chemistry A. 2013;**1**(15):4685-4692. DOI: 10.1039/C3TA00072A

[51] Li Y, Zhang J, Liu Z, Liu M, Lin H, Che R. Morphology-dominant microwave absorption enhancement and electron tomography characterization of CoO self-assembly 3D nano-flowers. Journal of Materials Chemistry C. 2014;**2**(26):5216-5222. DOI: 10.1039/C4TC00739E

[52] Lee CW, Seo S-D, Park HK, Park S, Song HJ, Kim D-W, et al. High-areal-capacity lithium storage of the kirkendall effect-driven hollow hierarchical NiSx nanoarchitecture.

Nanoscale. 2015;**7**(6):2790-2796. DOI: 10.1039/C4NR05942E

[53] Huang X, Willinger M-G, Fan H, Xie Z-l, Wang L, Klein-Hoffmann A, et al. Single crystalline wurtzite ZnO/ zinc blende ZnS coaxial heterojunctions and hollow zinc blende ZnS nanotubes: synthesis, structural characterization and optical properties. Nanoscale. 2014;**6**(15):8787-8795. DOI: 10.1039/ C4NR01575D

[54] Liao Y, Pan K, Pan Q, Wang G, Zhou W, Fu H. In situ synthesis of a NiS/Ni_3S_2 nanorod composite array on Ni foil as a FTO-free counter electrode for dye-sensitized solar cells. Nanoscale. 2015;**7**(5):1623-1626. DOI: 10.1039/C4NR06534D

[55] Wang Z, Li X, Yang Y, Cui Y, Pan H, Wang Z, et al. Highly dispersed [small beta]-NiS nanoparticles in porous carbon matrices by a template metal-organic framework method for lithium-ion cathode. Journal of Materials Chemistry A. 2014;**2**(21):7912-7916. DOI: 10.1039/C4TA00367E

[56] Xitao W, Rong L, Kang W. Synthesis of ZnO@ZnS-Bi_2S_3 core-shell nanorod grown on reduced graphene oxide sheets and its enhanced photocatalytic performance. Journal of Materials Chemistry A. 2014;**2**(22):8304-8313. DOI: 10.1039/C4TA00696H

[57] Zhu Y-P, Li J, Ma T-Y, Liu Y-P, Du G, Yuan Z-Y. Sonochemistry-assisted synthesis and optical properties of mesoporous ZnS nanomaterials. Journal of Materials Chemistry A. 2014;**2**(4):1093-1101. DOI: 10.1039/ C3TA13636A

[58] Zhang Q , Li C, Chen Y, Han Z, Wang H, Wang Z, et al. Effect of metal grain size on multiple microwave resonances of Fe/TiO_2 metal-semiconductor composite. Applied Physics Letters. 2010;**97**(13):133115. DOI: 10.1063/1.3496393

[59] Watts PCP, Hsu WK, Barnes A, Chambers B. High permittivity from defective multiwalled carbon nanotubes in the x-band. Advanced Materials. 2003;**15**(7-8):600-603. DOI: 10.1002/ adma.200304485

[60] Zhao B, Shao G, Fan B, Zhao W, Chen Y, Zhang R. Facile synthesis of crumpled ZnS net-wrapped Ni walnut spheres with enhanced microwave absorption properties. RSC Advances. 2015;**5**(13):9806-9814. DOI: 10.1039/ C4RA15411H

[61] Zhao B, Shao G, Fan B, Zhao W, Zhang S, Guan K, et al. In situ synthesis of novel urchin-like ZnS/Ni_3S_2@Ni composite with a core-shell structure for efficient electromagnetic absorption. Journal of Materials Chemistry C. 2015;**3**(41):10862-10869. DOI: 10.1039/C5TC02063H

[62] Zhao B, Shao G, Fan B, Zhao W, Zhang R. Fabrication and enhanced microwave absorption properties of Al_2O_3 nanoflake-coated Ni core-shell composite microspheres. RSC Advances. 2014;**4**(101):57424-57429. DOI: 10.1039/ C4RA10638E

[63] Liu X, Zhou G, Or SW, Sun Y. Fe/ amorphous SnO_2 core-shell structured nanocapsules for microwave absorptive and electrochemical performance. RSC Advances. 2014;**4**(93):51389-51394. DOI: 10.1039/C4RA08998G

[64] Wang Z, Bi H, Wang P, Wang M, Liu Z, shen L, et al. Magnetic and microwave absorption properties of self-assemblies composed of core-shell cobalt-cobalt oxide nanocrystals. Physical Chemistry Chemical Physics. 2015;**17**(5):3796-3801. DOI: 10.1039/ C4CP04985C

[65] Li G, Wang L, Li W, Ding R, Xu Y. $CoFe_2O_4$ and/or Co3Fe7 loaded porous activated carbon balls as a lightweight microwave absorbent. Physical Chemistry Chemical Physics.

2014;**16**(24):12385-12392. DOI: 10.1039/C4TA05718J

[66] Li X, Feng J, Du Y, Bai J, Fan H, Zhang H, et al. One-pot synthesis of $CoFe_2O_4$/graphene oxide hybrids and their conversion into FeCo/graphene hybrids for lightweight and highly efficient microwave absorber. Journal of Materials Chemistry A. 2015;**3**(10):5535-5546. DOI: 10.1039/C4TA05718J

[67] Tong G, Yuan J, Wu W, Hu Q, Qian H, Li L, et al. Flower-like Co superstructures: Morphology and phase evolution mechanism and novel microwave electromagnetic characteristics. CrystEngComm. 2012;**14**(6):2071-2079. DOI: 10.1039/C2CE05910J

[68] Jiang J, Li D, Li S, Wang Z, Wang Y, He J, et al. Electromagnetic wave absorption and dielectric-modulation of metallic perovskite lanthanum nickel oxide. RSC Advances. 2015;**19**(19):14584-14591. DOI: 10.1039/C5RA00139K

[69] Yang H-J, Cao W-Q, Zhang D-Q, Su T-J, Shi H-L, Wang W-Z, et al. Hierarchical nanorings on SiC: Enhancing relaxation to tune microwave absorption at elevated temperature. ACS Applied Materials & Interfaces. 2015;**7**(13):7073-7077. DOI: 10.1021/acsami.5b01122

[70] Liu Q, Zhang D, Fan T. Electromagnetic wave absorption properties of porous carbon/Co nanocomposites. Applied Physics Letters. 2008;**93**(1):013110. DOI: 10.1063/1.2957035

[71] Zhao B, Shao G, Fan B, Zhao W, Xie Y, Zhang R. Synthesis of flower-like CuS hollow microspheres based on nanoflakes self-assembly and their microwave absorption properties. Journal of Materials Chemistry A. 2015;**3**(19):10345-10352. DOI: 10.1039/C5TA00086F

[72] Wang Q, Lei Z, Chen Y, Ouyang Q, Gao P, Qi L, et al. Branched polyaniline/molybdenum oxide organic/inorganic heteronanostructures: Synthesis and electromagnetic absorption properties. Journal of Materials Chemistry A. 2013;**1**(38):11795-11801. DOI: 10.1039/C3TA11591G

10

Phase-Shift Transmission Line Method for Permittivity Measurement and its Potential in Sensor Applications

Vasa Radonic, Norbert Cselyuszka,
Vesna Crnojevic-Bengin and Goran Kitic

Abstract

This chapter offers a detailed insight into a dielectric characterization of the materials based on the phase-shift measurements of the transmission signal. The chapter will provide in-depth theoretical background of the phase-shift transmission line measurement in the microstrip architecture and determination of dielectric permittivity of design under test for several measurement configurations. Potential of the phase-shift method will be demonstrated through applications in the characterization of an unknown dielectric constant in multilayered structure, realization of the soil moisture sensor, and sensor for determination of the dielectric constant of a fluid in microfluidic channel. Moreover, specific techniques for increasing the phase shift based on the electromagnetic bandgap structure, the aperture in the ground plane and the left-handed effect will be presented. In the end, the realization of simple in-field detection device for determination of permittivity based on the phase-shift measurement will be demonstrated.

Keywords: phase-shift method, transmission line, microstrip, permittivity, sensors

1. Introduction

The measurement of dielectric properties of the materials found application in different fields, such as material science, absorber development, biomedical research, tissue engineering, wood industry, food quality control, etc. [1–5]. A number of methods have been developed over a time for characterization of the dielectric properties of the materials such as time domain method [6, 7], capacitive method [3, 8], transmission line (TL) methods [3, 9, 10], resonant method [3, 9], etc. The selection of the appropriate method depends on the measured frequency range, expected values of permittivity, measurement accuracy, form of material (solid, powder, and liquid), sample shape and size, temperature, etc. [8, 11]. Moreover, depending on some important aspects, the measurement methods can be divided into contact or noncontact methods (depends on whether the sample is touched or not), destructive or nondestructive (depends on whether sample can be destroyed or not), narrowband and wideband (depends on frequency range), etc. Since each method has its own advantages and limitations, the selection of the appropriate one depends on a particular application, required accuracy, sample, and other factors. There are a number of commercially available holders, kits, and probes that operate on different principles and allow measurement of the dielectric constant of the material in different forms on different temperatures and frequency ranges [8, 11]. However, the most of them are designed to be connected with expensive instruments such as network analyzers, LCR meters, or impedance analyzers.

One of the commonly used methods suitable for the material characterization in a wide frequency range, from around 10 MHz to 75 GHz, is the TL method. The TL method includes both measurements of the reflection and/or transmission characteristic [3–10]. This method can be used for characterization of permittivity as well as permeability of hard solid materials with medium losses. High-loss materials can be also characterized using this method, if the sample is kept relatively thin. The TL holders are usually made of a coaxial, a waveguide, or a microstrip line section. However, the specific design of the holder or specific multilayered configurations can be used for characterization of powders, liquid, or gases. Usually, the method requires initial sample preparation to fit into the section of the TL, typically the waveguide or the coaxial line. For accurate permittivity measurement, the sample has to be exposed to the maximal electric field, and therefore the position of the sample is very important. A typical measurement configuration of this method consists of the TL section with a sample placed inside, a vector network analyzer (VNA) used to measure the two ports complex scattering parameters (S-parameters), and a software that converts the measured S-parameters to the complex permittivity or permeability. In addition, the TL method requires initial calibration with various terminations before the measurement.

The measurement of the phase shift of the transmitted signal represents relatively fast and simple method for determination of the dielectric properties of the material. It is characterized by fast time response, and in comparison with other methods, it is less sensitive to the noise [10, 12]. Furthermore, this method allows characterization at a single frequency which simplifies the development of a supporting electronic, allows easy integration with sensor element, and allows realization of low-cost in-field sensing devices. Therefore, it found application in the realization of different types of sensors such as soil moisture sensors [13], microfluidic sensor for detention fluid mixture concentration [14], etc.

In this chapter, the phase-shift method will be explained on the example of a microstrip line configuration, and the permittivity of the materials will be determined by measuring the phase shift of the transmitted signal. Theoretical background of the phase-shift method and mathematical equations for determination of real and imaginary part of complex permittivity based on the phase of the transmitted signal will be presented in Section 2. The unknown permittivity of material will be determined for several measurement configurations in Section 3. Potential of the phase-shift method will be demonstrated through several applications in the characterization of an unknown dielectric constant in multilayered structure, a soil moisture sensor, and sensor for determination of fluid properties in microfluidic channel. Advance techniques for increasing the sensitivity of the phase-shift measurement will be presented in Section 4, while the simple in-field detection device for determination of the permittivity based on the phase measurement will be presented in Section 5. The conclusions are given in Section 6.

2. Phase-shift method

2.1 Determination of real part of the dielectric constant

Phase-shift method is based on the measurement of the phase shift of a sinusoidal signal that propagates along a transmission line.

Phase shift $\Delta\varphi$ is defined by velocity and frequency of the propagating signal as well as physical properties of the transmission line:

$$\Delta\varphi = \frac{\omega L_{TL}}{v_p}, \tag{1}$$

where ω is the angular frequency, v_p is the phase velocity, and L_{TL} is the length of transmission line.

In order to determine a phase velocity of electromagnetic wave, we will start with the expression for imaginary part of the complex propagation constant for lossy medium [15]:

$$\beta = \frac{\omega\sqrt{\mu\varepsilon}}{\sqrt{2}}\sqrt{1+\sqrt{1+\frac{\sigma^2}{\omega^2\varepsilon^2}}}, \tag{2}$$

where μ, ε, and σ are the real parts of the permeability, permittivity, and electrical conductivity of the medium through which the signal is propagating, respectively. If the imaginary part of the complex propagation constant is known, the phase velocity can be determined as

$$v_p = \frac{\omega}{\beta} = \frac{\sqrt{2}}{\sqrt{\mu\varepsilon}}\frac{1}{\sqrt{1+\sqrt{1+\frac{\sigma^2}{\omega^2\varepsilon^2}}}}. \tag{3}$$

Based on Eq. (3), it can be seen that phase velocity is dominantly influenced by permittivity, permeability and signal frequency, and then electrical conductivity.

The main advantage of the phase-shift method lies on the fact that on the frequencies high enough, the influence of conductivity can be neglected:

$$\frac{\sigma^2}{\omega^2\,\varepsilon^2} \ll 1; \tag{4}$$

therefore, expression for velocity on high frequencies can be reduced to

$$v_p = \frac{1}{\sqrt{\mu\varepsilon}}, \tag{5}$$

where phase velocity is determined by permeability and permittivity only. This is especially important for soil moisture sensor which will be discussed later since most of the materials in the soil are diamagnetic or paramagnetic. If we assume that electromagnetic wave propagates through a nonmagnetic medium, that is, the magnetic permeability is equal to $\mu_0 = 4\pi \bullet 10^{-7}$ H/m, the phase velocity and the phase shift are dependent on dielectric permittivity only:

$$\Delta\varphi = \omega L_{TL}\sqrt{\mu_0\varepsilon} = \omega L_{TL}\sqrt{\mu_0\varepsilon_0\varepsilon_r} = \frac{\omega L_{TL}}{c_0}\sqrt{\varepsilon_r}, \tag{6}$$

where c_0 is the speed of the light in the vacuum and ε_r is a relative dielectric constant of the medium.

Unlike the coaxial line or the waveguide, the microstrip structure is particularly interesting for a simple and low-cost fabrication of compact sensors based on dielectric permittivity change which can be easily integrated with supporting electronics. In this way, standard printed circuit board technology used for microstrip manufacturing offers fabrication of a complete solution of the sensor in a single substrate. Therefore, the concept of the phase-shift measurement will be explained on the example of microstrip line. In the microstrip, the influence of the change in permittivity is reflected in effective permittivity which is determined by permittivity of a medium above the microstrip line and permittivity of the dielectric substrate, **Figure 1**.

Effective permittivity of the microstrip line shown in **Figure 1** can be expressed as

$$\varepsilon_{eff} = \frac{\varepsilon_s + \varepsilon_m}{2} + \frac{\varepsilon_s - \varepsilon_m}{2} \frac{1}{\sqrt{1 + 12\frac{h}{w}}}, \tag{7}$$

where ε_r and ε_m are the permittivity of dielectric substrate and the medium above the line, respectively, h is the height of the substrate, and w is the width of the microstrip line [16].

It can be seen that a variation in the permittivity of the medium causes a variation in effective permittivity that results in the change of the phase velocity. The change in the phase velocity changes the phase shift of the signal. It is evident that phase shift is determined by the value of the permittivity of the medium above the microstrip line. Therefore, the real part of dielectric constant of unknown medium can be detected with simple measurement of the phase shift of the transmitted signal.

The range of the phase shift $\Delta\Phi$, is determined by the upper and lower values of the measured permittivity:

$$\Delta\Phi = \Delta\varphi_{max} - \Delta\varphi_{min} = \omega L_{TL}\sqrt{\mu_0\varepsilon_0}\left(\sqrt{\varepsilon_{eff\,max}} - \sqrt{\varepsilon_{eff\,min}}\right). \tag{8}$$

This parameter needs to be adjusted to the supporting electronics that measure the phase shift. From Eq. (8) it can be seen that with appropriate choice of the operating frequency and the proper optimization of the geometrical parameters (mostly the length of the transmission line), the range of the phase shift can be optimized to the maximal measurable value.

As stated above, the phase shift depends on the properties of the transmission line. Since the microstrip line can be characterized by inductance per unit length, L',

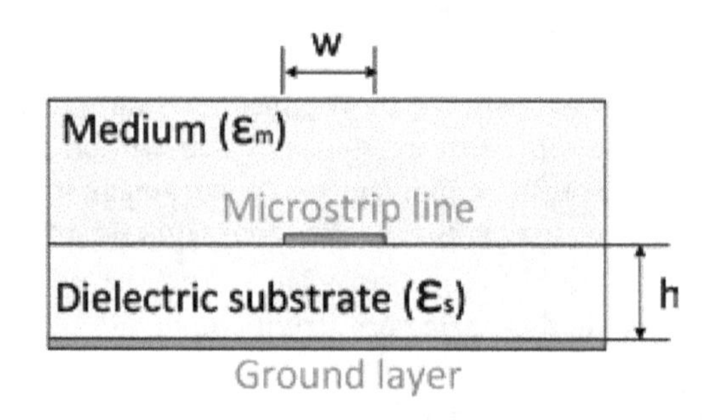

Figure 1.
Configuration of the microstrip line.

and capacitance per unit length, C', on high frequencies, the phase velocity of the signal that propagates along microstrip can be defined as [17]

$$v_p = \frac{1}{L'C'}. \tag{9}$$

Combining Eqs. (1) and (9), phase shift of the signal propagating along microstrip line with length L_{TL} can be expressed as

$$\Delta\varphi = \omega L_{TL}\sqrt{L'C'} = \omega\sqrt{LC}, \tag{10}$$

while the phase-shift range is

$$\Delta\Phi = \Delta\varphi_{max} - \Delta\varphi_{min} = \omega\left(\sqrt{L_{max}C_{max}} - \sqrt{L_{min}C_{min}}\right). \tag{11}$$

The capacitance can be written in the form of the vacuum capacitance C_0, which represents the capacitance of the microstrip line when both substrate and medium relative permittivities are equal to 1, and effective permittivity

$$C = C_0\varepsilon_{eff}. \tag{12}$$

In addition, if we assume that the electromagnetic wave propagates through a nonmagnetic medium, the inductivity of the transmission line does not depend on the sample under test ($L_{max} = L_{min} = L$). Therefore, the expression for the phase-shift range can be reduced to

$$\Delta\Phi = \omega\sqrt{LC_0}\left(\sqrt{\varepsilon_{eff\,max}} - \sqrt{\varepsilon_{eff\,min}}\right). \tag{13}$$

Based on Eq. (13), it can be concluded that the phase-shift range can be increased by performing the measurements on higher frequencies or by increasing the total microstrip line inductance or capacitance. Therefore, the phase-shift range can be optimized to the capabilities of the supporting electronics for measurement of the phase shift.

To validate the proposed method for the characterization of the material properties in wider frequency range, we compare the results of the dielectric constant calculated using proposed method with a predetermined value of dielectric constant, **Figure 2** . For the comparison, the microstrip transmission line was used where the unknown medium was placed above the microstrip line and dielectric substrate, as shown in **Figure 1**. The calculated dielectric constant was extracted from the phase shift of the transited signal using described procedure. The proposed result reveals that the phase-shift detection method provides high accuracy with the relative errors lower than 0.5% for the real parts of the dielectric constant in the measured frequency range.

2.2 Determination of the imaginary part of the dielectric constant

From the previous analysis, we demonstrate how the real part of the complex permittivity can be determined from the phase shift of the transmitted signal. However, the imaginary part of the complex permittivity cannot be directly calculated from the phase shift. It can be estimated using Kramers-Kronig (K-K) relation [18–22].

The real and imaginary part of the complex dielectric constant is correlated with K-K relation [18 – 20]. This relation is a direct consequence of the principle of

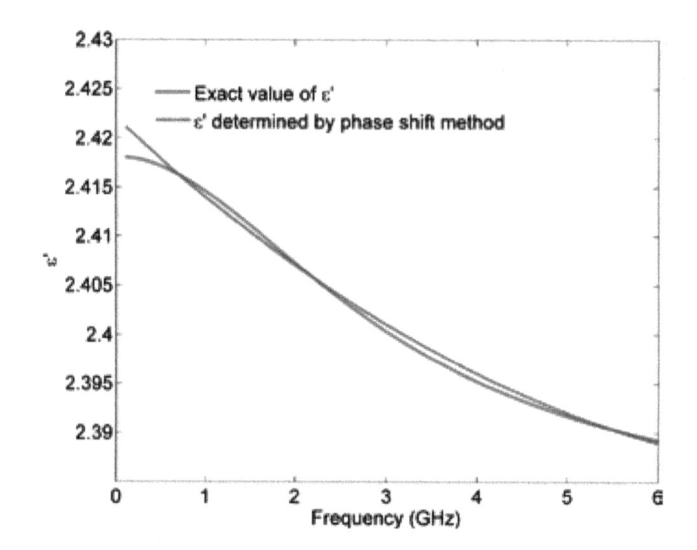

Figure 2.
Real part of the dielectric constant calculated using the phase-shift method compared with the predetermined value.

causality [21]. K-K relation describes a fundamental correlation between the real and imaginary part of the complex dielectric constant and allows us to retrieve imaginary part of the dielectric constant from the measured real part or vice versa.

The imaginary part of the complex dielectric constant can be calculated as

$$\varepsilon''(\omega) = -\frac{2\omega}{\pi}\wp\int_0^{\infty}\frac{\varepsilon'(\Omega)-1}{\Omega^2-\omega^2}d\Omega, \tag{14}$$

where ω is the angular frequency, ε' is the frequency-dependent real part of the dielectric constant, and $\wp$ is the Cauchy principal value [22].

The imaginary part of the complex dielectric constants retrieved using K-K relation from the real part of the dielectric constant (**Figure 2**) measured by phase-shift method is shown in **Figure 3**. The calculated imaginary part shows a good

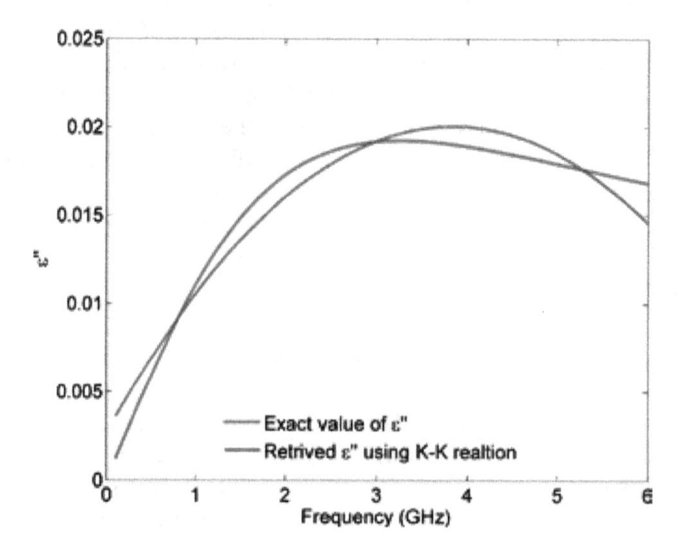

Figure 3.
Imaginary part of the complex dielectric constant retrieved by K-K relation from real part of the dielectric constant measured by phase-shift method.

agreement with the predetermined value of imaginary parts. However, the relative error is about 9% since the accuracy of Eq. (14) depends on the measured frequency range. It can be noted that the range from 0 to ∞ in the integral in Eq. (14) is not achievable in practice. Therefore, the limited range of frequencies affects the accuracy of the results and causes the error that can be observed in **Figure 3**.

3. Multilayered substrate configurations

In a previous section, we presented how the real and imaginary parts of the dielectric constant of the conventional microstrip configuration can be determined using the phase-shift method based on the phase of the transmitted signal. The same method can be applied for nonhomogeneous substrate such as multilayered or heterogeneous substrate. In this section we will analyze different microstrip multilayered substrate configurations interesting for the realization of different sensor topologies. The phase-shift method will be used for the calculation of the effective dielectric constant and determination of the real part of dielectric constant of individual layers. Similarly, the imaginary part of the complex permittivity can be determined from real part using procedure described in Section 2.2.

If we assume that the microstrip transmission line is realized on a multilayered substrate consisting of N layers with different dielectric constants, the effective dielectric constant, ε_s, of the multilayer substrate with N layers can be calculated using [14, 23]

$$\varepsilon_s = \frac{\sum_{i=1}^{N} |d_i|}{\sum_{i=1}^{N} \left|\frac{d_i}{\varepsilon_i}\right|}, \tag{15}$$

where N is the number of the layers, ε_i is the dielectric constant of the i-th layer, and d_i is a coefficient which can be calculated using the following equation:

$$d_i = \frac{K(k_i)}{K'(k_i)} - \frac{K(k_{i-1})}{K'(k_{i-1})} \cdots - \frac{K(k_1)}{K'(k_1)}, \tag{16}$$

where K and $K' = K(k_i)$ are the complete elliptical integrals of the first kind [24] and k_i is

$$k_i = \frac{1}{\cosh\left(\frac{\pi w}{4\sum_{i=1}^{N} h_i}\right)}, \tag{17}$$

where w is the width of the microstrip line and h_i is the thickness of the i-th layer. If all geometrical parameters are known as well as the dielectric constants of all layers except of one arbitrary layer, this unknown dielectric constant can be determined based on effective dielectric constant of the multilayered substrate. Previously, the effective dielectric constant has to be determined from the phase shift of the transmitted signal.

If the parameters of several layers are unknown, the unknown values of the dielectric constants can be found by solving system of equations. This procedure requires several measurements with different sets of geometrical parameters, typically the length of the microstrip line. In that case, the number of the essential measurements required for the determination of all parameter depends on the number of the unknown materials.

Beside conventional microstrip line, next three examples present typical microstrip configurations commonly used in the sensor design. Therefore, the determination of the dielectric constant will be explained on bilayered, tri-layered, and the embedded substrate configurations. In the following section, the practical applications of the analyzed configurations will be demonstrated.

The simplest case is a microstrip line realized on a bilayer substrate, shown in **Figure 4**, when the property of one layer is known (ε_{r1}) and the other one is unknown (ε_{r2}). This presents a typical configuration of the microstrip sensor with thin sensitive film deposited on the dielectric substrate.

The effective dielectric constant of the substrate combination bellow microstrip line (ε_s) can be obtained using procedure explained in Section 2, while the unknown dielectric constant in the bilayered configuration can be expressed as

$$\varepsilon_{r2} = d_2\left[\pm\left(\frac{|d_1| + |d_2|}{\varepsilon_s} - \frac{|d_1|}{|\varepsilon_{r1}|}\right)\right]^{-1}, \tag{18}$$

where coefficients d_i can be obtained using Eq. (16).

The example of the calculated unknown dielectric constant in the bilayered configuration is shown in **Figure 5**. The preset values for the dielectric constants were set to constant value of ε_{r1} = 3.6 and ε_{r2} = 2.4, while the geometrical parameters of the microstrip line were set to h_1 = 0.1 mm, h_2 = 0.2 mm, and w = 1.5 mm, and the length of the microstrip line was set to L = 10 mm. It can be seen that the

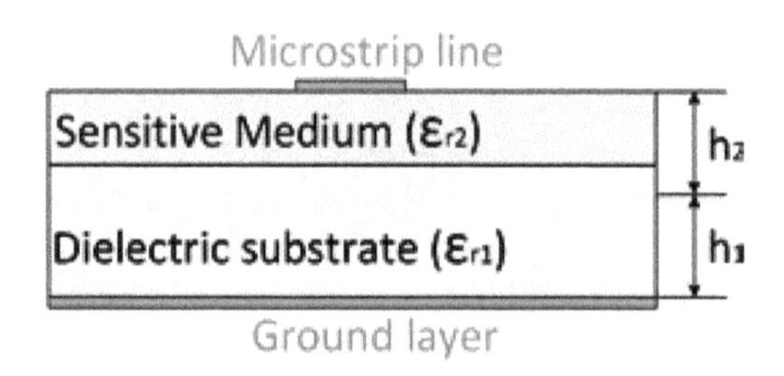

Figure 4.
Microstrip line on bilayered substrate.

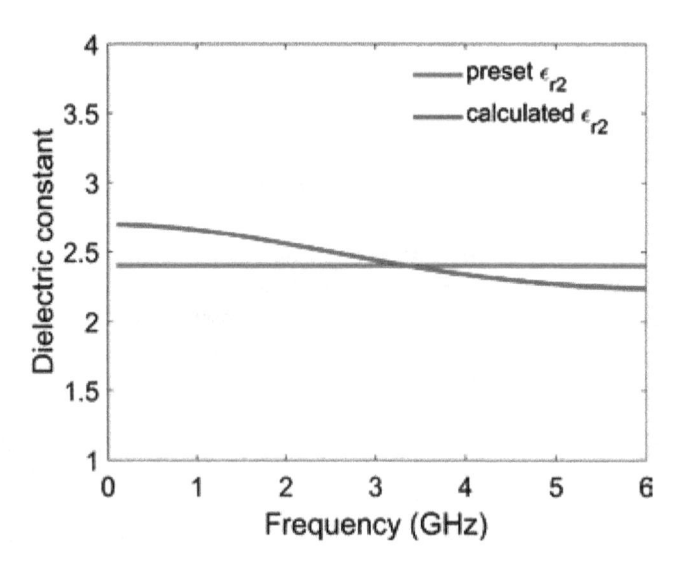

Figure 5.
Calculated unknown dielectric constant in the bilayered substrate combination.

calculated dielectric constant is in a good agreement with the preset value in the calculated frequency range.

Another example presents tri-layered substrate where middle layer is the one with an unknown dielectric constant, **Figure 6**. The bottom and top layers have the same known dielectric contestants, while the middle one is with an unknown material and with known geometrical parameters. This geometrical configuration is typical for the realization of the sensor for the characterization of a fluid or gasses in which the reservoir is placed between two dielectrics bellow microstrip line [25, 26].

Using above-described procedure, the effective dielectric constant of the tri-layer substrate (ε_s) can be obtained from the phase shift, while the unknown dielectric constant, ε_{r2}, can be calculated using the following equation:

$$\varepsilon_{r2} = d_2 \left[\pm \left(\frac{\sum_{i=1}^{3} |d_i|}{\varepsilon_s} - \frac{|d_1| + |d_3|}{|\varepsilon_{r1}|} \right) \right]^{-1}, \tag{19}$$

where coefficients d_i can be determined using Eq. (16).

The example of the calculated unknown dielectric constant in tri-layered configuration is shown in **Figure 7**. The dielectric constants used for this configuration

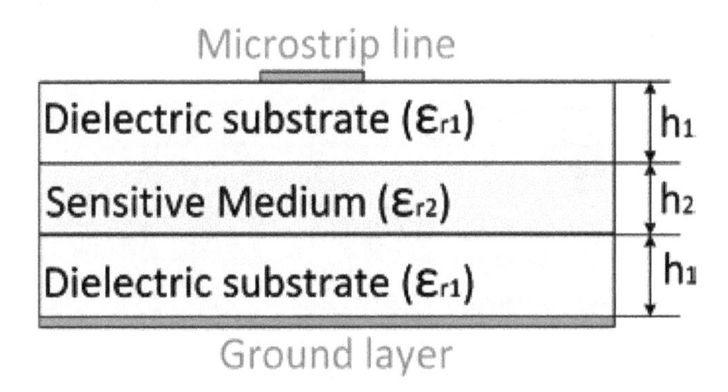

Figure 6.
Microstrip line on tri-layered substrate.

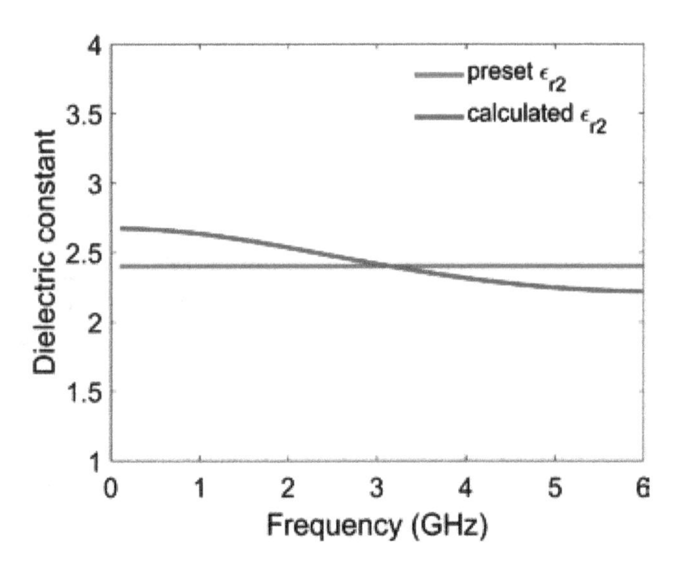

Figure 7.
Calculated unknown dielectric constant in the tri-layered substrate configuration.

were set to constant values of ε_{r1} = 3.6 and ε_{r2} = 2.4, while the geometrical parameters of the microstrip line were set to h_1 = 0.1 mm, h_2= 0.2 mm, w = 1.5 mm, and L = 10 mm. Good agreement is obtained with the maximal relative error of 10% in the observed frequency range.

The third configuration in which the substrate with an unknown dielectric constant is embedded into the substrate with known properties and dimensions is shown in **Figure 8**. This configuration is particularly interesting for the microfluidic applications, for the characterization of the fluid inside the channel [21].

This configuration of the substrate can be observed as a tri-layered substrate in which the middle layer is composed of two materials with different dielectric constants, ε_{r1} and ε_{r2}. The first and third layers are with known dielectric constant, ε_{r1}. If we assume that all geometrical parameters are known, the calculation of the effective dielectric constants of the middle layer ($\varepsilon_{r1_{eff}}$) can be calculated using Eq. (19). This formula does not determine the unknown dielectric constant, just the effective dielectric constant of the middle layer. Therefore, the unknown dielectric constant ε_{r2} can be calculated using Bruggeman formalism [27]:

$$\varepsilon_{r2e} = V\varepsilon_{r2} + (1 - V)\varepsilon_{r1}, \tag{20}$$

where V is the volumetric fraction of the microfluidic channel in the surrounding substrate. The dielec tric constant of the unk nown em bedded m aterial can be

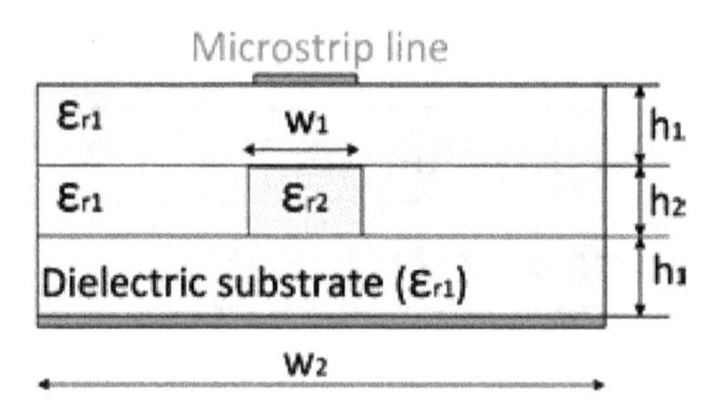

Figure 8.
Microstrip line on the substrate with embedded rectangular channel.

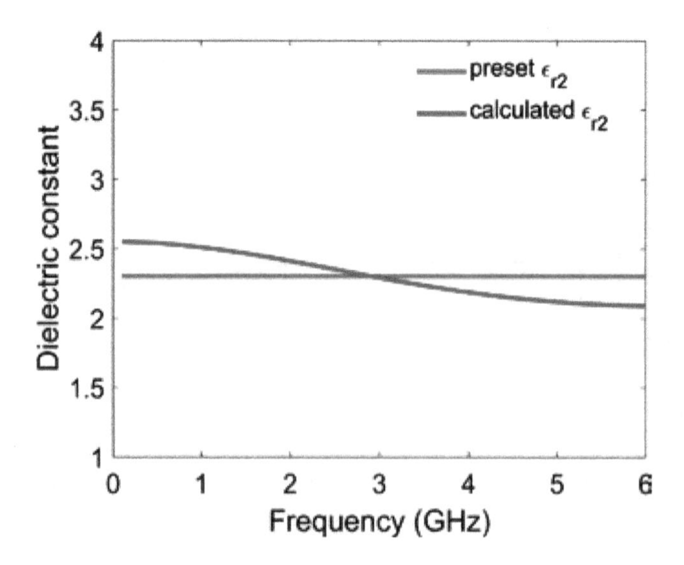

Figure 9.
Calculated unknown dielectric constant in substrate configuration with embedded channel.

determined using the procedure for the extraction of the effective dielectric constant of the tri-layered substrate from the phase shift combined with Eqs. (19) and (20).

The example of the calculated unknown dielectric constant in the embedded substrate configuration is shown in **Figure 9**. The preset values in this case were set to constant values of ε_{r1} = 3.6 and ε_{r2} = 2.35, while the geometrical parameters of the microstrip line were set to h_1 = 0.1 mm, h_2 = 0.2 mm, w = 1.5 mm, and L = 10 mm. The good agreement between the calculated dielectric constant and the preset value is obtained with the relative error lower than 8.5%.

From the previous examples, it can be seen that the phase-shift method has a potential for the characterization of the unknown dielectric materials in different configurations with a good accuracy, and therefore it presents good choice for the rapid characterization of material.

4. Techniques for increasing the phase shift

This section summarized various innovative techniques that can be used for improvement of the sensitivity of phase-shift measurement in the microstrip architecture and their advantages over conventional design.

The first technique is based on increasing the effective length of the transmission line, that is, increasing the inductance and capacitance of the microstrip line. However, this technique results in increased length and reduction in the compactness of the structure. To preserve compactness and satisfy the requirements for good performances, different shapes of the microstrip line have been recently used [28–31]. Since the space-filling property of a fractal offers high potentials for miniaturization of microwave circuits [28–31], the application of the fractal curves theoretically allows the design of infinite-length lines on finite substrate area. In that manner compact transmission line with improved phase response can be obtained using fractal curves. Based on that principle, we have designed a compact soil moisture sensor [28] that consists of two parallel fractal line segments, where each segment comprises two Hilbert fractal curves of the fourth order connected in serial, **Figure 10a**. The additional analyses confirm that the increase of the iteration of the Hilbert fractal curve

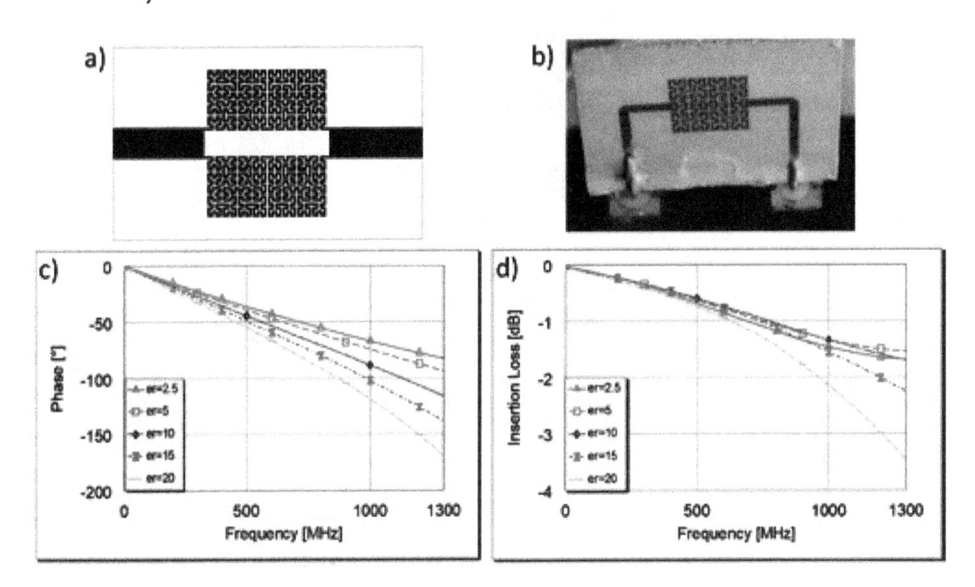

Figure 10.
Soil moisture sensor based on Hilbert fractal curve: (a) layout of the sensor, (b) fabricated sensor, (c) phase-shift response, and (d) insertion loss [28].

increases the range of phase shift [32], but accordingly the insertion loss too. Therefore, in the proposed configuration, two transmission lines are connected in parallel to reduce insertion losses. When Hilbert fractal curves are connected in parallel, the range of the phase shift is approximately the same, but insertion losses are reduced. The Hilbert curve itself is realized with the line width and the spacing between the lines of 100 μm. The results of the proposed sensor placed in the medium with different values of the dielectric constants are shown in **Figure 10**. The range of the phase shift for this configuration at the frequency of 1.2 GHz is 66.64°, while the insertion loss in the worst case is 2.98 dB. The consequence of the line modification is 0.4 dB greater insertion loss comparing to the conventional microstrip line. Although the insertion loss in this case is larger, it is still within acceptable range of 3 dB. However, for the same line length, the range of the phase shift is increased for more than three times. Stated results show that the usage of Hilbert fractal curve leads to more compact sensor characterized by higher sensitivity.

Another technique to increase the phase shift is based on an aperture in the ground plane, where the part of the ground plane, positioned under Hilbert curves of the sensor, shown in **Figure 11**, was removed. In this manner a certain passage for the lines of the electric field is made, so they can pass through it into the soil under the sensor and end up at the bottom side of the ground plane, **Figure 11**. In this manner, soil moisture has larger effect on the sensor characteristics. The results for the sensor with the aperture in the ground plane are presented in **Figure 11**. It can be seen that the range of the phase shift for the sensor with the aperture in the ground plane is 70.76° at the frequency of 1.2 GHz, and the insertion loss is 2.98 dB. Modification in ground plane improved the range of the phase shift for additional 6%, while insertion loss did not change. Quartz sand was used to validate the sensor performances, and the phase-shift measurements were performed for different moisture levels, at operating frequencies in the range of 500–2500 MHz [33, 34]. To investigate the influence of the conductivity, that is, of the soil type, on the calibration curves, the measurement procedure was repeated for the sand moisturized with water with two different salinities (conductivities): 29 and 70‰. If the operating frequency is sufficiently high, Eq. (4) will be satisfied. Therefore, the electrical conductivity can be neglected, and the expression for the phase velocity can be

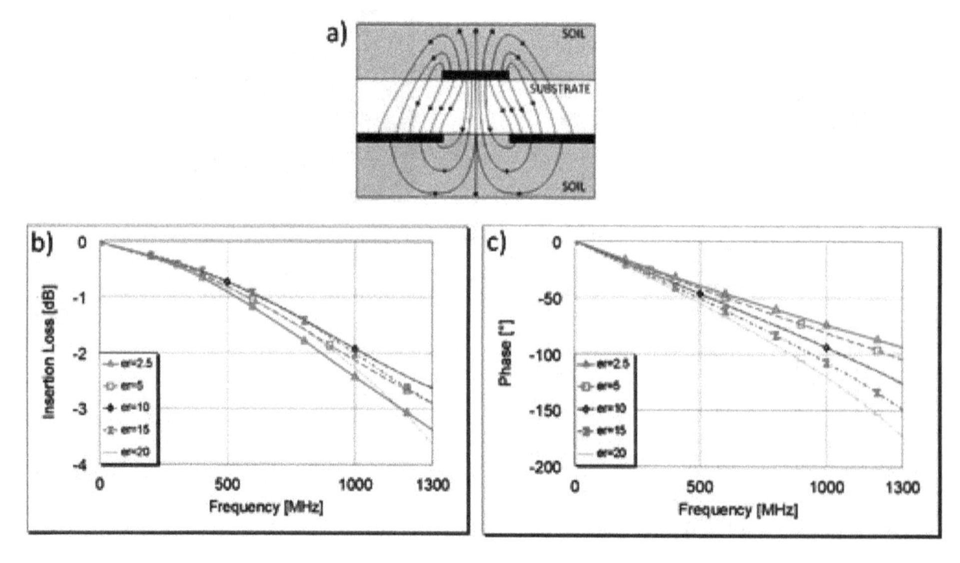

Figure 11.
Soil moisture sensor with the aperture in the ground plane: (a) cross section, (b) phase-shift response, and (c) insertion loss [28].

simplified to Eq. (5). In (**Figure 12**), fitted calibration curves (volumetric water content in the function of the phase shift) for the measurement performed at 500 and 2500 MHz are presented. A significant difference between the calibration curves is obtained at 500 MHz from differently treated samples. However, if the operating frequency is 2500 MHz, the phase shifts obtained from all three samples are almost identical. These results indicate that the Hilbert sensor with the aperture in the ground plane can be successfully used to measure soil moisture independently of the soil type (**Figure 12**).

Third technique to improve sensitivity is based on a defected electromagnetic bandgap (EBG) structure, periodical structure realized as a pattern in the microstrip ground plane. A concept to improve microstrip sensor sensitivity based on the EBG structure was firstly proposed in [35], where it is demonstrated that the sensor sensitivity can be increased by reducing the wave group velocity of the propagating signal. Illustration of this technique is shown in **Figure 13** in the realization of the 3D-printed microfluidic sensor for the determination of the characteristics of different fluids in the microfluidic channel [14]. The bottom layer of the sensor, shown in **Figure 13c**, represents the ground plane realized using defected EBG structure, periodical structure that consists of etched holes. The introduction of the uniform EBG structure in the ground plane forms a frequency region where propagation is forbidden, that is, bandgap in the transmission characteristic [36], while the defect in the EBG results in a resonance in the bandgap, which frequency is determined by the size of the defect. Introduction of the defected EBG structure improves phase shift of the microstrip line in comparison to the conventional microstrip line.
Moreover, in comparison with a conventional microstrip line, the intensity of the electric field is stronger in the vicinity of the defect in the EBG [14]. Therefore, the changes of the dielectric constant of the liquid that flows in the channel will have the highest impact to the phase response. In the proposed configuration, the EBG structure is designed to provide bandgap between 5 and 9 GHz, while the defect in the EBG causes the resonance at 6 GHz. By introducing defected EBG structure, the phase change significantly increases, especially at the frequencies that are close to the bandgap edges and at the resonance in the bandgap, due to decrease in the wave phase velocity. From the transmission characteristic, **Figure 14**, it can be seen that the resonant frequency of the defect in the bandgap slightly shifts due to the change of the dielectric constant of the fluid in microchannel. The effect of the EBG structure is predominant at the frequency of 6 GHz where the wave phase velocity is minimal. The results show that the change of the fluid permittivity from 1 (air) to

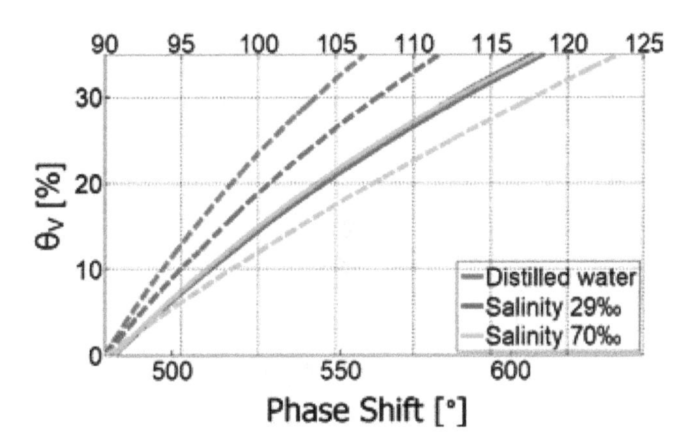

Figure 12.
Volumetric water content for 500 MHz (dashed lines, upper abscissa) and 2500 MHz (full lines, lower abscissa) [33, 34].

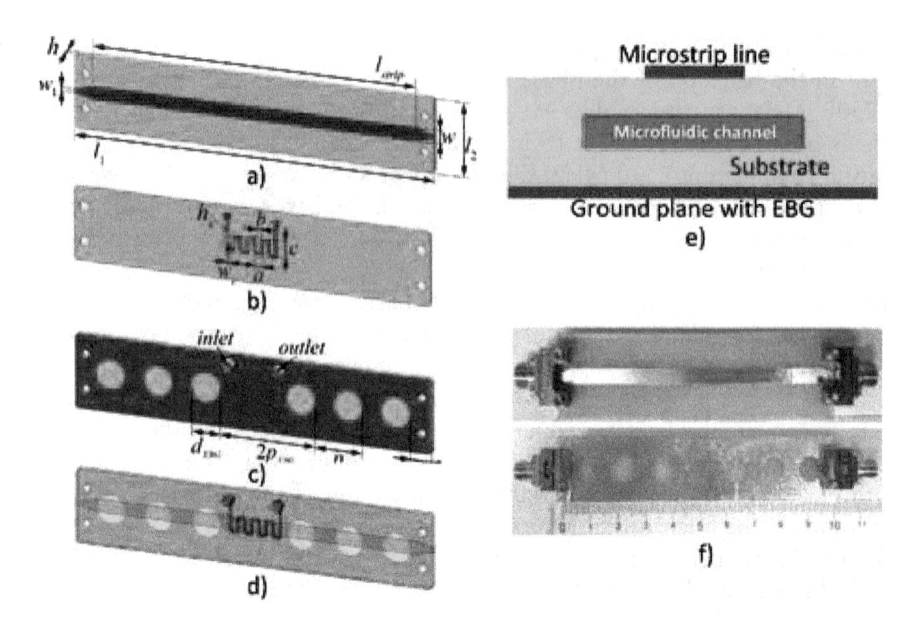

Figure 13.
Microfluidic EBG sensor: (a) top layer, (b) substrate with embedded channel, (c) bottom layer with defected EBG, (d) 3D view, (e) cross section, and (f) top and bottom side of the fabricated sensor [14].

81 (water) causes the phase-shift difference of 84°, **Figure 14**. Compared to the phase shift of the conventional microstrip line without EBG which is only 10.2° at 6 GHz, the proposed design shows eight times higher phase shift. Furthermore, the proposed sensor shows relatively high and almost linear dependence for the fluid materials with permittivity lower than 30. Therefore, the proposed sensor is characterized by relatively high sensitivity and linearity, which makes it a suitable candidate for monitoring small concentrations of a specific fluid in different mixtures. The potential application has been demonstrated in the realization of the microfluidic sensor for detection of toluene concentration in toluene-methanol mixture [14], **Figure 15**.

The fourth technique is based on metamaterials and a left-handed (LH) transmission line approach [37, 38]. Metamaterials are artificial structures that can be designed to exhibit specific electromagnetic properties not commonly found in nature. Metamaterials with simultaneously negative permittivity and

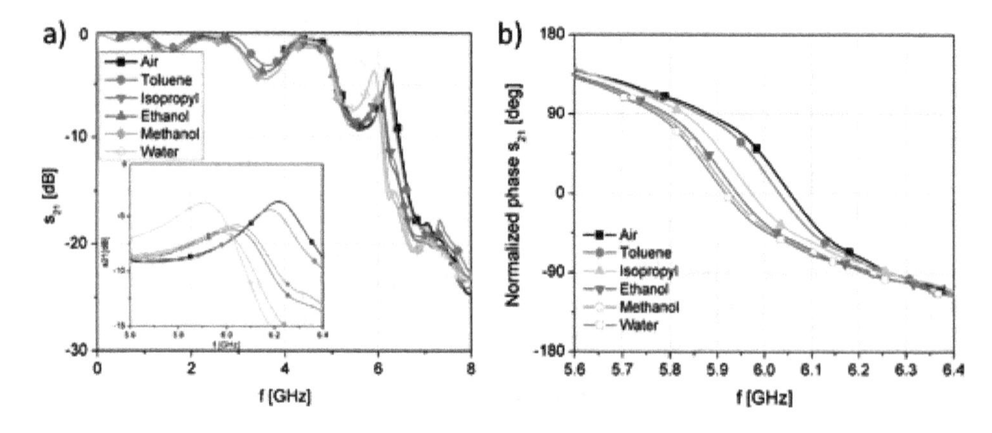

Figure 14.
Measured results of the microfluidic EBG sensor with different fluids inside the microfluidic channel: (a) transmission characteristic and (b) phase characteristic [14].

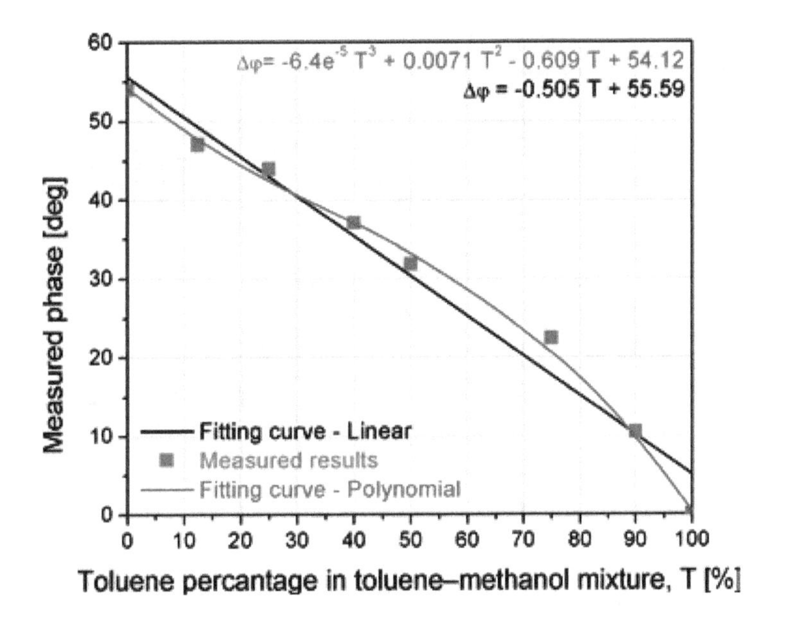

Figure 15.
Measurement of the toluene concentration in toluene-methanol mixture: sensor measured response (dots) and the corresponding fitting curves (lines) [14].

permeability, more commonly referred to as LH materials, have a propagation constant equal to zero at non-zero frequency [37]. Therefore, they can support electromagnetic waves with a group and phase velocities that are antiparallel, known as backward waves [39]. Consequently, energy will travel away from the source, while wave fronts travel backward toward the source. However, the ideal LH structure does not exist in the nature and can be formed as a combination of the LH section and conventional transmission line (RH). This structure known as a composite left-/right-handed (CLRH) transmission line can be formed using a capacitance in series with shunted inductance. In the case of CLRH due to the backward wave propagation, the phase "advance" occurs in the LH frequency range, while phase delay occurs in the RH frequency range [38]. This concept was used in the design of microfluidic sensor for the measurement of the characteristic of the fluid that flows in the microfluidic reservoir placed under CLRH transmission line. The conventional microstrip line in phase comparator was replaced with CRLH transmission line, **Figure 16**. Since the proposed line consists of one CRLH unit cell, it provides passband response at frequency of 1.9 GHz that is characterized with narrowband LH behavior. Therefore, the phase advance at Output 2 is obtained comparing to the signal that propagates true the conventional RH line (Output 1), **Figure 17b**. By changing the properties of the fluid that flows in the microfluidic channel, the central resonance of the LH band slightly shifts to the lower frequency, **Figure 17a**, while the slope of the phase characteristics intensely changes, **Figure 17b**. It can be mentioned that the level between two signals is lower than 8 dB in the worst case at 1.9 GHz; therefore the standard phase detector can be used for the phase-shift measurement. In the proposed configuration, the con-ventional RH line was used as a referent one, and therefore it was bent in the meander shape to provide maximal measurement phase-shift range at 1.9 GHz. The proposed sensor is characterized with maximal achievable sensitivity, good linearity, and design flexibility, since the maximum extent of the phase can be obtained for the arbitrary range of the dielectric constants by simple modification of the CLRH transmission line.

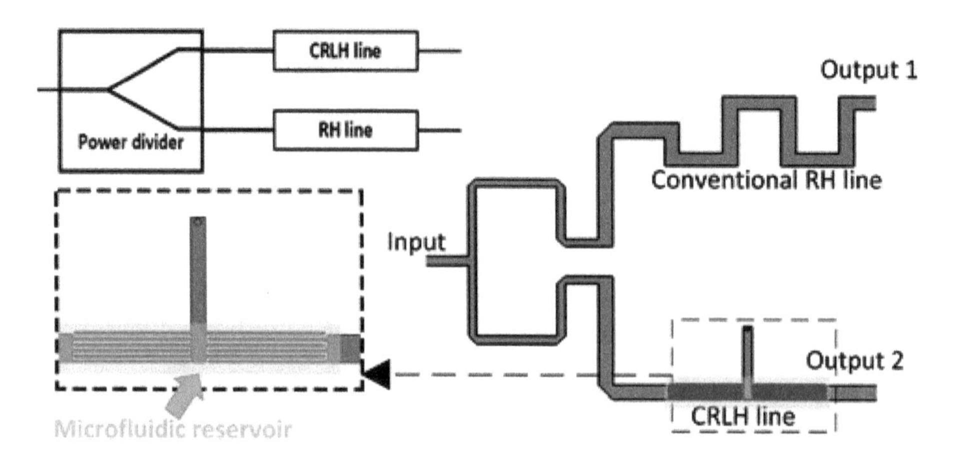

Figure 16.
Blok diagram and layout of the CRLH sensor.

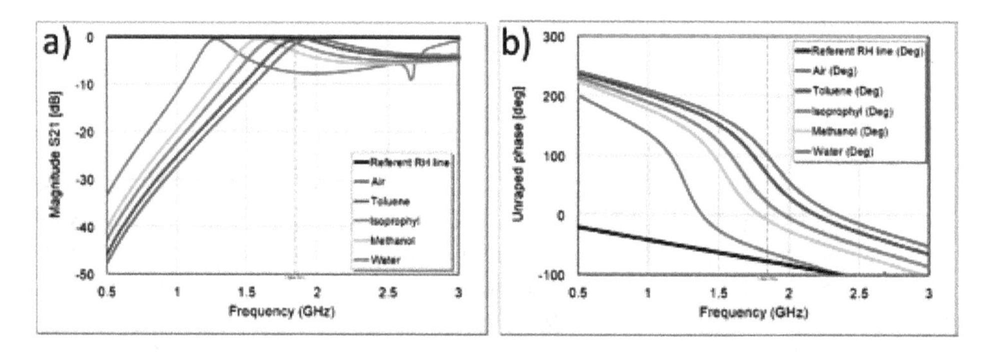

Figure 17.
Response of the CLRH sensor: (a) transmission characteristic and (b) phase characteristic.

5. Simple in-field detection circuit

The phase-shift method, as it was stated above, is the simple method that can be used for the determination of the complex permittivity. Therefore, the expensive instrument for the determination of permittivity can be replaced with a simple detection circuit. The block diagram, and detailed electrical scheme for measurement of the phase shift at single operating frequency, is presented in **Figure 18** together with the fabricated prototype realized in LTCC (low-temperature cofired ceramics) technology.

Detection circuit consists of a microwave oscillator, a quadrature hybrid, a sensor element, and a phase detector functional blocks. The operating principle can be described in the following way: sinusoidal signal generated by source, that is, microwave oscillator, is divided by quadrature hybrid on measurement and referent signal. The measurement signal propagates along sensor element where phase shift occurs according to the dielectric properties of the surrounding medium. The signal from the sensor element is fed into the input of the phase detector which compares the phase of the measurement and the referent signal. The output of the phase detector is the voltage proportional to the phase shift of the input signals which can be related to the measured dielectric constant. Referent signal propagates through the phase-shifter module which has a purpose to provide the calibration of the sensor and to overcome deviations of the nominal properties of the materials and

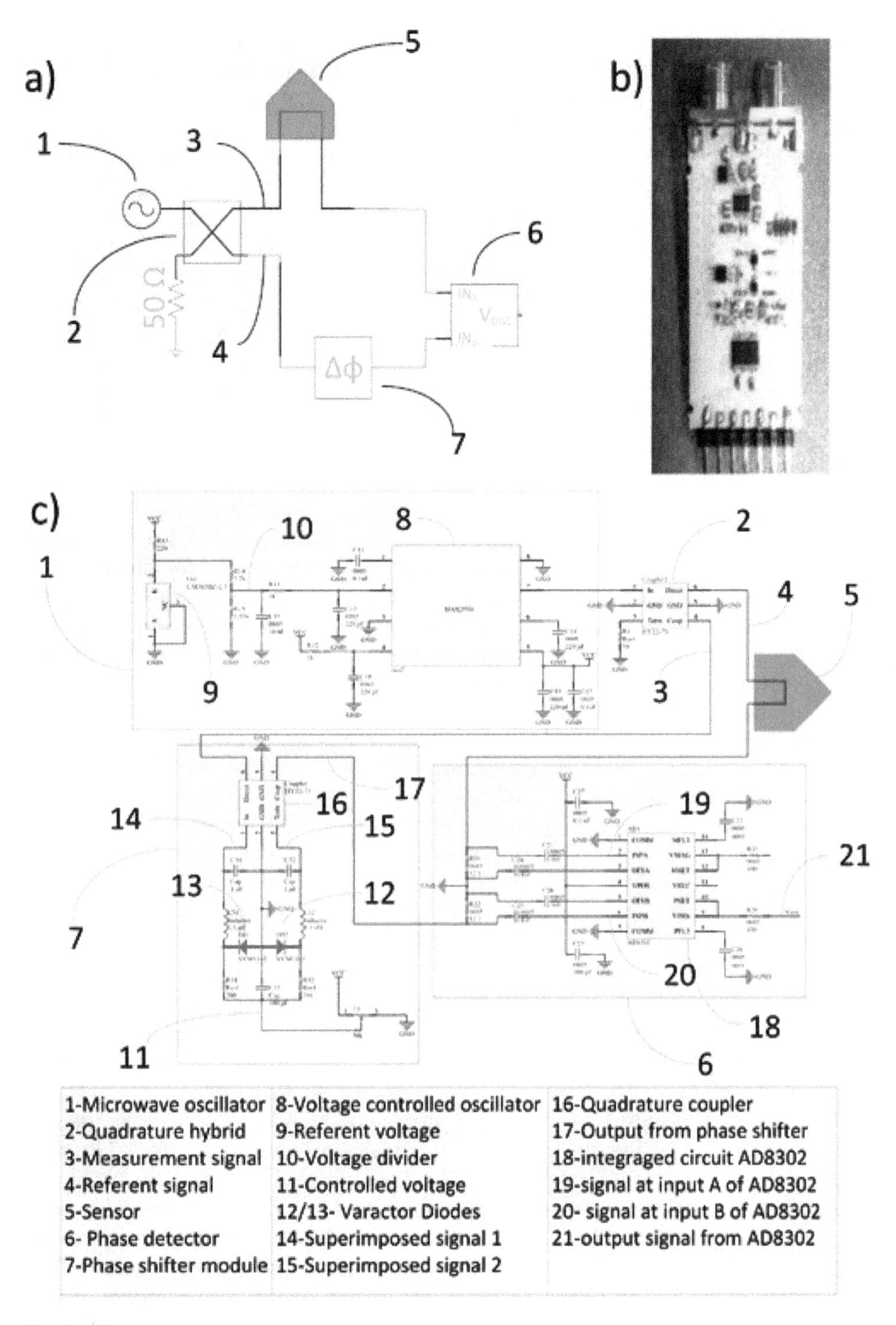

Figure 18.
Phase-shift measurement device: (a) block diagram, (b) fabricated circuit, and (c) detail electronic circuit of the phase-shift measurement device [13].

components used for sensor fabrication. In this way the repeatability of measurement of the different sensors will be secured.

The phase-shift measurement electrical circuit is detailly presented in **Figure 18c**, indicating abovementioned functional blocks of the circuit, **Figure 18a**. The oscillator is realized by the voltage-controlled oscillator MAX2751 (8), which is adjusted by the referent voltage (9) and the voltage divider (10) to operating frequency of 2.2 GHz. The signal from the oscillator is divided into measurement and referent signals using commercially available quadrature coupler circuit HY22-73 (2). The phase-shifter module allows control of the phase shift and enables fine-tuning of the difference in the phase between the measurement and referent signals. The phase shift can be achieved by varying the control voltage (11) that affects the capacitance of the varactor diodes (12 and 13). The change in capacitance affects the signals (14) and (15) which superimpose with the input referent signal (3) through the quadrature coupler (16) and in this way changes the phase of the resultant signal at the output of the module (17). The measurement of the phase shift between the measurement and referent signal is done using a phase detector module implemented with the integrated circuit AD8302 Analogue Devices (18). The phase detector circuit is set to the phase difference measurement mode according to the manufacturer's recommendation. Integrated circuit AD8302 on its output (21) gives a voltage signal that is proportional to the phase difference of the signal on its inputs (19) and (20).

The accuracy of the designed phase-shift measurement device was experimentally verified by comparing the results of the measurement of the phase shift of the signal, induced by phase-shift circuit (functional block (7), **Figure 18c**), in the range from 0 to 90° with the results of the measurement obtained using vector network analyzer (VNA). The results of the comparison are shown in **Figure 19**. It can be seen that the results agree well and the relative error in respect of the full-scale output is 5.56% which confirms performance of the designed phase-shift circuit.

Additional module for different sensor topologies can be designed to provide a conversion of the measured voltage to the corresponding value of the dielectric constant.

6. Conclusions

In this chapter dielectric characterization technique based on the phase-shift method has been presented. It has been shown that both the real and imaginary part of a complex permittivity can be calculated solely by using the phase shift of the transmitted signal. The exact formulas for the calculation of the unknown dielectric properties in the microstrip and multilayered substrate configurations such as bilayered, tri-layered, and embedded configurations were presented.

To enhance sensitivity of sensors that operate on phase-shift principle, various techniques for increasing the phase-shift range have been presented. The first technique uses space-filling property of Hilbert fractal curve to increase the phase-shift range and yet preserve the compactness of the structure. The technique that increases sensitivity of the sensor by introducing an aperture in the ground layer has been presented on the example of the soil moisture sensor. The third technique based on EBG effect has been illustrated on the example of 3D-printed microfluidic sensor for detection of toluene concentration in toluene-methanol mixture. Metamaterial and CRLH transmission line approach is used as the technique in the realization of microfluidic sensor with a maximum extent of the phase-shift range.

In the end a simple in-field detection circuit for determination of permittivity based on the phase-shift measurement on single operating frequency has been described. With the help of described concepts, a complete set of tools has been introduced which enables the design and optimization of the phase-shift-based sensors.

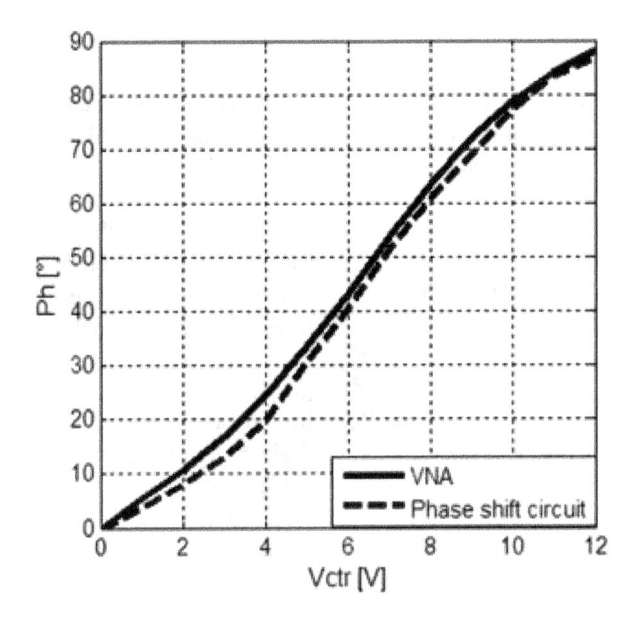

Figure 19.
Comparisons between VNA and designed phase-shift measurement circuit.

Acknowledgements

This result is part of a project that has received funding from the European Union's Horizon 2020 research and innovation programme under the grant agreement No. 664387.

Author details

Vasa Radonic*, Norbert Cselyuszka, Vesna Crnojevic-Bengin and Goran Kitic
University of Novi Sad, BioSense Institute–Research and Development Institute for Information Technologies in Biosystems, Novi Sad, Serbia

*Address all correspondence to: vasarad@uns.ac.rs

References

[1] Foster KR, Schwan HP. Dielectric properties of tissues and biological materials: A critical review. Critical Reviews in Biomedical Engineering. 1989, 1989;**17**(1):25-104

[2] Nelson SO. Fundamentals of dielectric properties measurements and agricultural applications. Journal of Microwave Power and Electromagnetic Energy. 2010;**44**(2): 98-113. DOI: 10.1080/08327823. 2010.11689778

[3] Jarvis JB, Janezic MD, Riddle BF, Johnk RT, Kabos P, Holloway CL, et al. Measuring the Permittivity and Permeability of Lossy Materials: Solids, Liquids, Metals, Building Materials, and Negative-Index Materials. NIST Technical Note 1536; 2005

[4] Krraoui H, Mejri F, Aguili T. Dielectric constant measurement of materials by a microwave technique: Application to the characterization of vegetation leaves. Journal of Electromagnetic Waves and Applications. 2016;**30**(12):1643-1660. DOI: 10.1080/09205071.2016.1208592

[5] Brodie G, Jacob MV, Farrell P. Techniques for measuring dielectric properties. In: Microwave and Radio-Frequency Technologies in Agriculture. 1st ed. Warsaw, Poland: Sciendo; 2015. pp. 52-77. DOI: 10.1515/ 9783110455403-007

[6] Nozaki R, Bose TK. Broadband complex permittivity measurements by time-domain spectroscopy. IEEE Transactions on Instrumentation and Measurement. 1990;**39**(6):945-951. DOI: 10.1109/19.65803

[7] Cataldo A, Tarricone L, Attivissimo F, Trotta A. A TDR method for real-time monitoring of liquids. IEEE Transactions on Instrumentation and Measurement. 2007;**56**(5):1616-1625. DOI: 10.1109/TIM.2007.903596

[8] Keysight Technologies. Basics of Measuring the Dielectric Properties of Materials, Application note, Literature number 5989-2589EN, 2017

[9] Stuchly MA, Stuchly SS. Coaxial line reflection methods for measuring dielectric properties of biological substances at radio and microwave frequencies—A review. IEEE Transaction of Instrumentation and Measurement. 1980;**M-29**:176-183. DOI: 10.1109/TIM.1980.4314902

[10] Chen L, Varadan VV, Ong CK, Neo CP. Microwave Electronics: Measurement and Materials Characterization. 1st ed. Chichester, England: John Wiley & Sons; 2005. DOI: 10.1002/0470020466

[11] Rode & Swartz. Measurement of Dielectric Material Properties. Application Note. RAC-0607-0019, 2012

[12] You KY. Effects of sample thickness for dielectric measurements using transmission phase-shift method. International Journal of Advances in Microwave Technology. 2016;**1**(3): 64-67

[13] Kitic G. Microwave soil moisture sensor based on phase shift method independent of electrical conductivity of the soil. The Intellectual Property Office Republic of Serbia; 2018. П-2018/ 0253 —March 2018

[14] Radonic V, Birgermajer S, Kitic G. Microfluidic EBG sensor based on phase-shift method realized using 3D Printing technology. Sensors. 2017; **17**(4):892. DOI: 10.3390/s17040892

[15] Balanis CA. Advanced Engineering Electromagnetics. New York: John Wiley & Sons, Publisher Inc; 1989

[16] Liu R, Zhang Z, Zhong R, Chen X, Li J. Nanotechnology synthesis study: Technical Report 0-5239-1; April 2007. pp. 76-79

[17] Pozar DM. Microwave Engineering. 4th ed. New Jersey: John Wiley and Sons; 2011

[18] Kuzuoglu M, Mittra R. Frequency dependence of the constitutive parameters of causal perfectly matched anisotropic absorbers. IEEE Microwave and Guided Wave Letters. 1996;**6**(12): 447-449. DOI: 10.1109/75.544545

[19] Steeman PAM, Turnhout JV. A numerical Kramers-Kronig transform for the calculation of dielectric relaxation losses free from Ohmic conduction losses. Colloid and Polymer Science. 1997;**275**(2):106-115. DOI: 10.1007/s003960050059

[20] Lucarini V, Saarinen JJ, Peiponen K-E, Vartiainen EM. Kramers–Kronig Relations in Optical Materials Research. Berlin/Heidelberg: Springer-Verlag; 2005. DOI: 10.1007/b138913

[21] Landau LD, Lifshitz EM. Electrodynamics of Continuous Media. Moscow, Russia: Pergamon Press; 1960

[22] Arfken GB, Weber HJ. Mathematical Methods for Physicists. 6th ed. Burlington, MA, USA: Elsevier Academic Press; 2005, 2005. DOI: 10.1016/C2009-0-30629-7

[23] Jha KR, Singh G. Terahertz Planar Antennas for Next Generation Communication. Cham, Switzerland: Springer International Publishing; 2014. DOI: 10.1007/978-3-319-02341-0

[24] Whittaker ET, Watson GN. A Course in Modern Analysis. 4th ed. Cambridge, England: Cambridge University Press; 1990. DOI: 10.1017/CBO9780511608759

[25] Lee K, Hassan A, Lee CH, Bae J. Microstrip patch sensor for salinity determination. Sensors. 2017;**17**(12): 2941. DOI: 10.3390/s17122941

[26] Jankovic N, Radonic V. A microwave microfluidic sensor based on a dual-mode resonator for dual-sensing applications. Sensors. 2017;**17**(12):2713. DOI: 10.3390/s17122713

[27] Lakhtakia A, Michel B, Weiglhofer WS. Bruggeman formalism for two models of uniaxial composite media; dielectric properties. Composites Science and Technology. 1997;**57**: 185-196. DOI: S0266353896001224

[28] Radonic V, Kitic G, Crnojevic-Bengin V. Novel Hilbert soil moisture sensor bassed on the phase shift method, In: Proceedings of Mediterranean Microwave Symposium (MMS 2010); 25–27 August 2010; Guzelyurt, Cyprus; 2010

[29] Crnojevic-Bengin V, Radonic V, Jokanovic B. Fractal geometries of split-ring resonators. IEEE Transactions of Microwave Theory and Techniques. 2008;**56**(10):2312-2321. DOI: 10.1109/TMTT.2008.2003522

[30] Jarry P, Beneat J. Design and Realizations of Miniaturized Fractal Microwave and RF Filters. 1st ed. Hoboken, NJ, USA: Wiley; 2009. ISBN: 978-0-470-48781-5

[31] Stojanović G, Radovanović M, Radonic V. A New Fractal-Based Design of Stacked Integrated Transformers, Active and Passive Electronic Components; 2008. Article ID 134805. DOI: 10.1155/2008/134805

[32] Kitic G. Microwave soil moisture sensors based on distributed elements [thesis]. University of Novi Sad, Faculty of Technical Sciences; 2015

[33] Will B, Crnojević-Bengin V, Kitić G. Microwave soil moisture sensors. In: Proceedings of the 43rd European Microwave Conference (EuMA); 6–10 October 2013; Nuremberg, Germany. IEEE; 2013. DOI: 10.23919/EuMC. 2013.6686793

[34] Kitić G, Crnojević Bengin V. The influence of conductivity on the microstrip soil moisture sensor. In: 3rd Global Workshop on Proximal Soil Sensing; 26–29 May 2013; Potsdam, Germany

[35] García-Baños B, Cuesta-Soto F, Griol A, Catalá-Civera JM, Pitarch J. Enhancement of sensitivity of microwave planar sensors with EBG structures. IEEE Sensors Journal. 2006; **6**:1518-1522. DOI: 10.1109/ JSEN.2006.884506

[36] Griol A, Mira D, Martinez A, Marti J. Multiple frequency photonic bandgap microstrip structures based on defects insertion. Microwave and Optical Technology Letters. 2003;**36**:479-481. DOI: 10.1002/mop.10795

[37] Christophe Caloz C, Itoh T. Electromagnetic Metamaterials: Transmission Line Theory and Microwave Applications. 1st ed. Wiley-IEEE Press; 2005. DOI: 10.1002/ 0471754323

[38] Lai A, Itoh T, Caloz C. Composite right/left-handed transmission line metamaterials. IEEE Microwave Magazine. 2004;**5**(3):34-50. DOI: 10.1109/MMW.2004.1337766

[39] Ramo S, Whinnery JR, Duzer TV. Fields and Waves in Communication Electronics. 2nd ed. New York, NY, USA: Wiley; 1984

11

Metal- and Dielectric-Loaded Waveguide: An Artificial Material for Tailoring the Waveguide Propagation Characteristics

Vishal Kesari

Abstract

In the present chapter a number of loaded structures are studied with circular cross-section to explore the deviation in their dispersion characteristics from their parent circular waveguide. The dielectric and/or metal loading to the waveguide tailors its dispersion characteristics. In general, the dielectric depresses and the metal elevates the dispersion characteristics from the characteristics of their parent circular waveguide. The axial periodicity results in periodic dispersion characteristics with a lower and an upper cut-off frequency (bandpass). However such a characteristic is not reported for the azimuthal periodic structures. The bandpass characteristic arises due to the shaping of the dispersion characteristics. Therefore the dispersion shaping is only possible with axial periodicity and not with the azimuthal periodicity. The sensitivity of the structure (geometry) parameters on the lower and upper cut-off frequencies, the extent of passband and the dispersion shaping are also included. In the axial periodic structures, the periodicity is found to be the most sensitive parameter for tailoring the dispersion characteristics and the disc-hole radius is the most sensitive parameter for shifting the dispersion characteristics over the frequency axis.

Keywords: periodically loaded waveguide, metal and dielectric loading, disc-loaded circular waveguide, dispersion characteristics, dispersion shaping

1. Introduction

Waveguides, due to their low insertion loss and high power handling capabilities at microwave and millimeter-wave frequencies, are the transmission line commonly used for transmitting or propagating or guiding the signals of these frequencies from one point to another. The propagation characteristics of a guiding structure are generally represented by its dispersion characteristics. The dispersion characteristics are the study of structure supported frequency for a given phase propagation constant, and most commonly being plotted as the supported angular frequency ω ($= 2\pi f, f$ being frequency) against the phase propagation constant β. Therefore, the dispersion characteristics are also known as $\omega - \beta$ characteristics.
The dispersion ($\omega - \beta$) characteristics of a waveguide is a hyperbola that has a cutoff frequency, more specifically a lower cutoff frequency. All the signals having

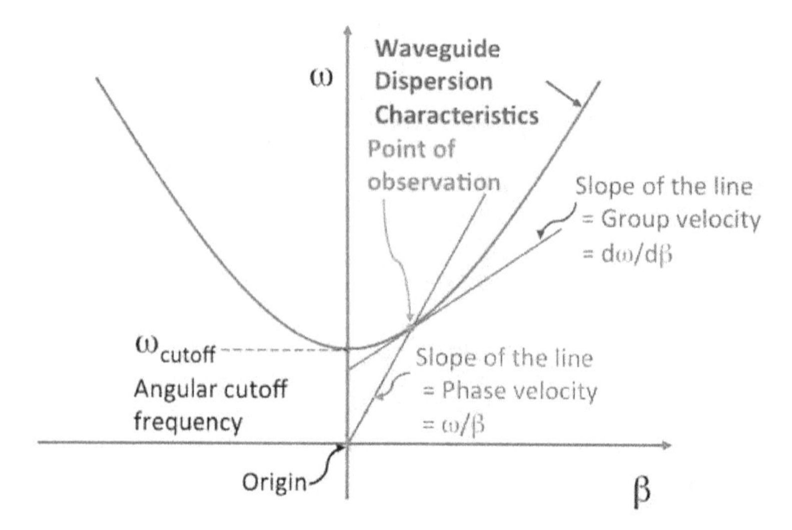

Figure 1.
$\omega - \beta$ dispersion characteristics of a circular waveguide showing the waveguide cutoff frequency and phase and group velocities.

frequencies above this lower cutoff frequency are allowed to propagate through the waveguide, and the signals having frequencies below this frequency will face a high reflection. Because of this characteristics a waveguide is inherently a high pass filter. The waveguide supports two kinds of velocities namely the phase velocity and the group velocity. The phase velocity at a chosen frequency is the one with which the signal of constant phase travels, which is represented by the slope of a line joining a chosen frequency point on $\omega - \beta$ dispersion characteristics to the origin (point representing zero frequency and zero phase propagation constant), i.e. mathematically given as ω/β. The group velocity at a chosen frequency point is the one with which the energy in signal travels, which is represented by the slope of the $\omega - \beta$ dispersion characteristics at the chosen frequency point, i.e. mathematically given as $d\omega/d\beta$ (**Figure 1**). Thus, one can control the supported phase and group veloci-ties in a waveguide by tailoring its dispersion characteristics. Such tailoring can be achieved by loading the waveguide by metal and/ or dielectric in to the smooth wall waveguide [1–5]. The characteristics (propagation or dispersion) of the conventional (smooth wall) waveguide changes with the metal and/or dielectric loading, and the same cannot be generated naturally. Therefore, the metal- and/or dielectric-loaded waveguide may be considered as artificially created material or artificial material. In part of the chapter to follow, a number of circular waveguide models containing various metal and/or dielectric loading are considered (Section 2). The electromagnetic boundary conditions (Section 3) and the dispersion relations (Section 4) of these loaded waveguides are outlined. Further, the dispersion characteristics of all the considered loaded waveguides are discussed with their sensitivity against variation in structure (geometrical) parameters (Section 5). Finally, the conclusion is drawn (Section 6).

2. Structure models

Although, the considered structures being a single conductor structure support TE ($E_z = 0$) as well as TM ($H_z = 0$) modes, they are being analyzed for the TE modes. The structures excited in these modes are of the interest for a specific class of vacuum electronic fast-wave devices, specifically the gyro-devices. In the category of gyro-devices, for the broadband amplifier namely the gyro-traveling-wave tube (gyro-TWT) the growth rate of the TM-mode vanishes at higher frequencies [6]. The models are also restricted to circular waveguide and the analyses are carried out in cylindrical (r, θ, z) system of coordinates.

2.1 Dielectric-loaded circular waveguide

In this section, we will explore two variants of dielectric-loaded structure: (i) the circular waveguide with dielectric lining on metal wall (**Figure 2**), and (ii) the circular waveguide with dielectric coaxial insert (**Figure 3**) for their dispersion characteristics and the tailoring of these characteristics with change of the relative permittivity of the dielectric material.

2.1.1 Circular waveguide with dielectric lining on metal wall

This model (model-1) includes a metallic circular waveguide of inner radius r_W, inner wall of which is containing a dielectric lining of inner radius r_L and relative permittivity ε_r for the full length of the waveguide [7]. (Here, the waveguide is considered to be infinitely long and there is no reflection of the traveling signals

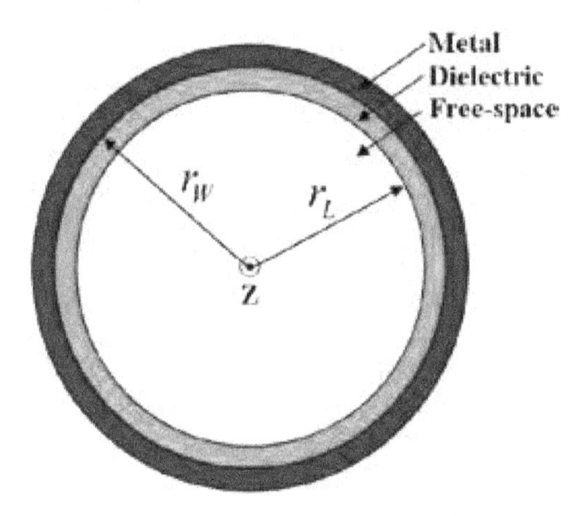

Figure 2.
Circular waveguide with dielectric lining on metal wall [7].

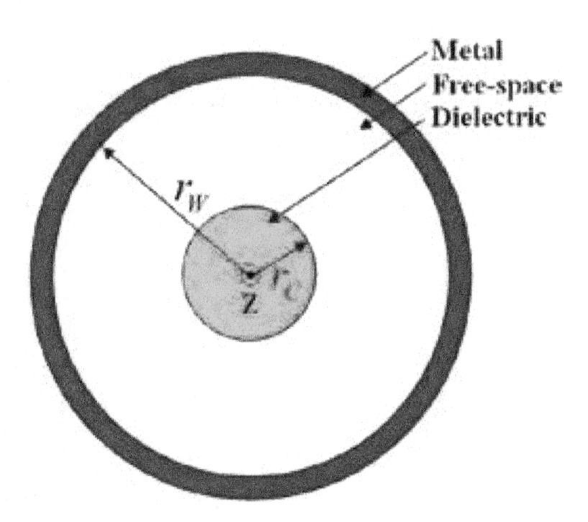

Figure 3.
Circular waveguide with dielectric coaxial insert [7].

from the waveguide extremes). Thus the radial thickness of the dielectric lining can be calculated as $r_W - r_L$. For the sake of analysis, the structure may be divided into two analytical regions, the central free-space (dielectric free) region I: $0 \leq r < r_L$, $0 \leq z < \infty$; and the dielectric filled region II: $r_L \leq r < r_W$, $0 \leq z < \infty$ (**Figure 2**). The relevant (axial magnetic and azimuthal electric) field intensity components may be written as [7]:

In region I:

$$H_z^I = \sum_{n=-\infty}^{+\infty} A_n^I J_0\{\gamma_n^I r\} \exp j(\omega t - \beta_n z) \quad (1)$$

$$E_\theta^I = j\omega\mu_0 \sum_{n=-\infty}^{+\infty} \frac{1}{\gamma_n^I} A_n^I J_0'\{\gamma_n^I r\} \exp j(\omega t - \beta_n z) \quad (2)$$

In region II:

$$H_z^{II} = \sum_{n=-\infty}^{+\infty} \left(A_n^{II} J_0\{\gamma_n^{II} r\} + B_n^{II} Y_0\{\gamma_n^{II} r\}\right) \exp j(\omega t - \beta_n z) \quad (3)$$

$$E_\theta^{II} = j\omega\mu_0 \sum_{n=-\infty}^{+\infty} \frac{1}{\gamma_n^{II}} \left(A_n^{II} J_0'\{\gamma_n^{II} r\} + B_n^{II} Y_0'\{\gamma_n^{II} r\}\right) \exp j(\omega t - \beta_n z) \quad (4)$$

where J_0 and Y_0 are the zeroth-order Bessel functions of the first and second kinds, respectively. Prime with a function represents the derivative with respect to its argument. A_n^I, A_n^{II} and B_n^{II} are the field constants, superscript identifying its value, in different analytical regions. $\gamma_n^I \left[= \left(k^2 - \beta_n^2\right)^{1/2}\right]$ and $\gamma_n^{II} \left[= \left(\varepsilon_r k^2 - \beta_n^2\right)^{1/2}\right]$ are the radial propagation constants in regions I and II, respectively. β_n and k are the phase and the free-space propagation constants, respectively [7].

2.1.2 Circular waveguide with dielectric coaxial insert

Similar to model-1, this model (model-2) also contains a metallic circular waveguide of inner radius r_W, inner wall of which is free from any dielectric. The model includes a coaxial dielectric insert of radius r_C and relative permittivity ε_r for the full length of the waveguide (**Figure 3**) [7].

For the sake of analysis, the structure may be divided into two analytical regions, the central dielectric filled region I: $0 \leq r < r_C$, $0 \leq z < \infty$; and the free-space (dielectric free) region II: $r_C \leq r < r_W$, $0 \leq z < \infty$. The relevant (axial magnetic and azimuthal electric) field intensity components may be written same as for model-1 (1)–(4), in which the radial propagation constants γ_n^I and γ_n^{II} are interpreted as: $\gamma_n^I = \left(\varepsilon_r k^2 - \beta_n^2\right)^{1/2}$ and $\gamma_n^{II} = \left(k^2 - \beta_n^2\right)^{1/2}$, respectively [7].

2.2 Metal-loaded circular waveguide

In Section 2.1, we have studied the dispersion characteristics of dielectric-loaded structures. In the present section, we will explore three variants of all metal-loaded structure: (i) the conventional annular metal disc-loaded circular waveguide (**Figure 4**), (ii) the interwoven-disc-loaded circular waveguide (**Figure 5**), and (iii) metal vane-loaded circular waveguide (**Figure 6**) for their dispersion characteristics and the effect of change of the geometry parameters on these characteristics.

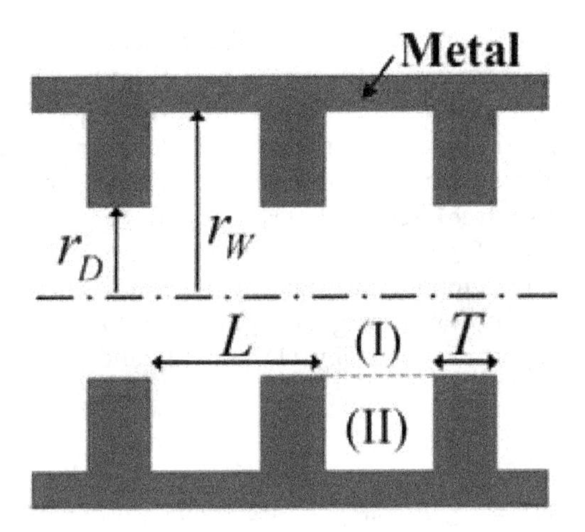

Figure 4.
Circular waveguide with annular metal discs [9–13].

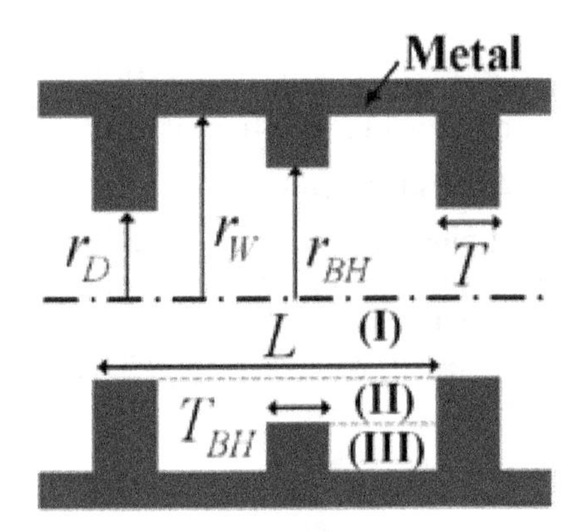

Figure 5.
Interwoven disc-loaded circular waveguides [2, 9, 14, 15].

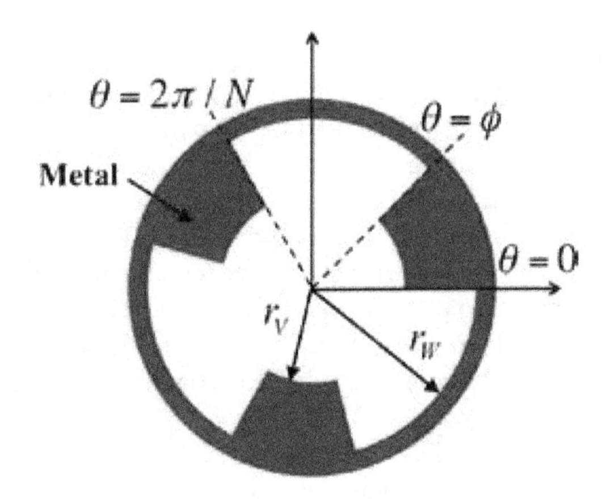

Figure 6.
Circular waveguide loaded with metal vanes [16–18].

2.2.1 Circular waveguide with annular metal discs

In this model (model-3) a circular metallic waveguide of inner radius r_W is considered in which annular disc of thickness T, inner radius r_D and outer radius r_W

are arranged periodically with periodicity L. The structure is commonly known as the disc-loaded circular waveguide (conventional) (**Figure 4**) [2, 5, 8–13]. As the structure is periodic, therefore one period of the structure coupled with Floquet's theorem is sufficient for the analysis of the infinitely long structure [1, 2, 5, 8, 9]. For the sake of analysis, the structure may be divided into two analytical regions, the central free-space (disc free) region I: $0 \le r < r_D$, $0 \le z < \infty$; and the disc occupied region II: $r_D \le r < r_W$, $0 \le z \le L - T$ (**Figure 4**). The disc free and disc occupied regions are assumed to support propagating (traveling) and stationary waves, respectively. The relevant (axial magnetic and azimuthal electric) field intensity components in the region I is given by (1) and (2) and in the region II may be written as [2, 5, 9–13]:

In region II:

$$H_z^{II} = \sum_{m=1}^{\infty} A_m^{II} Z_0\{\gamma_m^{II} r\} \exp(j\omega t) \sin(\beta_m z) \tag{5}$$

$$E_\theta^{II} = j\omega\mu_0 \sum_{m=1}^{\infty} \frac{1}{\gamma_m^{II}} A_m^{II} Z_0'\{\gamma_m^{II} r\} \exp(j\omega t) \sin(\beta_m z) \tag{6}$$

where $Z_0\{\gamma_m^{II} r\} = J_0\{\gamma_m^{II} r\} Y_0'\{\gamma_m^{II} r_W\} - J_0'\{\gamma_m^{II} r_W\} Y_0\{\gamma_m^{II} r\}$; A_m^{II} is the field constants, superscript identifying its value, in different analytical regions.

$\gamma_m^{II} \left[= \left(k^2 - \beta_m^2\right)^{1/2}\right]$ is the radial propagation constant in region II. β_n, defined as $\beta_n = \beta_0 + 2\pi n/L$, is the axial phase propagation constant in region I with β_0 as the axial phase propagation constant for fundamental space harmonic, and n [= 0, ±1, ±2, ±3, ...] as space harmonic number. β_m, defined as $\beta_m = m\pi/(L - T)$, is the axial propagation constants in region II with m (= 1, 2, 3, ...) as the modal harmonic numbers in region II [2, 5, 9–13].

2.2.2 Interwoven-disc-loaded circular waveguide

This model (model-4) differs from the conventional disc-loaded circular waveguide due to different additional disc included in between two identical con-secutive discs of conventional disc-loaded circular waveguide [2, 9, 14, 15]. Thus, this model is considered with a circular metallic waveguide of inner radius r_W in which annular disc of thickness T, inner radius r_D and outer radius r_W are arranged periodically with periodicity L. In addition, another annular disc of thickness T_{BH}, inner radius r_{BH} and outer radius r_W are also arranged periodically with periodicity L such that the disc of thickness T_{BH} is placed in middle of two identical consecutive discs of thickness T. The structure is known as the interwoven-disc-loaded circular waveguide. Similar to conventional disc-loaded circular waveguide, this structure is also periodic, therefore considering unit cell of the structure with Floquet's theorem suffices for the analysis of the infinitely long structure. The anal ytical regions of the m odel may be considered as: (i) region I: $0 \le r < r_D$, $0 \le z < \infty$; (ii) region II: $r_D \le r < r_{BH}$, $0 \le z \le L - T$; and (iii) region III: $r_{BH} \le r < r_W$, $0 \le z \le (L - T - T_{BH})/2$, where $L - T$ and $(L - T - T_{BH})/2$ represent the axial-gaps between two consecutive discs of smaller hole and between discs of bigger and smaller holes (**Figure 5**).

Similar to the conventional - disc loaded circular waveguide (model-3), it is assumed that the disc free (I) and disc occupied (II and III) regions, respectively, support propagating and standing waves. The relevant (axial magnetic and azimuthal electric) field intensity components in the region I is given by (1) and (2) and in the regions II and III may be written as [2, 9, 14, 15]:

In region II:

$$H_z^{II} = \sum_{m=1}^{\infty} \left[A_m^{II} J_0\{\gamma_m^{II} r\} + B_m^{II} Y_0\{\gamma_m^{II} r\}\right] \exp(j\omega t) \sin(\beta_m z) \tag{7}$$

$$E_\theta^{II} = j\omega\mu_0 \sum_{m=1}^{\infty} \frac{1}{\gamma_m^{II}} \left[A_m^{II} J_0'\{\gamma_m^{II} r\} + B_m^{II} Y_0'\{\gamma_m^{II} r\}\right] \exp(j\omega t) \sin(\beta_m z) \tag{8}$$

In region III:

$$H_z^{III} = \sum_{p=1}^{\infty} A_p^{III} Z_0\left\{\gamma_p^{III} r\right\} \exp(j\omega t) \sin\left(\beta_p z\right) \tag{9}$$

$$E_\theta^{III} = j\omega\mu_0 \sum_{p=1}^{\infty} \frac{1}{\gamma_p^{III}} A_p^{III} Z_0'\{\gamma_m^{III} r\} \exp(j\omega t) \sin\left(\beta_p z\right), \tag{10}$$

where $Z_0\left\{\gamma_p^{III} r\right\} = J_0\left\{\gamma_p^{III} r\right\} Y_0'\left\{\gamma_p^{III} r_W\right\} - J_0'\left\{\gamma_p^{III} r_W\right\} Y_0\left\{\gamma_p^{III} r\right\}$; A_m^{II}, B_m^{II} and A_p^{III} are the field constants, superscript identifying its value, in different analytical regions. $\gamma_m^{II}\left[= \left(k^2 - \beta_m^2\right)^{1/2}\right]$ and $\gamma_p^{III}\left[= \left(k^2 - \beta_p^2\right)^{1/2}\right]$ are the radial propagation constants in regions II and III, respectively. The axial phase propagation constants β_n in region I and β_m in region II are defined in the same manner as for model-3. β_p $[= 2p\pi/(L - T - T_{BH})]$ is the axial phase propagation constants in region III; here p is the modal harmonic number in region III [2, 9, 14, 15].

2.2.3 Circular waveguide loaded with metal vanes

This model (model-5) considers a circular waveguide of radius r_W and N number of metal vanes of vane-inner-tip radius r_V and vane angle ϕ extending axially over the length of the waveguide arranged on the waveguide wall to maintain the azimuthal periodicity (**Figure 6**) [16–18]. Clearly, the azimuthal periodicity is $2\pi/N$. For the analysis of the structure, it may be divided into two regions; (i) the central cylindrical vane-free free-space region I: $0 \le r < r_V$, $0 \le \theta < 2\pi$, and (ii) the free-space region II between two consecutive metal vanes $r_V \le r \le r_W$, $\phi < \theta < 2\pi/N$(**Figure 6**). The relevant (axial magnetic and azimuthal electric) field intensity components in the regions I and II may be written as [16–18]:

In region I:

$$H_z^I = \sum_{q=-\infty}^{+\infty} A_q^I J_q\{\gamma^I r\} \exp(-jq\theta) \tag{11}$$

$$E_\theta^I = \frac{j\omega\mu_0}{\gamma^I} \sum_{q=-\infty}^{+\infty} A_q^I J_q'\{\gamma^I r\} \exp(-jq\theta) \tag{12}$$

In region II:

$$H_z^{II} = \sum_{v=0}^{+\infty} \left[A_v^{II} J_v\{\gamma^{II} r\} + B_v^{II} Y_v\{\gamma^{II} r\}\right] \cos\left(\frac{v\pi(\theta - \phi)}{2\pi/N - \phi}\right) \tag{13}$$

$$E_\theta^{II} = \frac{j\omega\mu_0}{\gamma^{II}} \sum_{v=0}^{+\infty} \left[A_v^{II} J_v'\{\gamma^{II} r\} + B_v^{II} Y_v'\{\gamma^{II} r\}\right] \cos\left(\frac{v\pi(\theta - \phi)}{2\pi/N - \phi}\right) \tag{14}$$

where A_q^I, A_v^{II} and B_v^{II} are the field constants; J and Y are the ordinary Bessel function of first and second kinds, respectively, with their primes representing the derivatives with respect to their arguments. q is an integer; and v is a non -negative integer. $\gamma^{I} = \gamma^{II}\left(= \left(k^2 - \beta^2\right)^{1/2}\right)$ and β are the radial and the axial phase propagation constants, respectively. In order to include the effect of azimuthal harmonics due to angular periodicity of the structure, the azimuthal dependence is considered as $\exp(-jv\theta)$, such that $v = s + qN$, where s is also an integer [16–18].

2.3 Metal- and dielectric-loaded circular waveguide

In Sections 2.1 and 2.2, we have respectively explored the independent dielectric- and metal-loaded structures. However, in this section we will study the metal as well as dielectric loading in the circular waveguide.

2.3.1 Circular waveguide loaded with dielectric and metal discs

This model (model-6) is formed by alternatively stacking the metal and dielectric discs each of same disc hole radii r_D. This is similar to conventional disc-loaded circular waveguide in which the volume between two consecutive metal discs is filled with dielectric of relative permittivity ε_r. Similar to conventional disc-loaded circular waveguide for the sake of analysis, one may divide the structure into two regions: central disc free region I: $0 \le r < r_D$, $0 < z < \infty$, and disc occupied region II: $r_D \le r < r_W$, $0 < z < L - T$ (**Figure** 7). The relevant (axial magnetic and azimuthal electric) field intensity components in the regions I and II may be given by (1), (2), (5) and (6). In (5) and (6), the radial propagation constant in region II is interpreted as $\gamma_m^{II} = \left(\varepsilon_r k^2 - \beta_m^2\right)^{1/2}$ [19].

2.3.2 Circular waveguide loaded with dielectric and metal discs having different hole radius

This model (model-7) is similar to that of model-6, which has a circular waveguide consisting of alternate dielectric and metal discs, such that the hole-radius of metal discs is lesser than that of dielectric discs [20]. For the purpose of analysis, one may divide the structure into three regions: i) the central disc free region I: $0 \le r < r_D$, $0 < z < \infty$; ii) the disc occupied free space region II: $r_D \le r < r_{DD}$, $0 < z < L - T$; and iii) the dielectric filled disc occupied region III: $r_{DD} \le r < r_W$, $0 < z < L - T$; where r_{DD} is the hole radius of dielectric disc. It is considered that the region I (disc free region) supports propagating and regions II and III (disc occupied regions) support standing waves [20].

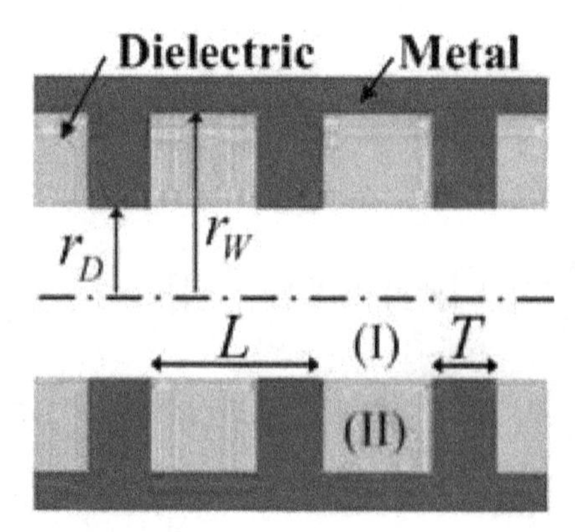

Figure 7.
Circular waveguide loaded with dielectric and metal discs [19].

The relevant (axial magnetic and azimuthal electric) field intensity components in the region I may be given by (1) and (2), and in regions II and III are written as [20]:

In region II:

$$H_z^{II} = \sum_{m=1}^{\infty} \left[A_m^{II} J_0\{\gamma_m^{II} r\} + B_m^{II} Y_0\{\gamma_m^{II} r\}\right] \exp(j\omega t)\ \sin(\beta_m z) \tag{15}$$

$$E_\theta^{II} = j\omega\mu_0 \sum_{m=1}^{\infty} \frac{1}{\gamma_m^{II}} \left[A_m^{II} J_0'\{\gamma_m^{II} r\} + B_m^{II} Y_0'\{\gamma_m^{II} r\}\right] \exp(j\omega t)\ \sin(\beta_m z) \tag{16}$$

In region III:

$$H_z^{III} = \sum_{m=1}^{\infty} A_m^{III} Z_0\{\gamma_m^{III} r\}\ \exp(j\omega t)\ \sin(\beta_m z) \tag{17}$$

$$E_\theta^{III} = j\omega\mu_0 \sum_{m=1}^{\infty} \frac{1}{\gamma_m^{III}} A_m^{III} Z_0'\{\gamma_m^{III} r\}\ \exp(j\omega t)\ \sin(\beta_m z) \tag{18}$$

where $Z_0\{\gamma_m^{III} r\} = J_0\{\gamma_m^{III} r\} Y_0'\{\gamma_m^{III} r_W\} - J_0'\{\gamma_m^{III} r_W\} Y_0\{\gamma_m^{III} r\}$; A_m^{II}, B_m^{II} and A_m^{III} are the field constants in different analytical regions, identified by given superscript, respectively. $\gamma_n^I \left[= \left(k^2 - \beta_n^2\right)^{1/2}\right]$, $\gamma_m^{II} \left[= \left(k^2 - \beta_m^2\right)^{1/2}\right]$, and $\gamma_m^{III} \left[= \left(\varepsilon_r k^2 - \beta_m^2\right)^{1/2}\right]$ are the radial propagation constants in regions I, II, and III, respectively (**Figure 8**). The axial phase propagation constants β_n in region I and β_m in regions II and III are defined in the same manner as for model-3 [20].

2.3.3 Circular waveguide loaded with alternate dielectric and metal vanes

This model (model-8) is similar to model-5 except the region II filled with dielectric of relative permittivity ε_r between the two consecutive metal vanes [21]. For the sake of analysis, the structure may be divided into two regions; (i) the central cylindrical vane-free free-space region I: $0 \le r < r_V$, $0 \le \theta < 2\pi$, and (ii) the dielectric filled region II between two consecutive metal vanes: $r_V \le r \le r_W$, $< \theta < 2\pi/N$ (**Figure 9**). The relevant (axial magnetic and azimuthal electric) field intensity components in the regions I and II may be given by (11)–(14), in which the radial propagation constant is given as: $\gamma^{II} = \left(\varepsilon_r k^2 - \beta^2\right)^{1/2}$ [21].

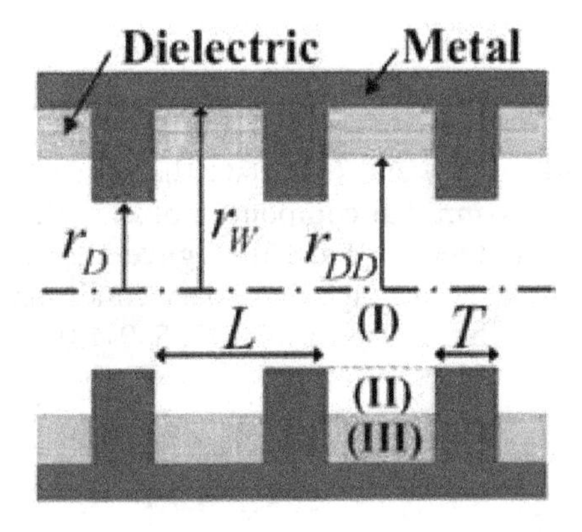

Figure 8.
Circular waveguide loaded with dielectric and metal discs with the hole radius of metal discs lesser than that of dielectric discs [20].

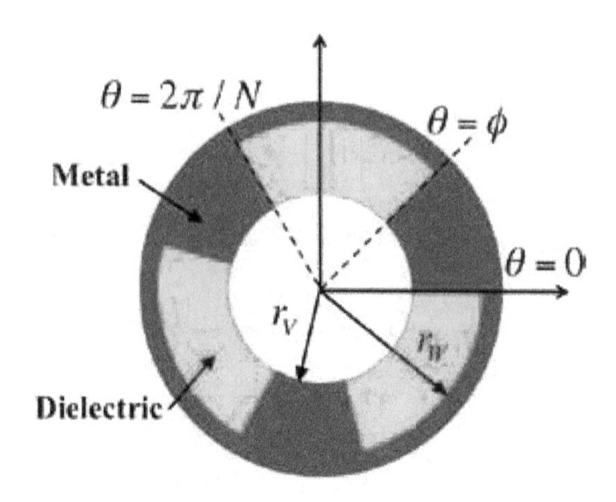

Figure 9.
Circular waveguide loaded with alternate dielectric and metal vanes [21].

3. Boundary conditions

One of the general boundary conditions comprising all the considered models is due to the null tangential electric field intensity at the inner surface of the metallic waveguide that is represented as [2, 5, 9–21]:

$$E_\theta = 0|_{r=r_W} \quad 0 < z < \infty \tag{19}$$

Model-1: The relevant electromagnetic boundary conditions for model-1 may be written in the mathematical form as [7]:

$$H_z^I = H_z^{II}|_{r=r_L} \quad 0 < z < \infty \tag{20}$$

$$E_\theta^I = E_\theta^{II}|_{r=r_L} \quad 0 < z < \infty \tag{21}$$

The boundary conditions (20) and (21) state the continuity of the axial component of magnetic and the azimuthal component of electric field intensities at the interface, $r = r_L$, between the regions I and II (**Figure 2**) [7].

Model-2: The relevant electromagnetic boundary conditions for model-2 (**Figure 3**) may be written in the mathematical form as [7]:

$$H_z^I = H_z^{II}, \quad 0 < z < \infty|_{r=r_C} \tag{22}$$

$$E_\theta^I = E_\theta^{II}, \quad 0 < z < \infty|_{r=r_C} \tag{23}$$

The boundary conditions (22) and (23) state the continuity of the axial component of magnetic and the azimuthal component of electric field intensities at the interface, $r = r_C$, between the regions I and II (**Figure 3**) [7].

Model-3: The relevant electromagnetic boundary conditions for model-3 (**Figure 4**) may be written in the mathematical form as [2, 5, 9–13]:

$$H_z^I = H_z^{II} \quad 0 < z < L - T|_{r=r_D} \tag{24}$$

$$E_\theta^I = \left\{ \begin{array}{ll} E_\theta^{II} & 0 < z < L - T \\ 0 & L - T \le z \le L \end{array} \right|_{r=r_D} \tag{25}$$

The boundary conditions (24) and (25) state the continuity of the axial component of magnetic and the azimuthal component of electric field intensities at the interface, $r = r_D$, between the regions I and II, and the null azimuthal component of electric field intensity at the disc hole metallic surface (**Figure 4**) [2, 5, 9–13].

Model-4: The boundary conditions (24) and (25) are also true for the model-4 (**Figure 5**) at the interface, $r = r_D$, between the regions I and II. The additional boundary conditions at the interface, $r = r_{BH}$, between the regions II and III may be written as [2, 9, 14, 15]:

$$H_z^{II} = H_z^{III} \quad 0 \le z \le (L - T - T_{BH})/2\big|_{r=r_{BH}} \tag{26}$$

$$E_\theta^{II} = \begin{cases} E_\theta^{III} & 0 \le z \le (L - T - T_{BH})/2 \\ 0 & (L - T - T_{BH})/2 \le z \le (L - T + T_{BH})/2 \end{cases}\Bigg|_{r=r_{BH}} \tag{27}$$

The boundary conditions (26) and (27) state the continuity of the axial component of magnetic and the azimuthal component of electric field intensities at the interface, $r = r_{BH}$, between the regions II and III, and the null azimuthal component of electric field intensity at the disc hole metallic surface (**Figure 5**) [2, 9, 14, 15].

Model-5: The relevant electromagnetic boundary conditions for model-5 (**Figure 6**) may be written in the mathematical form as [16–18]:

$$H_z^I = H_z^{II} \qquad \phi \le \theta \le 2\pi/N\big|_{r=r_V} \tag{28}$$

$$E_\theta^I = \begin{cases} 0 & 0 \le \theta < \phi \\ E_\theta^{II} & \phi \le \theta \le 2\pi/N \end{cases}\Bigg|_{r=r_V} \tag{29}$$

The boundary conditions (28) and (29) state the continuity of the axial component of magnetic and the azimuthal component of electric field intensities at the interface, $r = r_V$, between the regions I and II, and the null azimuthal component of electric field intensity at the vane tip metallic surface (**Figure 6**) [16–18].

Model-6: The relevant electromagnetic boundary conditions for model-6 (**Figure 7**) are given by (24) and (25), same as for model-3 (**Figure 4**) [19].

Model-7: The relevant electromagnetic boundary conditions (24) and (25) are also true for the model-7 (**Figure 8**) at the interface, $r = r_D$, between the regions I and II. The additional boundary conditions at the interface, $r = r_{DD}$, between the regions II and III may be written as [20]:

$$H_z^{II} = H_z^{III} \qquad 0 < z < L - T\big|_{r=r_{DD}} \tag{30}$$

$$E_\theta^{II} = E_\theta^{III} \qquad 0 < z < L - T\big|_{r=r_{DD}} \tag{31}$$

The boundary conditions (30) and (31), respectively, state the continuity of the axial magnetic and the azimuthal electric field intensities at the interface, $r = r_{DD}$, between the disc occupied free space region II and disc-occupied dielectric region III (**Figure 8**) [20].

Model-8: The relevant electromagnetic boundary conditions for model-8 (**Figure 9**) are given by (28) and (29), same as for model-5 (**Figure 6**) [21].

4. Dispersion relations

In general, the field intensity components (**1**)–(18) contain unknown field constants. In order to establish relations between these unknown field constants, the relevant field intensity components are substituted into the respective boundary conditions. Further, the algebraic manipulations of the obtained relations between these field constants eliminate all the field constants, and it results in a characteristic relation of the model known as the dispersion relation. The dispersion relations of various considered models are:

Model-1 [7]:

$$\begin{aligned}&\gamma^{II}J_0'\{\gamma^I r_L\}\left(J_0\{\gamma^{II}r_L\}Y_0'\{\gamma^{II}r_W\}-Y_0\{\gamma^{II}r_L\}J_0'\{\gamma^{II}r_W\}\right)\\&-\gamma^I J_0\{\gamma^I r_L\}\left(J_0'\{\gamma^{II}r_L\}Y_0'\{\gamma^{II}r_W\}-Y_0'\{\gamma^{II}r_L\}J_0'\{\gamma^{II}r_W\}\right)=0\end{aligned} \tag{32}$$

Model-2 [7]:

$$\begin{aligned}&\gamma^{II}J_0'\{\gamma^I r_C\}\left(J_0\{\gamma^{II}r_C\}Y_0'\{\gamma^{II}r_W\}-Y_0\{\gamma^{II}r_C\}J_0'\{\gamma^{II}r_W\}\right)\\&-\gamma^I J_0\{\gamma^I r_C\}\left(J_0'\{\gamma^{II}r_C\}Y_0'\{\gamma^{II}r_W\}-Y_0'\{\gamma^{II}r_C\}J_0'\{\gamma^{II}r_W\}\right)=0\end{aligned} \tag{33}$$

Model-3 [2, 5, 9–13]:

$$\det\left|M_{nm}J_0\{\gamma_n^I r_D\}Z_0'\{\gamma_m^{II}r_D\}-Z_0\{\gamma_m^{II}r_D\}J_0'\{\gamma_n^I r_D\}\right|=0 \tag{34}$$

where

$$M_{nm}=\frac{\gamma_n^I\beta_m^{II}\left(1-(-1)^m\exp\left[-j\beta_n^I(L-T)\right]\right)}{\gamma_m^{II}\left[\beta_m^{II}-\exp\left(-j\beta_0^I L\right)\left(\beta_m^{II}\cos\left(\beta_m^{II}L\right)+j\beta_n^I\sin\left(\beta_m^{II}L\right)\right)\right]}. \tag{35}$$

Model-4 [2, 9, 14, 15]:

$$\det\left|M_{nm}J_0\{\gamma_n^I r_D\}\left[J_0'\{\gamma_m^{II}r_D\}+\xi Y_0'\{\gamma_m^{II}r_D\}\right]-J_0'\{\gamma_n^I r_D\}\left[J_0\{\gamma_m^{II}r_D\}+\xi Y_0\{\gamma_m^{II}r_D\}\right]\right|=0 \tag{36}$$

where

$$\xi=\frac{\gamma_p^{III}J_0'\{\gamma_m^{II}r_{BH}\}Z_0\{\gamma_m^{III}r_{BH}\}-\gamma_m^{II}J_0\{\gamma_m^{II}r_{BH}\}Z_0'\{\gamma_m^{III}r_{BH}\}}{\gamma_m^{II}Y_0\{\gamma_m^{II}r_{BH}\}Z_0'\{\gamma_m^{III}r_{BH}\}-\gamma_m^{III}Y_0'\{\gamma_m^{II}r_{BH}\}Z_0\{\gamma_m^{III}r_{BH}\}} \tag{37}$$

Model-5 [16–18]:

$$\begin{vmatrix}P_{g-1} & Q_{g-1,g} & Q_{g-1,g+1}\\ Q_{g,g-1} & P_g & Q_{g,g+1}\\ Q_{g+1,g-1} & Q_{g+1,g} & P_{g+1}\end{vmatrix}=0 \tag{38}$$

where

$$\begin{aligned}P_{v'}=&\left(J_{v'}'\{\gamma^{II}r_V\}-\frac{J_{v'}'\{\gamma^{II}r_W\}}{Y_{v'}'\{\gamma^{II}r_W\}}Y_{v'}'\{\gamma^{II}r_V\}\right)\left(\frac{2\pi}{N}-\varphi\right)\\&-\frac{2\pi}{N}\frac{\gamma^{II}}{\gamma^I}\frac{J_{v'}'\{\gamma^I r_V\}}{J_{v'}\{\gamma^I r_V\}}\left(J_{v'}\{\gamma^{II}r_V\}-\frac{J_{v'}'\{\gamma^{II}r_W\}}{Y_{v'}'\{\gamma^{II}r_W\}}Y_{v'}\{\gamma^{II}r_V\}\right)\end{aligned} \tag{39}$$

$$Q_{v',v}=\left(J_v'\{\gamma^{II}r_V\}-\frac{J_v'\{\gamma^{II}r_W\}}{Y_v'\{\gamma^{II}r_W\}}Y_v'\{\gamma^{II}r_V\}\right)\frac{1-\exp j(v'-v)\varphi}{j(v'-v)} \tag{40}$$

Model-6 [19]:

The dispersion relation for model-6 (**Figure 7**) is same as that for model-3 (**Figure 4**) and is given by (34) through (35) with interpretation of radial propagation constant in dielectric filled region as $\gamma_m^{II} = \left(\varepsilon_r k^2 - \beta_m^2\right)^{1/2}$ [19].

Model-7 [20]:

$$\det\left|M_{nm} J_0\{\gamma_n^I r_{MD}\} \left[J_0'\{\gamma_m^{II} r_D\} + \xi Y_0'\{\gamma_m^{II} r_D\}\right] - J_0'\{\gamma_n^I r_D\}\left[J_0\{\gamma_m^{II} r_D\} + \xi Y_0\{\gamma_m^{II} r_D\}\right]\right| = 0 \tag{41}$$

$$\xi = \frac{\gamma_m^{III} J_0'\{\gamma_m^{II} r_{DD}\} Z_0\{\gamma_m^{III} r_{DD}\} - \gamma_m^{II} J_0\{\gamma_m^{II} r_{DD}\} Z_0'\{\gamma_m^{III} r_{DD}\}}{\gamma_m^{II} Y_0\{\gamma_m^{II} r_{DD}\} Z_0'\{\gamma_m^{III} r_{DD}\} - \gamma_m^{III} Y_0'\{\gamma_m^{II} r_{DD}\} Z_0\{\gamma_m^{III} r_{DD}\}} \tag{42}$$

Model-8 [21]:

The dispersion relation for model-8 (**Figure 9**) is same as that for model-5 (**Figure 6**) and is given by (38) through (39) and (40) with interpretation of radial propagation constant in dielectric filled region as $\gamma^{II} = \left(\varepsilon_r k^2 - \beta^2\right)^{1/2}$ [21].

5. Dispersion characteristics

One may clearly observe that the shape of dispersion characteristics of the model-1 (**Figure 2**) and model-2 (**Figure 3**) change with change in relative permittivity of the dielectric material (**Figures 10** and **11**). The cutoff frequency decreases with increase of the relative permittivity of the dielectric material. The increase of relative permittivity of the dielectric material depresses the dispersion characteristics of the model-1 and model-2 and more at higher value of phase propagation constant. The analytical dispersion characteristics are found within 3% of that obtained using HFSS (**Figures 10** and **11**).

Periodic loading a circular waveguide by the metal annular discs (model-3) brings out alternate pass and stop bands with their respective higher and lower

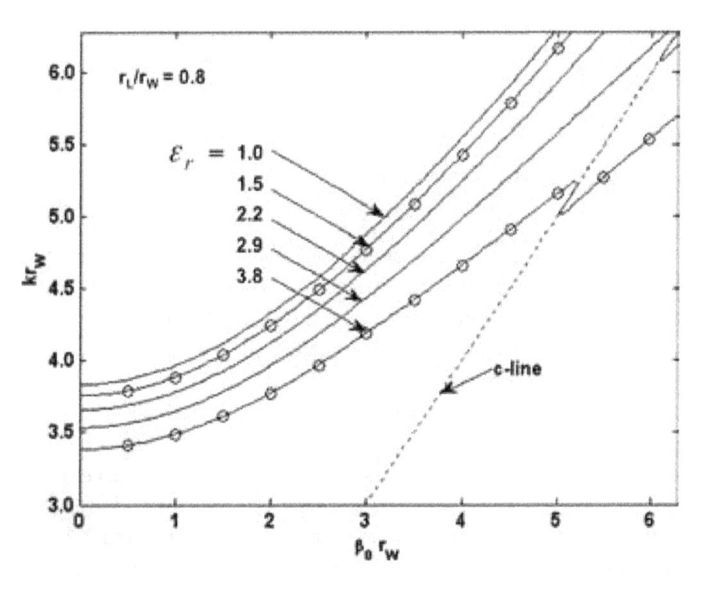

Figure 10.
TE_{01}-mode dispersion characteristics of circular waveguide with dielectric lining on metal wall taking ε_r as the parameter. The characteristics with $\varepsilon_r = 1$ (special case) represents the dispersion characteristics of conventional circular waveguide. Circles represent results obtained using HFSS.

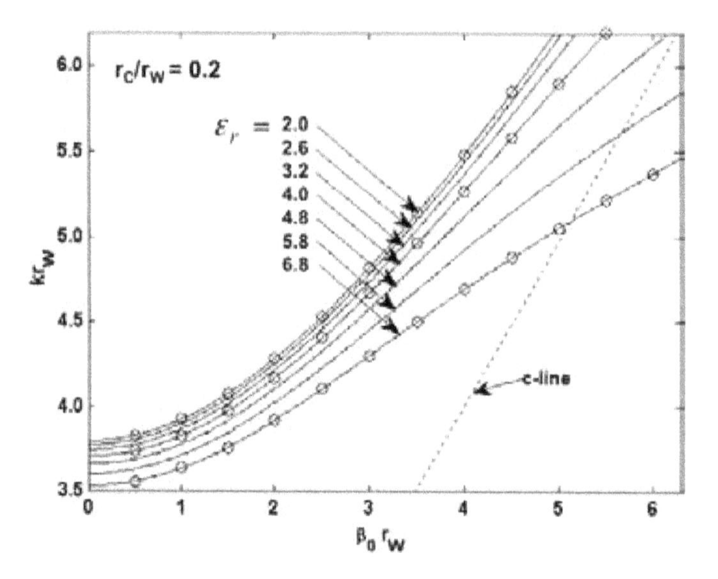

Figure 11.
TE_{01}-mode dispersion characteristics of circular waveguide with coaxial dielectric rod and taking ε_r as the parameter. Circles represent results obtained using HFSS.

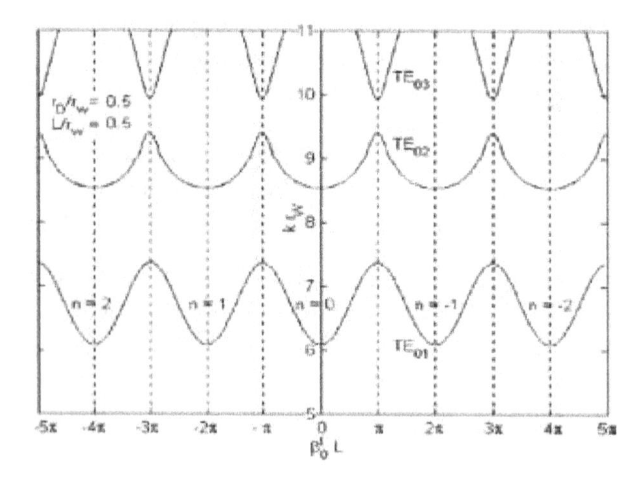

Figure 12.
Pass and stop band characteristics of the infinitesimally thin metal disc-loaded circular waveguide (including higher order harmonics $n = 0, \pm 1, \pm 2, \pm 3, \pm 4, \pm 5$; $m = 1, 2, 3, 4, 5, 6, 7, 8, 9, 10, 11$) [9, 10].

cutoff frequencies [1–3, 5, 9]. The dispersion characteristics taking horizontal axis as normalized phase propagation constant become periodic with the periodicity of $\beta^I{}_0 L = 2\pi$ for a given mode (TE_{01}, TE_{02} and TE_{03}). The normalized passband (kr_W scale) for the TE_{02} mode is narrower than that of the TE_{01} mode. Similarly, the normalized stopband above the TE_{02} mode is narrower than that above the TE_{01} mode (**Figure 12**). The RF group velocity (slopes of dispersion plot) is positive (fundamental forward wave mode, $n = 0$) for the TE01 or TE_{02} mode, it is negative (fundamental backward wave mode) for the TE_{03} mode (**Figure 12**). The dependencies of the structure dispersion characteristics, for typical mode TE_{01}, on the disc-hole radius (**Figure 13**), the structure periodicity (**Figure 14**) and the finite disc thickness (**Figure 15**) are studied. Further, with the increase of either of the parameters, namely, the disc-hole radius and the structure periodicity, the lower and upper edge frequencies of the passband of the dispersion characteristics both decrease, though not equally. This lead to decrease or increase of the passband

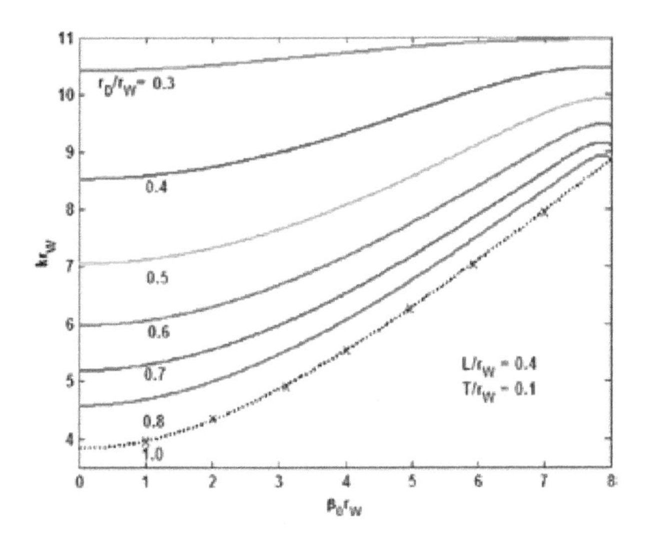

Figure 13.
TE_{01}-mode dispersion characteristics of the conventional disc-loaded circular waveguide (solid curve) taking the disc-hole radius as the parameter. The broken curve with crosses refers to a smooth-wall circular waveguide [2, 5, 11].

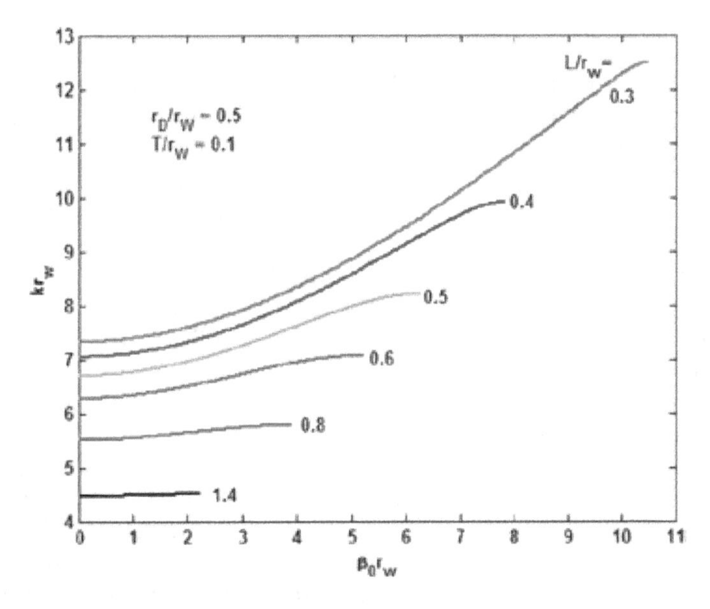

Figure 14.
TE_{01}-mode dispersion characteristics of the conventional disc-loaded circular waveguide (solid curve) taking the structure periodicity as the parameter [2, 5, 11].

according as the disc-hole radius decreases or the structure periodicity decreases, with the shift of the mid-band frequency of the passband to a higher value for the decrease of both the parameters (**Figures 13** and **14**) [2, 5, 11].

Although the disc-hole radius and the structure periodicity tailor the dispersion characteristics, the later one found to be more effective that the former one for widening the frequency range of the straight-line section of the characteristics. Reducing the structure periodicity can increase the frequency range of the straight-line section, however it accompanies with shift in waveguide cutoff (**Figure 14**). This wider straight-line section of the dispersion characteristics may be utilized for wideband coalescence with cyclotron wave (beam mode line) to result a wideband performance of a gyro-TWT. Thus, reducing the structure periodicity (**Figure 14**)

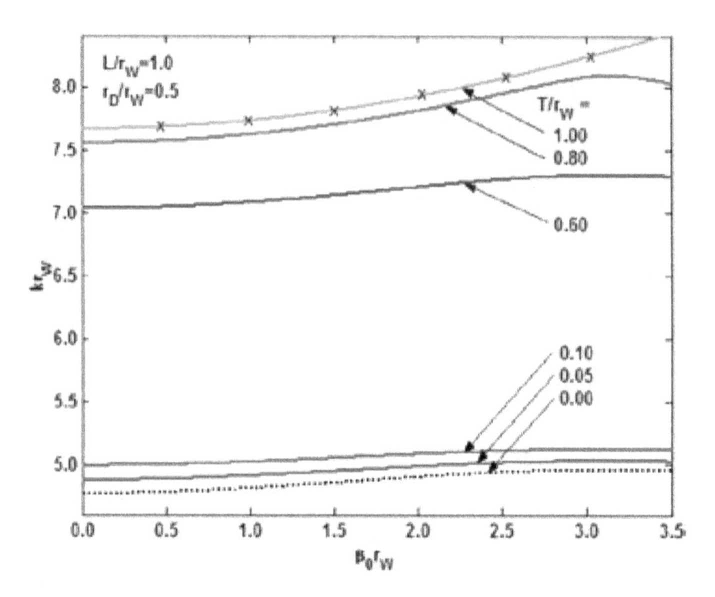

Figure 15.
TE_{01}-mode dispersion characteristics of the conventional disc-loaded circular waveguide taking the disc thickness as the parameter. The broken curve refers to the infinitesimally thin metal disc-loaded circular waveguide [2, 5, 11].

and increasing the disc-hole radius (**Figure 13**) may widen the device bandwidth. However, such broadbanding of coalescence is accompanied by the reduction of the bandwidth of the passband of the structure as well (**Figures 13** and **14**) [2, 5, 11]. The decrease of the disc thickness decreases both the lower and upper edge frequencies of the passband such that the passband first decreases, attains a minima and then increases; and the mid-band frequency of the passband as well as the start frequency of the straight-line section of the dispersion characteristics reduces (**Figure 15**). The shape of the dispersion characteristics depends on the disc thickness, though not as much as it does on the disc-hole radius or the structure periodicity (**Figures 13-15**) [2, 5, 11].

Similar to the conventional disc-loaded circular waveguide (model-3), both the hole-radii (bigger and smaller) of the interwoven-disc-loaded circular (model-4, **Figure 5**) waveguide tailor the dispersion characteristics. The lower- and the upper-cutoff frequencies decrease with increase in hole-radii, such that the passband increases and decreases with decrease of bigger and smaller hole-radii, respectively (**Figures 16** and **17**). Similar to the conventional disc-loaded circular waveguide (model-3), the structure periodicity of the interwoven-disc-loaded circular waveguide is the most effective for the increasing the passband and tailoring the dispersion characteristics (**Figure 18**). Neither, the extent of passband changes nor dispersion tailors with variation of disc-thickness of bigger-hole-disc, however, the mid frequency of the passband shifts to higher frequency with increase of disc-thickness of bigger-hole-disc (**Figure 19**). This nature may be used for shifting the operation band in the passive components or in order to optimizing the beam-wave interaction in designing a gyro-TWT with the interwoven-disc-loaded circular waveguide. In addition to tailoring the dispersion characteristics, required for designing a broadband gyro-TWT, the model-4 shows an interesting characteristic. The passband increases with increase as well as with decrease of disc-thickness of smaller- hole-disc with reference to that of bigger-hole-disc, however, the shift of the passband occur towards higher and lower frequency side, respectively, with increase and decrease of disc thickness of smaller-hole-disc with reference to that of bigger-hole-disc (**Figure 20**). Thus, the structure periodicity (**Figure 18**) and the

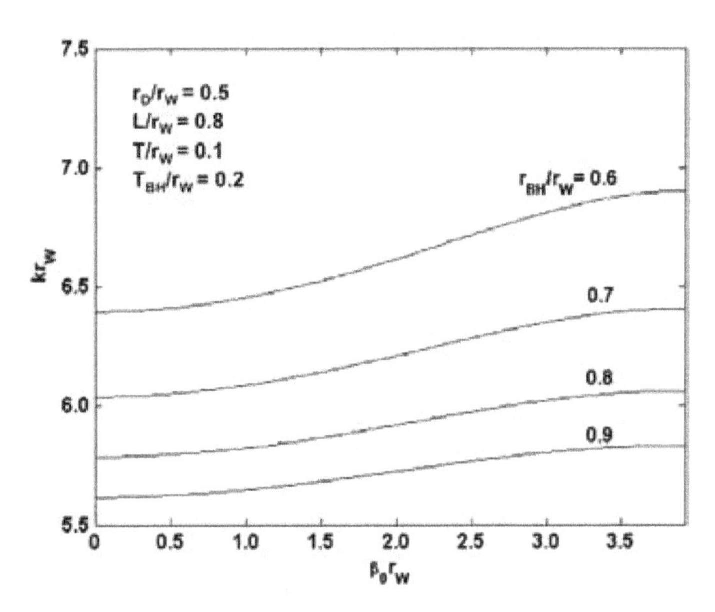

Figure 16.
TE_{01}-mode dispersion characteristics of the interwoven-disc-loaded circular waveguide taking the bigger disc-hole-radius as the parameter [9, 14, 15].

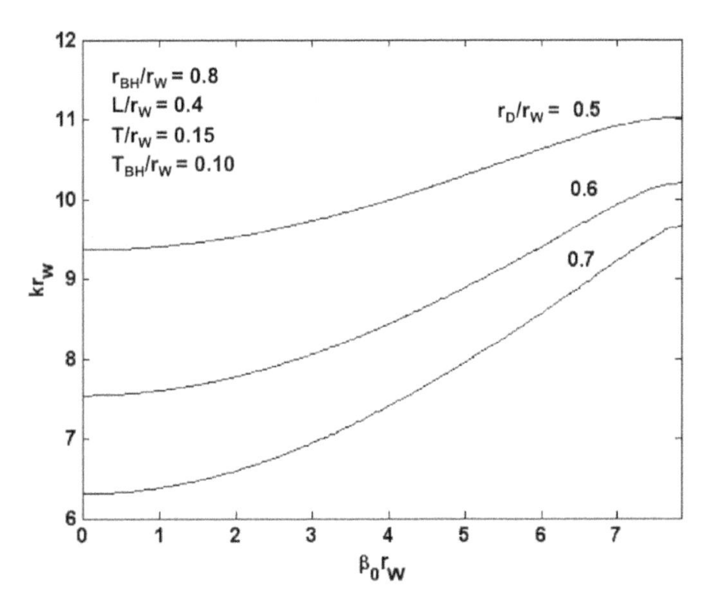

Figure 17.
TE_{01}-mode dispersion characteristics of the interwoven-disc-loaded circular waveguide taking the smaller disc-hole-radius as the parameter [9, 14, 15].

disc-thickness (**Figure 19**) of bigger-hole-disc of the interwoven-disc-loaded circular waveguide are, respectively, the most and the least sensitive parameter for controlling the passband as well as shape of the dispersion characteristics [14, 15].

Axial metal vane loading to a smooth-wall circular waveguide (model-5, **Figure 6**) forms an azimuthally periodic structure, which does not shape its disper-sion characteristics, however the insertion of the metal vanes in to the circular waveguide shifts the waveguide cutoff frequency to a higher value [16–18] (**Figures 21–23**). Specifically, the increase of either of the vane angle (**Figure 22**) and the number of metal vanes (**Figure 23**) and the decrease of vane-inner-tip

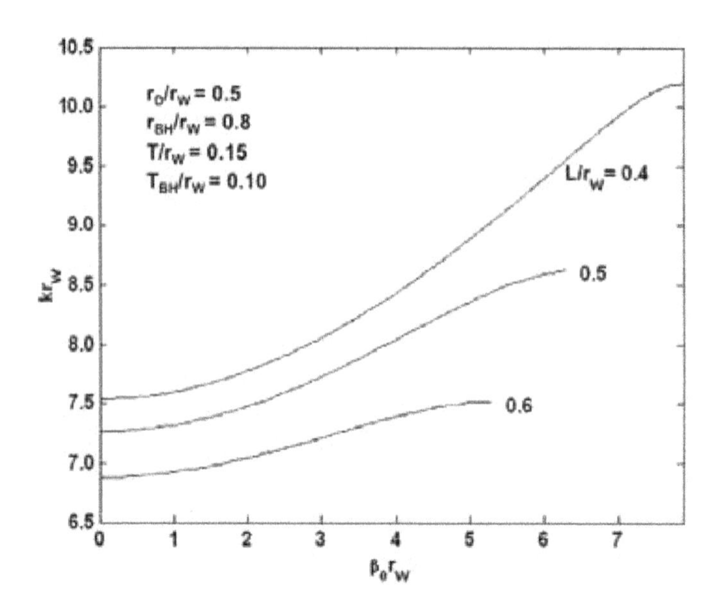

Figure 18.
TE_{01}-mode dispersion characteristics of the interwoven-disc-loaded circular waveguide taking the structure periodicity as the parameter [9, 14, 15].

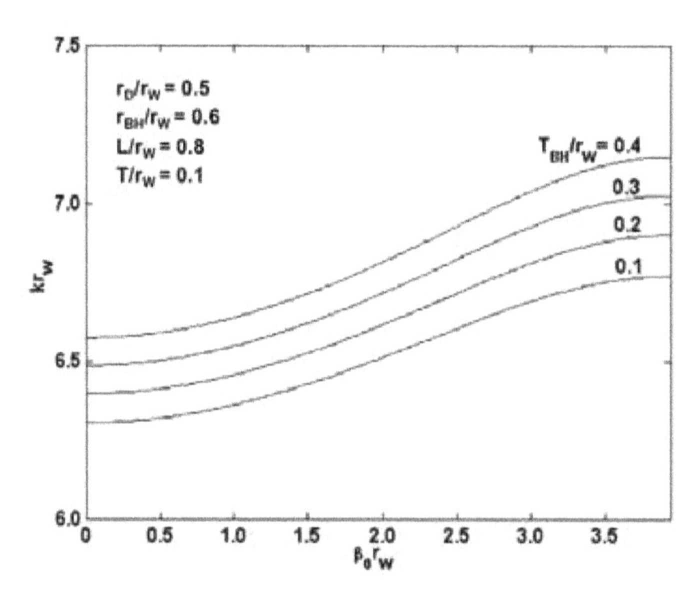

Figure 19.
TE_{01}-mode dispersion characteristics of the interwoven-disc-loaded circular waveguide taking the disc-thickness of bigger-hole-disc as the parameter [9, 14, 15].

radius (**Figure 21**) increases the waveguide cutoff frequency, and none of the parameters tailors the dispersion characteristics.

For the model-6, the variation of relative permittivity of the dielectric discs changes the lower and upper cutoff frequencies of the passband (**Figure 24**). Two lowest order azimuthally symmetric (TE_{01} and TE_{02}) modes are typically considered to study the performance of this model. With the increase of relative permittivity, the passband continuously decreases for the TE_{01} mode, and first decreases and then increases for the TE_{02} mode (**Figure 24**). Also, the variation in relative permittivity shapes of the dispersion characteristics of the structure, for both the

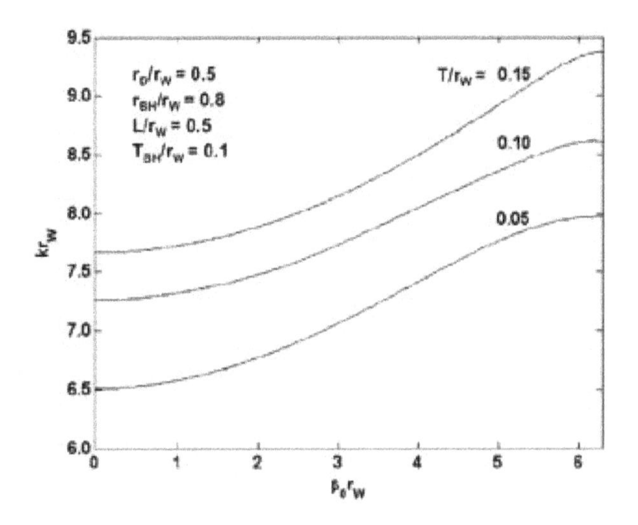

Figure 20.
TE_{01}-mode dispersion characteristics of the interwoven-disc-loaded circular waveguide taking the disc thickness of smaller-hole-disc as the parameter [9, 14, 15].

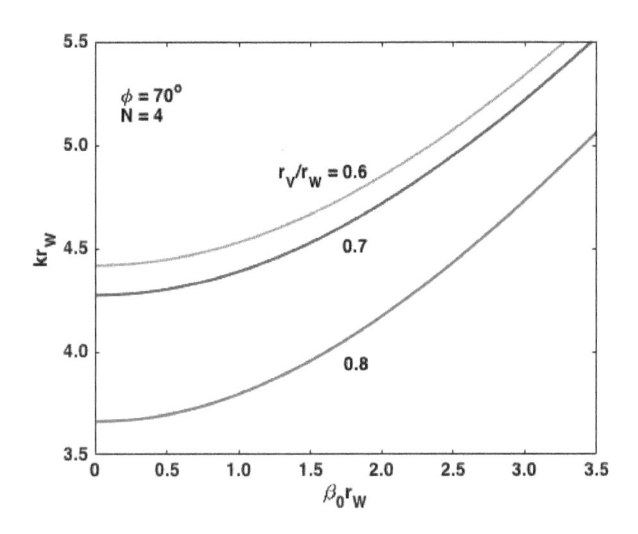

Figure 21.
TE_{01}-mode dispersion characteristics of the metal-vane-loaded circular waveguide taking the vane-inner-tip radius as the parameter [16–18].

modes TE_{01} and TE_{02}, however more significantly for the latter mode (**Figure 24**). The TE_{01} mode of the structure.

Exhibits fundamental forward wave (positive) dispersion characteristics irrespective of the value of relative permittivity (**Figure 24(a)**), however, the TE_{02} mode exhibits fundamental forward (positive) and backward (negative) wave dispersion characteristics, respectively, at higher and lower values of relative permittivity. This suggests that an appropriate selection of the value of relative permittivity in this structure would yield a straightened TE_{02} mode $\omega - \beta$ dispersion characteristics near low value of phase propagation constant for wideband coalescence with the beam-mode dispersion line and consequent wideband gyro-TWT performance (**Figure 24(b)**). Thus the introduction of the dielectric discs between metal discs in the a conventional metal disc-loaded waveguide, with lower values of relative permittivity for the TE_{01} mode and with higher values of relative permittivity for the TE_{02} mode enhances the frequency range of the straight line portion of

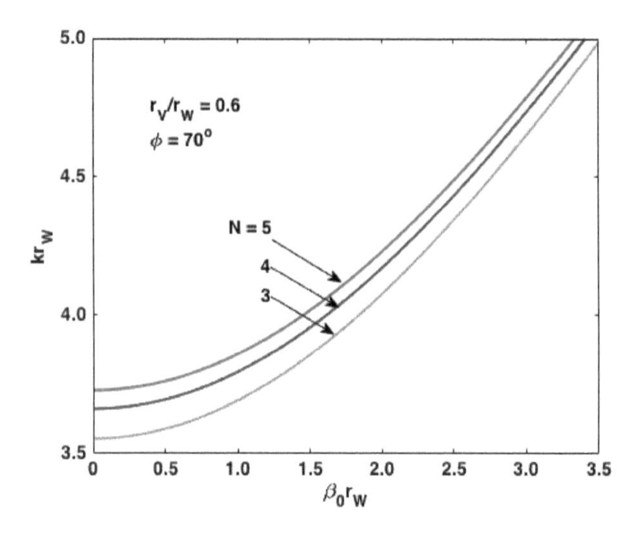

Figure 22.
TE_{01}-mode dispersion characteristics of the metal-vane-loaded circular waveguide taking the vane angle as the parameter [16–18].

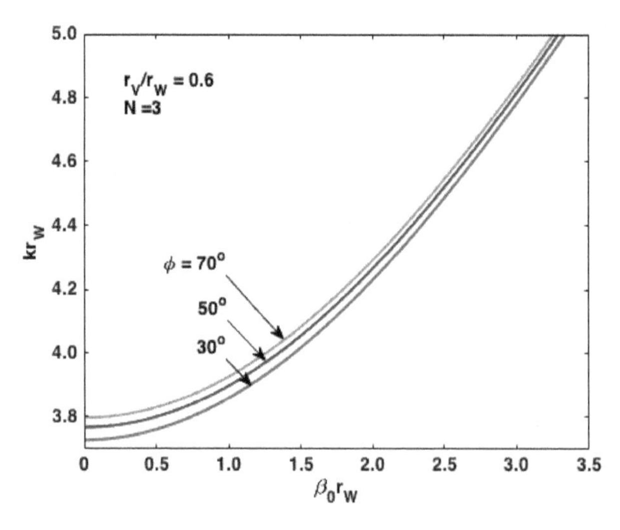

Figure 23.
TE_{01}-mode dispersion characteristics of the metal-vane-loaded circular waveguide taking the number of metal vanes as the parameter [16–18].

the $\omega - \beta$ dispersion characteristics, desired for wideband gyro-TWT performance (**Figure** 25) [19].

The lower and upper cutoff frequencies of the passband vary with thickness of dielectric disc $L - T()/r_W$ taking structure periodicity constant such that the pass-band decreases with an increase in thickness of dielectric disc for both the TE_{01} and TE_{02} modes. The thickness of dielectric disc tailors the dispersion characteristics, however, more for the TE_{02} than for the TE_{01} mode, and the control is more prominent for thinner dielectric disc (**Figure 26**) [19]. The less effective parameters, the disc-hole radius, in tailoring the dispersion characteristics of a conventional disc-loaded waveguide [2, 5, 9–13], effectively controls the shape of the characteristics after introducing the dielectric discs between the metal discs, while, the control is more for the TE_{02} (**Figure 27(b)**) than for the TE_{01} (**Figure 27(a)**) mode, however the characteristics is little irregular for higher disc-hole radius

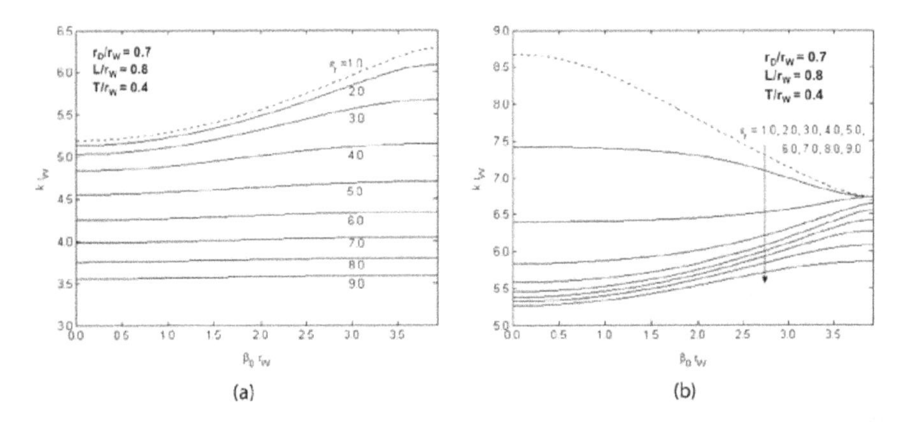

Figure 24.
TE_{01} (a) and TE_{02} (b) mode dispersion characteristics of a circular waveguide loaded with alternate dielectric and metal annular discs (model-6) taking relative permittivity as the parameter. The broken curves refer to a conventional metal disc-loaded circular waveguide [19].

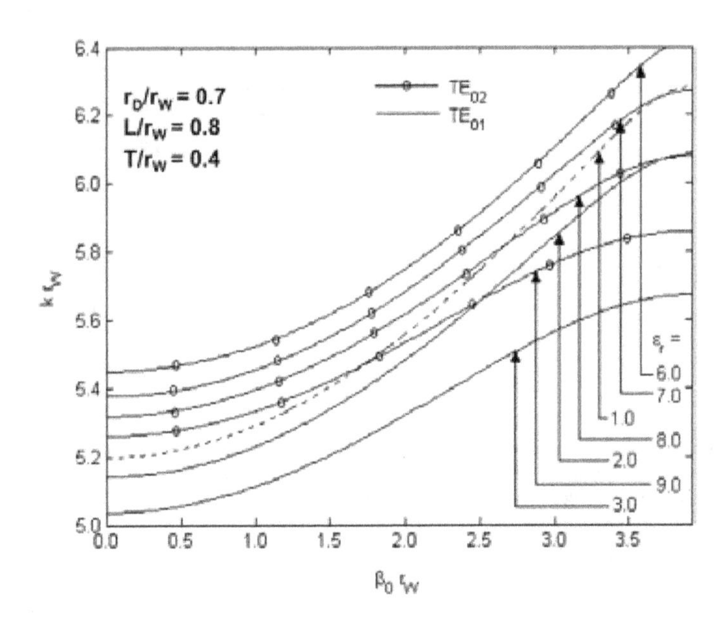

Figure 25.
Dispersion characteristics of a circular waveguide loaded with alternate dielectric and metal annular discs (model-6) for the selected dielectric disc relative permittivity values for the sake of comparison between the modes TE_{01} and TE_{02} with respect to the control of the shape of the dispersion characteristics. The broken curves referring to a conventional metal disc-loaded circular waveguide [19].

(**Figure 27**) [19]. Similar to a conventional all-metal disc-loaded waveguide (model-4), the structure periodicity is the most effective parameter for tailoring the dispersion characteristics of the structure with dielectric discs between the metal discs (model-6), for the TE_{01} and TE_{02} modes, more for the latter. The control of the structure periodicity in straightening the dispersion characteristics, as required for the desired wideband gyro-TWT performance, is enhanced by introducing the dielectric discs in the conventional disc-loaded waveguide, though not enhancing the frequency range of the straight line portion of the dispersion characteristics (**Figure 28**). In this model, a serious care is required while selecting the dielectric material because a heavily dielectric-loaded structure depresses the dispersion characteristics to the slow-wave region (below the velocity of light line in $\omega - \beta$ characteristics).

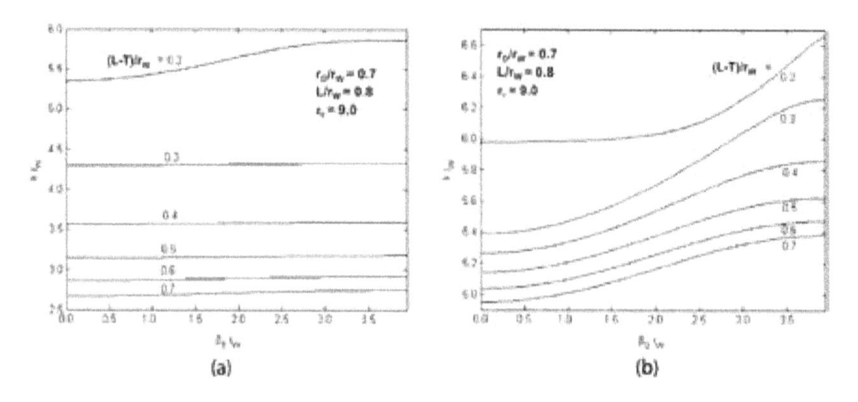

Figure 26.
TE_{01} (a) and TE_{02} (b) mode dispersion characteristics of a circular waveguide loaded with alternate dielectric and metal annular discs (model-6) taking dielectric disc thickness as the parameter for a constant structure periodicity [19].

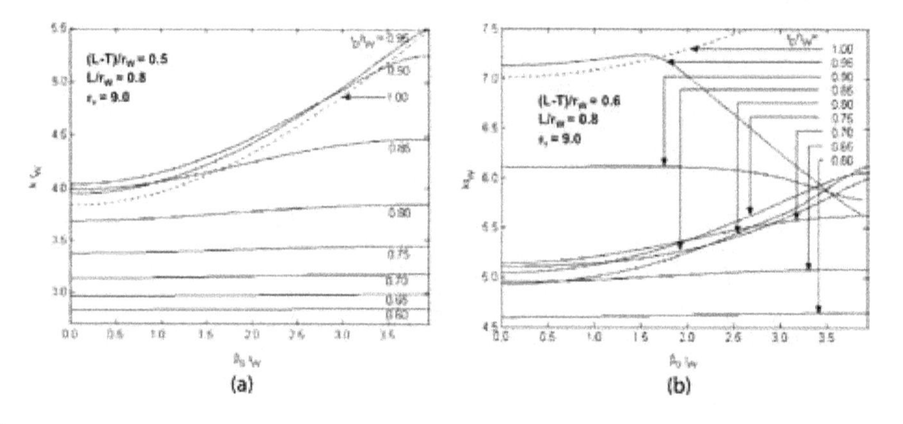

Figure 27.
TE_{01} (a) and TE_{02} (b) mode dispersion characteristics of a circular waveguide loaded with alternate dielectric and metal annular discs (model-6) taking the disc-hole radius as the parameter. The broken curves refer to the special case of a conventional smooth wall circular waveguide [19].

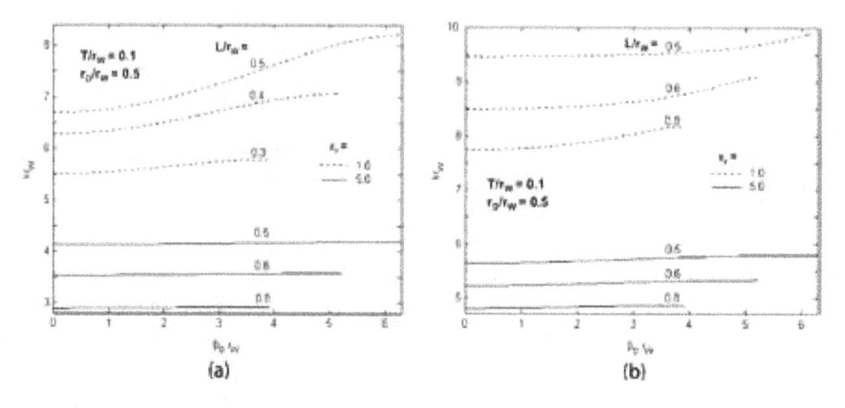

Figure 28.
TE_{01} (a) and TE_{02} (b) mode dispersion characteristics of a circular waveguide loaded with alternate dielectric and metal annular discs (model-6) taking structure periodicity as the parameter. The broken curves refer to a conventional metal disc-loaded circular waveguide [19].

We choose first three lowest order azimuthally symmetric modes in the composite (dielectric and metal) loaded structures for exploring the effect of structure parameter on dispersion characteristics while we chose only the lowest order azimuthally symmetric mode in all-metal variants of the axially periodic structure. The increase of the relative permittivity of the dielectric discs in model-7 reduces the lower and upper cutoff frequencies, however not equally therefore shortens the passband for the TE_{01} mode with shift of the mid-frequency of the passband towards lower value (**Figure 29(a)**). With the increase of the relative permittivity of the dielectric discs, the lower and upper cutoff frequencies shift to lower value for the TE_{01} and TE_{02} modes. The quantitatively the shift in upper cutoff frequency is higher than that of lower cutoff frequency for the TE_{01}, which in turn shortens the passband (**Figure 29(a)**), however, the shift in lower and upper cutoff frequencies are almost equal for the TE_{02} mode, effectively the passband does not change (**Figure 29(b)**). Interestingly for the TE_{02} mode the introduction of dielectric discs converts the fundamental backward mode (the zero group velocity follows to take negative values then again zero and further positive) into fundamental forward mode (the zero group velocity follows to take positive values then again zero and further negative) (**Figure 29(b)**). Thus, introduction of dielectric discs into the conventional disc-loaded waveguide turns the negative dispersion into positive. In absence as well as in presence of the dielectric discs, the TE_{03} mode dispersion characteristics of the disc-loaded waveguide represents the fundamental backward mode, in which the increase of the relative permittivity of the dielectric discs shifts the lower cutoff frequency more than that of upper cutoff frequency, and thus widens the passband for lower relative permittivity. For higher relative permittivity value the lower cutoff frequency remains unchanged and upper cutoff frequency shifts to lower value with the increase of the relative permittivity of the dielectric discs and shortens the passband (**Figure 29(c)**) [20, 21].

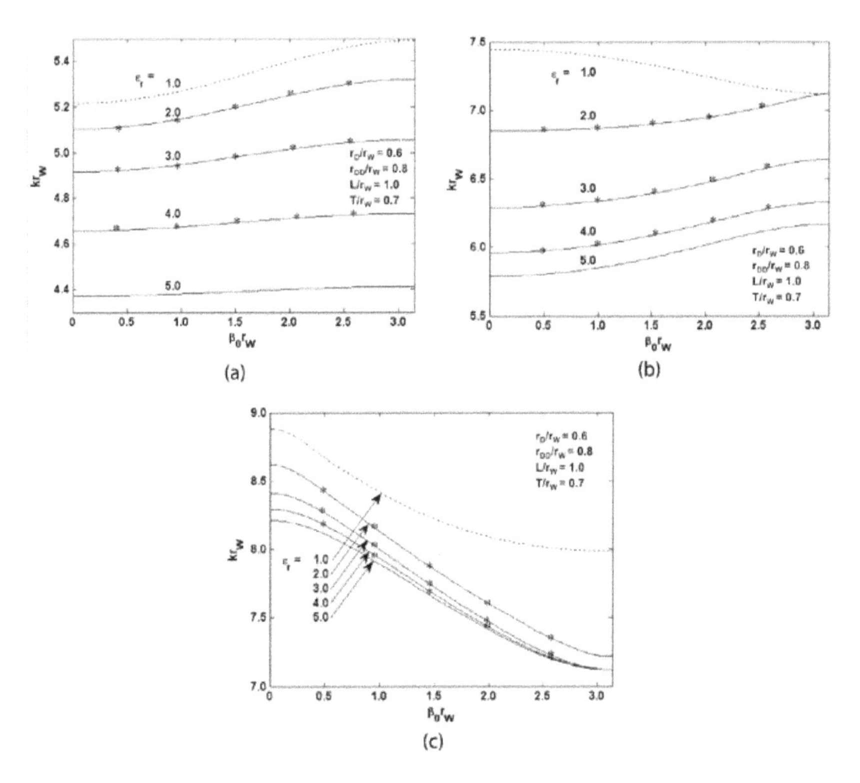

Figure 29.
TE_{01} (a), TE_{02} (b) and TE_{03} (c) mode dispersion characteristics of the alternate dielectric and metal disc-loaded circular waveguide taking relative permittivity of dielectric disc as the parameter [20, 21]. The broken curve refers to the conventional disc-loaded circular waveguide [2, 5, 9–13].

The increase of dielectric disc radius in the model-7 taking constant metal disc radius shifts the lower and upper cutoff frequencies of the TE_{01} mode up, and the passband increases due to lesser the shift in lower cutoff frequency (**Figure 30**). Similarly, the passband of the TE_{02} mode increases with the increase of dielectric disc radius. Here, it is interesting to note that for the taken structure parameters ($r_D/r_W = 0.6$, $L/r_W = 1.0$, $T_{DD}/r_W = 0.3$ and $\varepsilon_r = 5.0$) the frequency shift is maximum for r_{DD}/r_W equal to 0.8–0.9, and minimum for 0.7–0.8 (**Figure 30(b)**). For the lower and higher values of inner dielectric disc radius, the TE_{03} mode dispersion characteristics of the model-7 are fundamental forward (positive) and backward (negative) modes respectively. Thus, there is a possibility of getting straight-line dispersion characteristics parallel to phase propagation constant axis, i.e., zero group velocity line (**Figure 30(c)**) [20, 21]. The increase of periodicity of the alternate dielectric and metal disc-loaded circular waveguide (model-7) reduces the passband and both the lower and upper cutoff frequencies with higher relative reduction in upper cutoff frequency than that of lower (**Figure 31**) for the chosen three azimuthally symmetric modes TE_{01}, TE_{02} and TE_{03}. The TE_{01} and TE_{02} modes are fundamental forward and the TE_{03} is fundamental backward [20, 21]. The change of dielectric disc thickness does not much tailor the dispersion characteris-tics and the passband (**Figure 32**), however, the decrease of the dielectric disc thickness or the increase of metal disc thickness shifts the passband to lower fre-quency side for the TE_{01} and TE_{02} modes (**Figure 31(a)** and **(b)**) and least change occur to the TE_{03} mode. In very precise observation, the lower cutoff frequency of the TE_{03} mode is insensitive and the upper cutoff frequency first decreases and then increases with decrease of dielectric disc thickness or with increase of metal disc thickness (**Figure 32**) [20, 21].

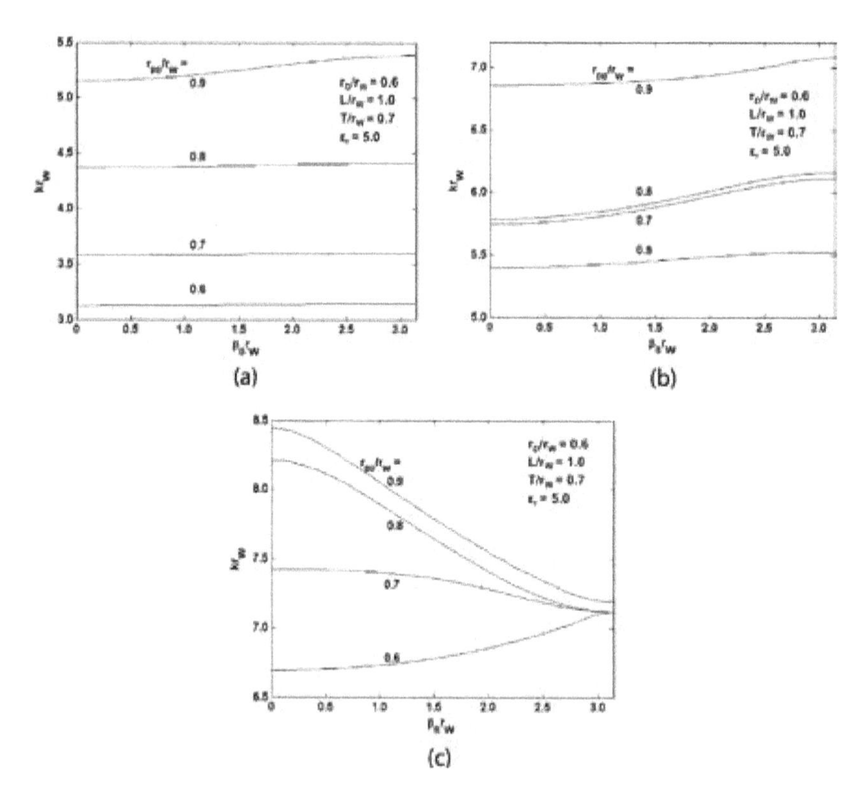

Figure 30.
TE_{01} (a), TE_{02} (b) and TE_{03} (c) mode dispersion characteristics of the alternate dielectric- and metal disc-loaded circular waveguide (model-7) taking inner radius of dielectric disc as the parameter [20, 21].

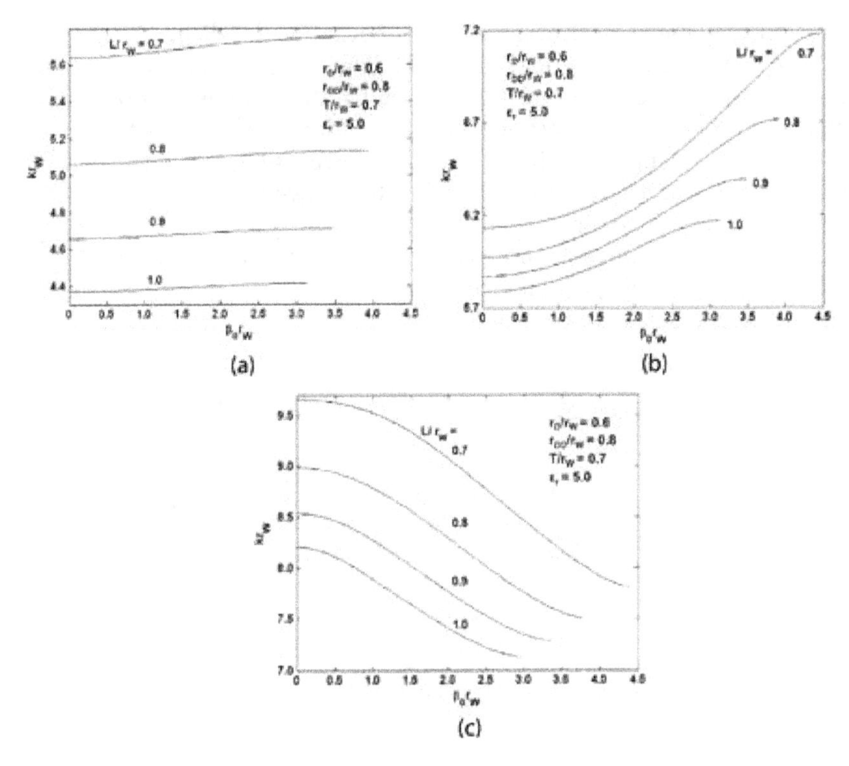

Figure 31.
TE_{01} (a), TE_{02} (b) and TE_{03} (c) mode dispersion characteristics of the alternate dielectric and metal disc-loaded circular waveguide (model-7) taking structure periodicity as the parameter [20, 21].

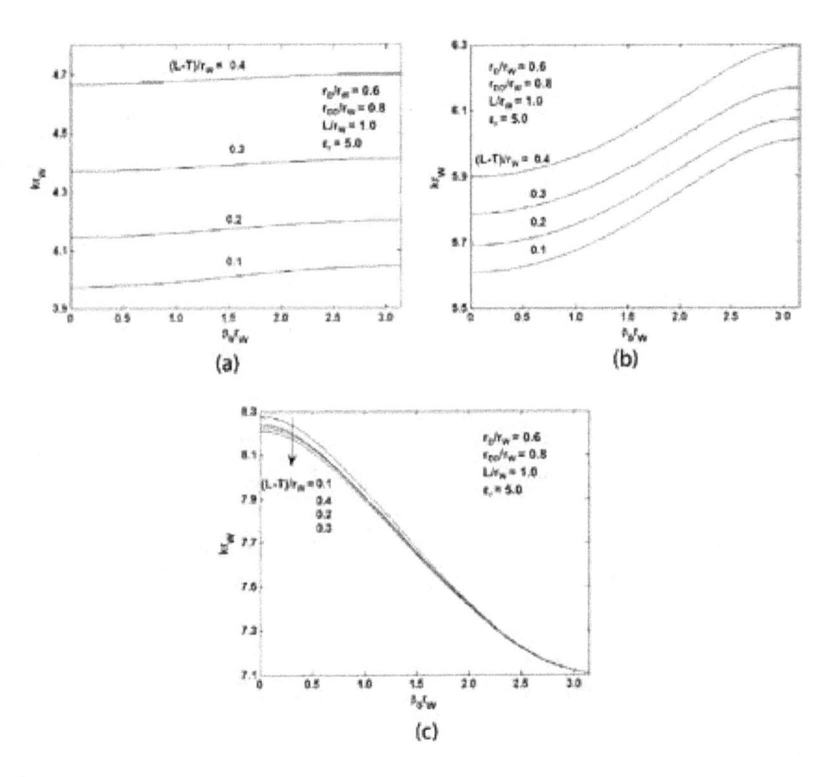

Figure 32.
TE_{01} (a), TE_{02} (b) and TE_{03} (c) mode dispersion characteristics of the alternate dielectric and metal disc-loaded circular waveguide (model-7) taking thickness of dielectric disc as the parameter [20, 21].

Although the geometrical parameters do not tailor the dispersion characteristics of the only metal vane-loaded waveguide [16–18], however, it does for composite-loaded structure (model-8). The radial dimensions (**Figure 33**) are less sensitive in tailoring the dispersion characteristics of the dielectric and metal vanes than their angul ar dimensions (**Figure 34**), relative permittivity (**Figure 35**) and number of vanes (**Figure 36**). However, the waveguide cutoff (eigenvalue) of the structure depends on all these parameters (**Figures 33-36**). Thus, one may choose angular dimensions, relative perm ittivity and number of vanes for tailoring the dispersion characteristics and the radial parameter to control the waveguide cutoff frequency [21].

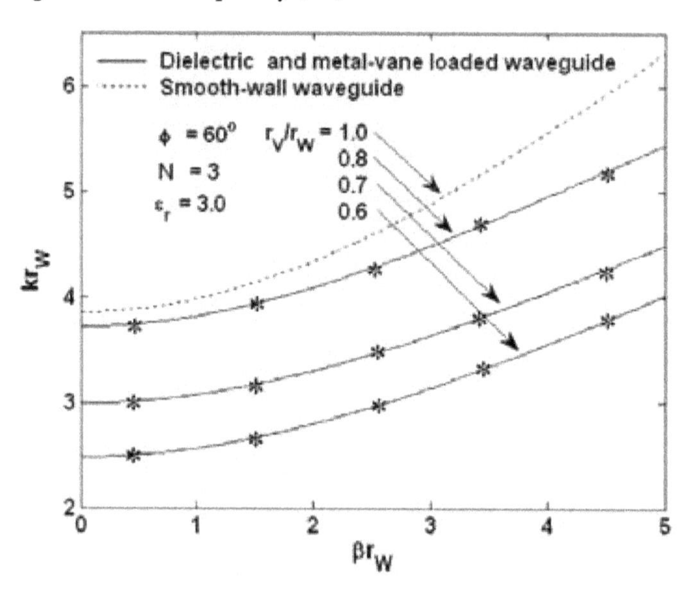

Figure 33.
TE_{01}-mode dispersion characteristics of a circular waveguide loaded with composite alternate dielectric and metal vanes taking the vane inner-tip radius as the parameters. The broken curve refers to a smooth-wall waveguide (free from dielectric and metal vanes) and the star (*) marker refers representative points obtained using HFSS [21].

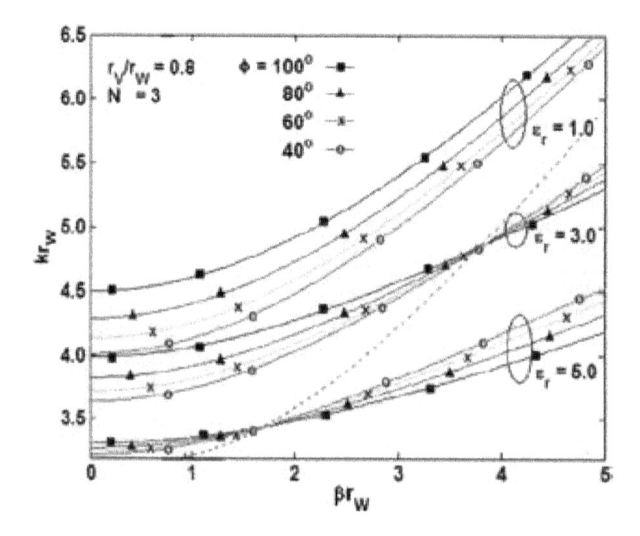

Figure 34.
TE_{01}-mode dispersion characteristics of a circular waveguide loaded with dielectric and metal vanes taking metal vane angle as the parameter. The broken curve represents the locus of the crossover point, which is same as the dispersion characteristics of the smooth-wall circular waveguide of radius r_V [21].

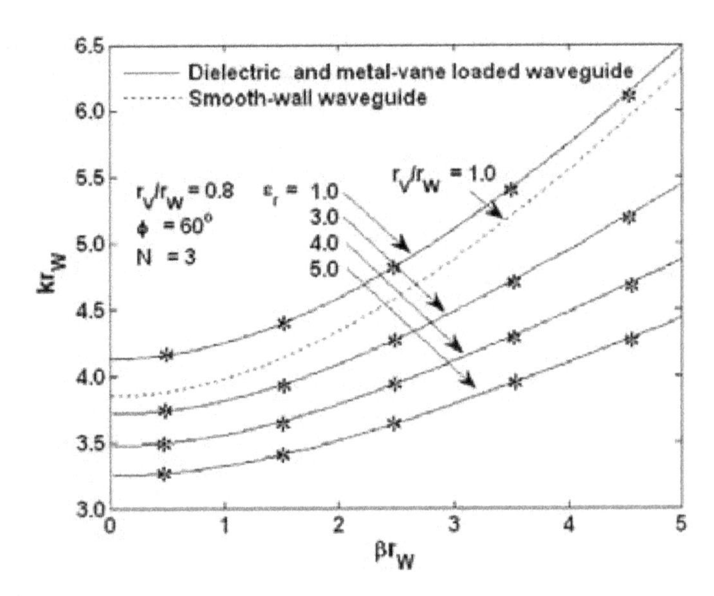

Figure 35.
TE_{01}-mode dispersion characteristics of a circular waveguide loaded with composite alternate dielectric and metal vanes, taking the relative permittivity of dielectric vanes as the parameters, along with the corresponding characteristics of a smooth-wall waveguide (free from dielectric and metal vanes) (broken curve) and the typical representative points of the characteristics obtained by simulation (HFSS) ((star) marker) [21]. $\varepsilon_r = 1$ represents the characteristics of a waveguide loaded with metal vanes alone (**Figure 6**) [16–18].*

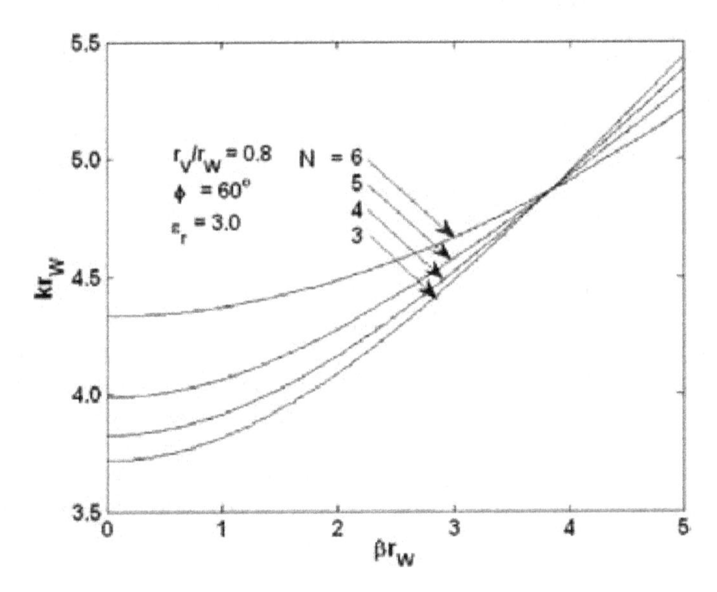

Figure 36.
TE_{01}-mode dispersion characteristics of a circular waveguide loaded with dielectric and metal vanes taking number of vanes as the parameter [21].

A crossover point in the dispersion characteristics appears with varying metal vane angle (**Figure 34**) or varying number of vanes (**Figure 36**). This crossover point sifts to another location for another value of relative permittivity. The locus of such crossover points for varying relative permittivity overlaps with the dispersion plot of a smooth-wall circular waveguide of radius equal to the vane tip radius (**Figure 34**). Below the crossover point, the increase of metal cross-section (either by increasing metal vane angle or by increasing number of vanes) in the cross-section of the waveguide elevates and that of dielectric depresses the dispersion characteristics, and vice versa above the crossover point. This clearly represents the behavior of model-8 as a smooth-wall waveguide of wall radius equal to the vane tip radius at the crossover point and change of metal cross-section (either by changing metal vane angle or by changing number of vanes) in the cross-section of the waveguide affects the dispersion characteristics away from the crossover (**Figures 34** and **36**) [21].

6. Conclusion

The waveguide is inherently a high pass filter with a lower cut-off frequency, a signal of above this frequency will be allowed to propagate or travel through the waveguide. Different kind of loading, such as metal and/or dielectric, has been suggested in the published literature for altering the propagation (dispersion) characteristics. In the present chapter a number of loaded structures are studied with circular cross-sections. All the structure parameters are varied to explore the sensitivity over the dispersion characteristics. In general, the axial periodicity results in periodic dispersion characteristics with a lower and an upper cut-off frequency, which makes the guiding structure a bandpass structure. However, such a characteristic is not reported in literature for the azimuthal periodic structures. The bandpass characteristic arises due to the shaping of the dispersion characteristics. Therefore, the dispersion shaping is only possible with axial periodicity and not with the azimuthal periodicity in all metal structure. However, the azimuthal periodicity in all metal waveguide structure shifts the cut-off frequency over the frequency scale. The dispersion characteristics of various loaded structures have been explored. It has been the interest to study the change in the lower and upper cut-off frequencies and in the passband. The sensitivity of the structure (geometry) parameters on the lower and upper cut-off frequencies and the extent of passband are also included. In case of conventional disc-loaded structure, the periodicity is found to be the most sensitive parameter for dispersion shaping and the disc-hole radius is the most sensitive parameter for shifting the dispersion characteristics over the frequency axis. In interwoven-disc-loaded circular waveguide, the structure periodicity and the disc-thickness of bigger-hole-disc are, respectively, the most and the least sensitive parameter for controlling the passband as well as shape of the dispersion characteristics. Insertion of metal vanes to a circular waveguide does not shape the dispersion characteristics but increases the waveguide cutoff frequency. On the other hand, introduction of dielectric discs into the conventional disc-loaded waveguide turns the negative dispersion into positive. In case of composite vane-loaded structure, the angular dimensions, relative permittivity and number of vanes tailor the dispersion characteristics and the radial parameter controls the waveguide cutoff frequency.

Author details

Vishal Kesari
Microwave Tube Research and Development Centre, Defence Research and Development Organisation, Bangalore, India

*Address all correspondence to: vishal_kesari@rediffmail.com

References

[1] Watkins DA. Topics in Electromagnetic Theory. New York: John Wiley; 1958

[2] Kesari V, Basu BN. High Power Microwave Tubes: Basics and Trends. San Rafael (CA)/Bristol: IOP Concise Physics, Morgan & Claypool Publishers/ IOP Publishing; 2018

[3] Collin RE. Foundations for Microwave Engineering. 2nd ed. USA: Wiley-IEEE Press; 2001

[4] Basu BN. Electromagnetic Theory and Applications in Beam-Wave Electronics. Singapore: World Scientific; 1996

[5] Kesari V. Analysis of Disc-Loaded Circular Waveguides for Wideband Gyro-TWTs. Koln: Lambert Academic Publishing AG & Co.; 2009

[6] Fliflet AW. Linear and nonlinear theory of the Doppler-shifted cyclotron resonance maser based on TE and TM waveguide modes. International Journal of Electronics. 1986;**61**(6):1049-1080

[7] Rao SJ, Jain PK, Basu BN. Broadbanding of gyro-TWT by dispersion shaping through dielectric loading. IEEE Transactions on Electron Devices. 1996;**43**:2290-2299

[8] Choe JY, Uhm HS. Theory of gyrotron amplifiers in disc or helix-loaded waveguides. International Journal of Electronics. 1982;**53**:729-741

[9] Kesari V, Basu BN. Analysis of some periodic structures of microwave tubes: Part II: Analysis of disc-loaded fast-wave structures of gyro-traveling-wave tubes. Journal of Electromagnetic Waves and Applications. 2018;**32**(4): 1-36

[10] Kesari V, Jain PK, Basu BN. Approaches to the analysis of a disc loaded cylindrical waveguide for potential application in wideband gyro-TWTs. IEEE Transactions on Plasma Science. 2004;**32**(5):2144-2151

[11] Kesari V, Jain PK, Basu BN. Analysis of a circular waveguide loaded with thick annular metal discs for wideband gyro-TWTs. IEEE Transactions on Plasma Science. 2005;**33**(4):1358-1365

[12] Kesari V, Jain PK, Basu BN. Analysis of a disc-loaded circular waveguide for interaction impedance of a gyrotron amplifier. International Journal of Infrared and Millimeter Waves. 2005; **26**(8):1093-1110

[13] Kesari V, Jain PK, Basu BN. Modal analysis of a corrugated circular waveguide for wideband potential in gyro-devices. International Journal of Microwave and Optical Technology. 2007;**2**(2):147-152

[14] Kesari V, Keshari JP. Interwoven-disc-loaded circular waveguide for a wideband gyro-traveling-wave tube. IEEE Transaction on Plasma Science. 2013;**41**(3):456-460

[15] Kesari V, Keshari JP. Propagation characteristics of a variant of disc-loaded circular waveguide. Progress in Electromagnetic Research. 2012;**26**: 23-37

[16] Singh G, Ravi Chandra SMS, Bhaskar PV, Jain PK, Basu BN. Analysis of dispersion and interaction impedance characteristics of an azimuthally-periodic vane-loaded cylindrical waveguide for a gyro-TWT. International Journal of Electronics. 1999;**86**:1463-1479

[17] Agrawal M, Singh G, Jain PK, Basu BN. Analysis of a tapered vane loaded broad-band gyro-TWT. IEEE Transactions on Plasma Science. 2001; **29**:439-444

[18] Shrivastava UA. Small-signal theories of harmonic gyrotron and peniotron amplifiers and oscillators [thesis]. Utah: University of Utah; 1985

[19] Kesari V, Jain PK, Basu BN. Modeling of axially periodic circular waveguide with combined dielectric and metal loading. Journal of Physics D: Applied Physics. 2005;**38**:3523-3529

[20] Kesari V, Keshari JP. Analysis of a circular waveguide loaded with dielectric and metal discs. Progress in Electromagnetic Research. 2011;**111**: 253-269

[21] Kesari V. Analysis of alternate dielectric and metal vane loaded circular waveguide for a wideband gyro-TWT. IEEE Transactions on Electron Devices. 2014;**61**(3):915-920

Permissions

All chapters in this book were first published by InTech Open; hereby published with permission under the Creative Commons Attribution License or equivalent. Every chapter published in this book has been scrutinized by our experts. Their significance has been extensively debated. The topics covered herein carry significant findings which will fuel the growth of the discipline. They may even be implemented as practical applications or may be referred to as a beginning point for another development.

The contributors of this book come from diverse backgrounds, making this book a truly international effort. This book will bring forth new frontiers with its revolutionizing research information and detailed analysis of the nascent developments around the world.

We would like to thank all the contributing authors for lending their expertise to make the book truly unique. They have played a crucial role in the development of this book. Without their invaluable contributions this book wouldn't have been possible. They have made vital efforts to compile up to date information on the varied aspects of this subject to make this book a valuable addition to the collection of many professionals and students.

This book was conceptualized with the vision of imparting up-to-date information and advanced data in this field. To ensure the same, a matchless editorial board was set up. Every individual on the board went through rigorous rounds of assessment to prove their worth. After which they invested a large part of their time researching and compiling the most relevant data for our readers.

The editorial board has been involved in producing this book since its inception. They have spent rigorous hours researching and exploring the diverse topics which have resulted in the successful publishing of this book. They have passed on their knowledge of decades through this book. To expedite this challenging task, the publisher supported the team at every step. A small team of assistant editors was also appointed to further simplify the editing procedure and attain best results for the readers.

Apart from the editorial board, the designing team has also invested a significant amount of their time in understanding the subject and creating the most relevant covers. They scrutinized every image to scout for the most suitable representation of the subject and create an appropriate cover for the book.

The publishing team has been an ardent support to the editorial, designing and production team. Their endless efforts to recruit the best for this project, has resulted in the accomplishment of this book. They are a veteran in the field of academics and their pool of knowledge is as vast as their experience in printing. Their expertise and guidance has proved useful at every step. Their uncompromising quality standards have made this book an exceptional effort. Their encouragement from time to time has been an inspiration for everyone.

The publisher and the editorial board hope that this book will prove to be a valuable piece of knowledge for researchers, students, practitioners and scholars across the globe.

List of Contributors

Zion Menachem
Department of Electrical Engineering, Sami Shamoon College of Engineering, Beer Sheva, Israel

Hitoshi Ohsato
Microelectronics Research Unit, Faculty of Information Technology and Electrical Engineering, University of Oulu, Oulu, Finland
Department of Research, Nagoya Industrial Science Research Institute, Nagoya, Japan

Jobin Varghese and Heli Jantunen
Microelectronics Research Unit, Faculty of Information Technology and Electrical Engineering, University of Oulu, Oulu, Finland

Vasa Radonic, Norbert Cselyuszka, Vesna Crnojevic-Bengin and Goran Kitic
University of Novi Sad, BioSense Institute–Research and Development Institute for Information Technologies in Biosystems, Novi Sad, Serbia

Heng Luo
School of Physics and Electronics, Central South University, Changsha, China
State Key Laboratory of Powder Metallurgy, Central South University, Changsha, China

Lianwen Deng
School of Physics and Electronics, Central South University, Changsha, China

Peng Xiao
State Key Laboratory of Powder Metallurgy, Central South University, Changsha, China

Tsun-Hsu Chang
Department of Physics, National Tsing Hua University, Hsinchu, Taiwan

Ravinder Kumar Kotnala
CSIR-National Physical Laboratory, New Delhi, India

Biao Zhao
Henan Key Laboratory of Aeronautical Materials and Application Technology, School of Material Science and Engineering, Zhengzhou University of Aeronautics, Zhengzhou, Henan, China
Department of Mechanical and Industrial Engineering, University of Toronto, Toronto, Ontario, Canada

Rui Zhang
Henan Key Laboratory of Aeronautical Materials and Application Technology, School of Material Science and Engineering, Zhengzhou University of Aeronautics, Zhengzhou, Henan, China
School of Material Science and Engineering, Zhengzhou University, Zhengzhou, Henan, China

Kuldeep Chand Verma, Ashish Sharma and Navdeep Goyal
Department of Physics, Panjab University, Chandigarh, India

Mangui Han
School of Materials and Energy, University of Electronic Science and Technology of China, Chengdu, China

Li Zhuang, Cao Rui, Tao Xiaohui, Jiang Lihui and Rong Dawei
Key Lab of Aperture Array and Space Applications, He Fei, China

Licinius Dimitri Sá de Alcantara
Cyberspatial Institute, Federal Rural University of Amazon, Belém, Brazil

Vishal Kesari
Microwave Tube Research and Development Centre, Defence Research and Development Organisation, Bangalore, India

Index

S

T

W

Printed in the USA
CPSIA information can be obtained
at www.ICGtesting.com
JSHW052311231023
50683JS00006BA/62

9 781647 284510